Progress in Physics
Volume 17

William E. Baylis

Electrodynamics
A Modern Geometric Approach

Birkhäuser
Boston • Basel • Berlin

William E. Baylis
Department of Physics
University of Windsor
Windsor, Ontario N9B 3P4
Canada

Library of Congress Cataloging-in-Publication Data

Baylis, William E. (William Eric), 1939–
Electrodynamics: A Modern Geometric Approach / William E. Baylis.
p. cm. — (Progress in physics)
Includes bibliographical references and index.
ISBN 0-8176-4025-8 (alk. paper)
1. Electrodynamics. I. Title II. Series: Progress in physics (New York, N.Y.)
QC631 .B333 1998
537.6—dc21 98-31155
CIP

AMS Subject Classifications: 00A79, 11E88, 15A66, 78A25, 78A30, 78A35, 78A40, 78-01, 81V10, 83Cxx, 83C22, 83C50, 83C60

Printed on acid-free paper.
 Birkhäuser 

ISBN 0-8176-4025-8
ISBN 3-7643-4025-8

Typeset by the author in LaTeX.
Printed and bound by Hamilton Printing Co., Castleton, NY.
Printed in the United States of America.

9 8 7 6 5 4 3 2 1

Contents

Preface **xi**

1 Introduction **1**

1.1 Units 2
1.2 Electromagnetic Theory and Vector Algebra 6
1.3 An Introduction to Clifford Algebras 8
1.3.1 Associative Vector Product 8
1.3.2 Overview 10
1.3.3 Matrix Representations 10
1.4 Geometric Interpretations 12
1.4.1 Bivectors 13
1.4.2 Higher-Grade Multivectors 14
1.4.3 Bivectors as Generators of Rotation 15
1.4.4 Rotations in n Dimensions 18
1.4.5 Other Clifford Algebras 20
1.4.6 Volume Element and Hodge Duals 21
1.4.7 Paravectors 22
1.4.8 Involutions of $C\ell_n$ 23
1.4.9 Inverses and the Metric of Paravector Space 25
1.5 Problems 28

2 Special Relativity **33**

2.1 The Spacetime Manifold 33
2.1.1 Proper Time and Velocity 35
2.2 Lorentz Transformations 40
2.2.1 Active and Passive Transformations 42
2.2.2 Linearity 44
2.3 Spacetime Planes 45

2.3.1 Biparavector Basis 46
2.3.2 Simple and Compound Biparavectors 49
2.4 Spacetime Rotations 51
2.4.1 Generators 52
2.5 Proper Velocity 54
2.6 Tetrads and Matrix Representations 59
2.6.1 Covariant vs. Invariant 63
2.6.2 Reflections in Spacetime 65
2.7 Spacetime Diagrams 67
2.8 Electromagnetic Field 69
2.9 Problems 72

3 Maxwell's Equation(s) 79
3.1 Covariant Form 81
3.1.1 Symmetries 83
3.1.2 Component Form 86
3.1.3 Geometrical Theorems 86
3.1.4 Interpretations of the Fields 91
3.2 Conservation of Charge 91
3.3 Wave Equation 93
3.4 Magnetic Monopoles 93
3.5 Electromagnetic Potential 95
3.6 Gauge Invariance 97
3.7 Energy and Momentum Density 98
3.8 Problems 102

4 Lorentz-Force Equation 108
4.1 Covariant Form 108
4.2 Eigenspinors 110
4.3 Invariant Planes and Directions 113
4.3.1 Transformations of Transformations 114
4.4 The Group SL(2,C) 117
4.4.1 SL(2,C) Diagrams 118
4.5 Time Evolution of Eigenspinor 120
4.6 Thomas Precession 121
4.7 Spinorial Form of Lorentz-Force Equation 122
4.8 Motion in Constant Fields 123
4.8.1 Pure Electric Field 124

4.8.2 Pure Magnetic Field 125
4.8.3 Electric and Magnetic Fields 126
4.8.4 Simple Fields 127
4.8.5 Compound Fields 127
4.9 Motion in Varying Fields 129
4.9.1 Coulomb Field 130
4.10 Field Momenta 133
4.11 Problems 137

5 Electromagnetic Waves in 1-D 143
5.1 Wave Equation 143
5.2 Waves in a String 145
5.3 Separation of Variables 146
5.4 Fourier Series 147
5.5 Fourier Integrals 150
5.5.1 Dirac Delta Function 151
5.5.2 Group Velocity 153
5.5.3 Frequency Distributions 154
5.6 Transmission Lines 156
5.7 Impedance and Reflections 158
5.8 Problems 160

6 Electromagnetic Waves in Vacuum 166
6.1 Maxwell's Equation in a Vacuum 167
6.1.1 Directed Plane-Wave Solutions 167
6.1.2 Gauge Choices 170
6.2 Projectors 171
6.2.1 Pacwoman Property 173
6.3 Directed Plane Waves and Flags 174
6.4 Monochromatic Plane Waves 177
6.5 Motion of Charges in Plane Waves 180
6.6 Monochromatic Standing Waves 185
6.7 Problems 188

7 Polarization 194
7.1 Polarization of Monochromatic Waves 195
7.2 Poincaré Spinors 197
7.3 Stokes Parameters 198

7.4 Polarizers and Phase Shifters 201
7.5 Coherent and Incoherent Superpositions 202
7.6 Problems 203

8 Waves in Media 205
8.1 Maxwell's Macroscopic Equations 205
8.2 Linear, Isotropic Media 207
8.3 Wave Equation in Dielectric Media 209
8.4 Directed Plane Waves 209
8.5 Waves in Conducting Media 212
8.5.1 Ohm's Law 212
8.5.2 Differential Equation for Conductors 213
8.5.3 Physical and Unphysical Solutions 213
8.5.4 Wave/Diffusion Equation 214
8.5.5 Penetration Depth 215
8.5.6 Reflection 217
8.6 Problems 218

9 Waves at Boundaries 220
9.1 Nonconducting Boundaries 220
9.1.1 Continuity Conditions 222
9.1.2 Snell's Law and Total Internal Reflection 223
9.1.3 Total Internal Reflection 225
9.1.4 Matching Conditions 226
9.2 Normal Incidence 227
9.3 Fresnel Equations 228
9.4 Conducting Surfaces 231
9.4.1 Conducting Screen 231
9.4.2 Real Conductors 234
9.5 Wave Guides 238
9.6 TE and TM Modes 240
9.7 Problems 245

10 The Field of a Moving Charge 248
10.1 Liénard-Wiechert Potentials 249
10.2 The Electromagnetic Field 251
10.2.1 Covariant Coulomb Field 253
10.2.2 Lab-Frame Expression of $\mathbf{F}_c$ 254

10.2.3 Covariant Radiation Field 257
10.2.4 Lab-Frame Expression of $\mathbf{F}_r$ 258
10.3 Example: Uniformly Accelerated Charge 259
10.4 Radiated Power 262
10.4.1 Thomson Scattering 266
10.5 Green Functions 268
10.5.1 Potentials of Static Charge Distributions 268
10.5.2 Potentials from Currents 270
10.5.3 Coulomb Gauge 271
10.5.4 Dipole Radiation 272
10.6 Fields from Currents 275
10.7 Virtual-Photon Streams 276
10.8 Review Questions 280
10.9 Problems 281

11 The Model Atom and Other Applications **286**
11.1 Static Polarizability 286
11.2 Atoms in Anisotropic Media 287
11.3 Scattering of Radiation by Bound Charge 289
11.4 Blue Sky 291
11.5 Scattering in a Magnetic Field 293
11.6 Drude Model of Van der Waals Forces 295
11.7 Diffraction 297
11.7.1 Babinet's and Huygen's Principles 298
11.8 Linear Antenna 299
11.9 Problems 301

12 The Lorentz-Dirac Equation **304**
12.1 Introduction 304
12.2 Self Interaction 305
12.2.1 Renormalized Mass 307
12.2.2 Avoiding Mass Renormalization 308
12.3 The Lorentz-Dirac Equation 309
12.4 Pathological Features 310
12.4.1 One-Dimensional Motion; Runaways 311
12.4.2 Preacceleration and Nonuniqueness 312
12.4.3 Potential Barrier 314
12.4.4 Backward Time Integration 317

12.5 Eigenspinor Form . . . 317
12.6 Conclusions . . . 319
12.7 Problems . . . 320

13 Rotations and Spherical Harmonics **322**
13.1 Spherical Harmonics . . . 322
13.2 Ladder Operators . . . 324
13.2.1 Generating the Spherical Harmonics . . . 327
13.3 Rotations and Matrix Representations . . . 332
13.3.1 Product Representations and Their Reduction . . . 335
13.3.2 Clebsch-Gordan Coefficients . . . 337
13.4 Infinitesimal Generators . . . 340
13.5 Multipoles of Charge Distributions . . . 342
13.6 Problems . . . 346

14 Radiation Multipoles **348**
14.1 Wave Equation . . . 348
14.2 Vector Spherical Harmonics . . . 349
14.3 Spherical Bessel Functions . . . 352
14.4 Electromagnetic Multipoles . . . 354
14.5 Problems . . . 355

A Appendix: Other Coordinate Systems **357**
A.1 Coordinate Systems in Physical Space . . . 357
A.2 Gradient Operator . . . 360
A.3 Spherical Coordinates . . . 361
A.4 Integrals . . . 364
A.5 Cylindrical Coordinates . . . 364

B Appendix: Maple Worksheet rad3.mws **365**

Bibliography **367**

Index **370**

Preface

The subject of this text is classical electromagnetic theory and electrodynamics, that is, dynamical solutions to Maxwell's equations and to the equation for the Lorentz force. The approach is essentially relativistic. The natural appearance of the Minkowski spacetime metric in the paravector space of Clifford's geometric algebra is used to formulate a covariant treatment in special relativity that seamlessly connects spacetime concepts to the spatial vector treatments common in undergraduate texts. The approach allows a component-free notation that fosters geometrical interpretation with powerful tools such as spinors and projectors while avoiding the clutter of indices required in tensorial treatments. Even the experienced professional should be pleasantly surprised at the number of fresh geometrical insights and simplified derivations.

The text begins with a discussion of electromagnetic units and an explanation of how the SI system adopted can be readily converted to the Gaussian or natural Heaviside-Lorentz systems. It continues with introductions to geometric algebra and the paravector model of spacetime and to special relativity. Treatments of Maxwell's equation(s), the Lorentz-force law, Fresnel equations, electromagnetic waves and polarization, wave guides, radiation from accelerating charges and time-dependent currents, Liénard-Wiechert potentials, and radiation reaction all benefit from the covariant algebraic approach. Geometric algebra achieves new efficiencies not only of notation but also of thought, and the text emphasizes these efficiencies in the concepts and fundamental principles rather than in the details of approximate calculations. For example, the reader will find a full discussion of the behavior of electromagnetic fields and Maxwell's equation(s) under Lorentz transformations, including time reversal, parity inversion, and reflection, but only brief treatments of diffraction and antenna theory. The text does not cover circuit theory and also has rather little coverage of potential theory, a subject sometimes taught in intermediate-level courses on electromagnetic theory or differential equations.

Numerous worked examples and exercises dispersed throughout the text help the reader understand new concepts and facilitate self-study of the material. Each chapter concludes with a set of problems, many with answers. Complete solutions are also available from the publisher for instructors who adopt the text.

The text is designed for upper-level undergraduate and beginning graduate courses in mathematics and physics. It should also be of interest to practicing physicists and electrical engineers who want a deeper geometrical appreciation of electrodynamics and access to powerful new calculational tools for its application. Mathematicians will find here an introduction to geometric methods with paravectors in Clifford algebras and their applications in relativistic physics. It is assumed that the reader has previously studied electro- and magnetostatics at a level common to first-year university physics texts and has completed a course in vector calculus. No prior study is required of relativistic dynamics or Clifford algebras.

Field theory is the foundation of modern physics. It is modeled on the theory of electromagnetic fields, the highly successful description of the interaction of moving charges, based on Maxwell's equations and the Lorentz-force law. Electrodynamics is also highly important in its own right. Of the four known basic forces, namely strong, weak, electromagnetic, and gravitational, it is the electromagnetic one that almost exclusively runs our machines and generally governs our daily lives. For the engineer, a solid understanding of electromagnetic theory is essential for solving practical problems in the real world. But for the physicist, its importance extends well beyond its immediate practical relevance: it is a basic paradigm by which we try to understand the natural world. Through the study of electromagnetic theory, the student is introduced to modern theoretical and mathematical physics.

For many practical applications, a nonrelativistic approach to electrodynamics is usually sufficient and is often preferred because it describes phenomena in terms of vector quantities from our familiar three-dimensional world. Typical undergraduate texts introduce a relativistically covariant formulation of the theory late in the course, usually in terms of Minkowski-space tensors, and if practical applications follow this introduction, the authors engage in what must appear to the student as a schizophrenic jumping between incompatible notations of relativistic covariance and of vectors with practical geometrical significance.

Yet, electromagnetic theory is essentially relativistic. Its beauty and much of its geometrical significance can only be properly appreciated in a covariant formulation. For example, the important concept of the electromagnetic field as a (para)bivector acting in a spacetime plane requires a relativistic approach. Certainly the use of electromagnetic theory as an introduction to the field theories of modern physics calls for a covariant formulation, but even practical applications can benefit from the increased understanding afforded by a spacetime approach.

To bridge the traditional gap between the vector and covariant approaches, physicists have increasingly turned to Clifford's geometric algebras. They are compatible with differential forms and naturally include many of their important geometrical concepts, such as bivectors, trivectors, and their Hodge duals. Most physical applications of Clifford algebras to relativistic phenomena start with the algebra $\mathcal{C}\ell_{1,3}$ (or $\mathcal{C}\ell_{3,1}$) based on Minkowski spacetime (*i.e.*, Hestenes' "space-time algebra"[1] or on its complexification, the algebra of Dirac matrices with complex coefficients). The choice of these algebras is evidently made on the reasonable (but actually mistaken) impression that one of them (or a larger algebra) is required for a covariant description of the phenomena, but they are larger than required and entail complications such as a volume element that acts as a fifth dimension rather than as a scalar. The cost has been a layer of abstraction from familiar physical quantities and their interrelationships that has impeded its adoption by the broader physics community.

Probably the simplest covariant formulation is one that represents spacetime vectors as paravectors in the Pauli algebra, the Clifford algebra $\mathcal{C}\ell_3$ of real three-dimensional Euclidean space, $\mathbb{E}^3$. It shares with higher-dimensional Clifford algebras the geometrical facility to treat restricted Lorentz transformations with ease as rotations in spacetime planes. But it is unique in the way it relates covariant quantities to the familiar scalar and vector quantities measured by any observer, as well as in the simple understanding it provides of complex quantities and of the origin of the Minkowski spacetime metric. It is the approach adopted here.

The algebra $\mathcal{C}\ell_3$ models well, for example, our understanding of the proper time τ of a free particle: it is both a Lorentz-invariant scalar and also the time coordinate of an inertial frame. In $\mathcal{C}\ell_3$ it is simply τ, but in $\mathcal{C}\ell_{1,3}$ one distinguishes between τ and $\tau\gamma_0$. Similarly, the mass of the particle, which (within factors of c) is both a Lorentz invariant and the energy component of its spacetime momentum in the commoving inertial frame, requires distinct symbols in $\mathcal{C}\ell_{1,3}$ but only one in $\mathcal{C}\ell_3$. The electromagnetic field is also modeled effectively in $\mathcal{C}\ell_3$, where a single symbol $\mathbf{F}$ is both a spacetime plane and a sum $\mathbf{F} = \mathbf{E} + ic\mathbf{B}$ of a spatial vector $\mathbf{E}$ and bivector $ic\mathbf{B}$ in the observer's inertial frame. Other approaches generally require different notations for the field, depending on whether we think of it covariantly or in terms of the spatial vectors measured in the laboratory.

[1] D. Hestenes (1966). See also W. E. Baylis (1996). See the Bibliography at the end of the text for complete references.

The goal of representing relativistic physics in a formalism that emphasizes the geometry but does not depend on an explicit choice of the coordinate frame was also the driving force behind the approaches both of differential forms[2] and of spinors.[3] The formalism of geometric algebra simplifies and unifies both of these approaches. As the recent book by Snygg[4] points out, differential forms were developed to work in metric-free spaces, and they maintain a formal distinction between n-forms and n-vectors that is not always helpful to physicists. For example, that momentum is sometimes best thought of as a vector but at other times appears more appropriately viewed as a 1-form, is a reflection of the fact that the spacetime in which it acts is a metric space. The metric establishes an isomorphism and hence an equivalence between dual spaces that is simply realized in geometric algebra but largely ignored in the formalism of differential forms. Thus, whereas in the formalism of differential forms, we are to picture the electromagnetic field as a spacetime "eggcrate" distinct from its dual 2-vector, and both distinct from their *Hodge* duals, it is sufficient to view the electromagnetic field as a spacetime bivector in geometric algebra. We can then concentrate on the physical action of the field as a generator of rotations and neglect its mathematical role as a 2-form that maps vectors to 1-forms. Furthermore, whereas Maxwell theory requires separate equations for the electromagnetic 2-form field and for its Hodge dual in differential forms, these are neatly expressed as a *single* relation in geometric algebra.

Unlike the old spacetime-tensor approach to Lorentz transformations, in the geometric algebra, the transformations appear in their spinorial form as elements of $SL(2,\mathbb{C})$. A particular spinor of importance in particle dynamics is the Lorentz transformation we call the particle eigenspinor: it relates the instantaneous rest frame of the particle with the lab frame and has the simple transformation properties of a spinor element while avoiding the messy bookkeeping of spinor indices.

These are some of the considerations that led to the choice of a covariant form of the geometric algebra $\mathcal{C}\ell_3$ for the formalism used in this text, but the full impact of the choice will only become clear as one works through the material. I hope readers will delight in the beauty of electrodynamics as revealed with the help of the formalism of Clifford's geometric algebra. Readers should be alert to new potential economies of thought and notation; I am sure there are many beautiful insights still to be discovered.

---

[2]See, for example, C. W. Misner *et al.* (1973).

[3]See, for example, R. Penrose and W. Rindler (1984 and 1986).

[4]J. Snygg (1997).

Efforts have been made to relate the formalism clearly to the more traditional tensor notation. Indeed, such efforts are responsible for much of the bulk and many of the more complicated parts of the text. However, they are not central to the algebraic development and its applications and can be skipped by students not concerned about the connection to older methods.

The exercises and problems, on the other hand, should *not* be skipped. Practice is invaluable for building understanding and confidence in unfamiliar territory. The exercises mainly illustrate concepts in the text with specific examples, whereas the problems frequently extend the development to new areas. The alert student will discover many additional opportunities to test ideas and general relations with simple examples.

Acknowledgments
Thanks are owed to many, but I especially want to thank several undergraduate and graduate classes of students in courses at the University of Windsor on electromagnetic waves and classical electrodynamics, who helped me refine and improve the presentation, and my colleague, John Huschilt, who provided invaluable support and proofreading expertise.

Chapter 1

Introduction

The theory of electromagnetic fields is a mature, well-developed area of physics. It deserves to be thoroughly understood in a modern notation because it is a model for modern field theories. The electromagnetic field equations and their solutions, particularly their wave solutions, together with the motion of charges and currents in such fields, are the subject matter of the present text.

When Albert Einstein (1879–1955) was working out the foundations of the theory of relativity during the first years of the twentieth century, he had to choose between two conflicting but highly successful theories. Isaac Newton (1642–1727) had developed laws based implicitly on the principle of universal time, and although they worked well in describing the dynamics of macroscopic neutral bodies, they gave Galilean transformations that conflicted with Maxwell's laws of electromagnetic phenomena (see problem 1). Einstein decided that nature sided more with James Clerk Maxwell (1831–1879), who happened to have died in the year of Einstein's birth. Another key person who died during that year was the 33-year-old British mathematician and philosopher, William Kingdon Clifford (1845–1879). His prescient work on algebraic descriptions of geometry helped open doors to the branch of mathematics now known as Clifford (or geometric) algebras.

Electromagnetic theory is inherently relativistic, but it is usually introduced in a nonrelativistic formalism. The common vector notation used was developed for vectors in 3-dimensional physical space rather than for the 4-dimensional space-time vectors of relativity. Advanced relativistic treatments fall largely into two groups: those that emphasize the physical content of familiar spatial vectors in a frame-dependent (noncovariant) approach, and those that sacrifice some intuition in a covariant indexed-tensor approach. A schizophrenic mix of both approaches is also common.

The view taken here is that much of the power, beauty, and symmetry of electromagnetic theory can only be fully appreciated in a covariant relativistic formulation, and that with the simple, geometric formulation of relativity provided by Clifford algebra, physical intuition can be enhanced, indexed tensors can be avoided, and relativity no longer poses a barrier to learning.

In this chapter, we prepare the ground for our study of classical electrodynamics by first discussing the perennially troublesome issue of electromagnetic units, and then by introducing the essential ideas of Clifford's geometric algebra, with emphasis in particular on the Pauli algebra, $\mathcal{C}\ell_3$.

1.1 Units

One of the practical difficulties with electromagnetic theory is the multiplicity of unit systems that are commonly used. The internationally accepted standard system is the SI (système international) or MKSA system, based on meters, kilograms, seconds, and amperes. It has been almost universally adopted in modern elementary physics texts. The older Gaussian system, based on the cgs units of centimeters, grams, seconds, and statcoulombs, was once popular in advanced courses on electromagnetic theory, but it now seems to be losing ground. Many research theoreticians prefer natural Heaviside-Lorentz units, which can be simply derived from SI units.

Most systems are directly related to each other by a single constant ϵ_0 (or $\mu_0 = c^2/\epsilon_0$ or $Z_0 = (c\epsilon_0)^{-1}$), provided we set the *speed of light* $c = 299\,792\,458$ m/s = 1. If speeds are to be dimensionless, then the units of time and distance are no longer independent but should, in fact, be the same. Thus, we might chose to measure time in seconds and distances in light-seconds (about 300 Mm), or we might prefer to measure distances in meters and time in "light-meters," meaning the time, roughly 3.33 564 ns, for light to travel one meter in vacuum. Such choices are supported by the fact that since 1983[1], the speed of light in a vacuum is now a *defined* quantity rather than something measured.

On the other hand, some unit systems are incompatible with the choice of $c = 1$. For example, in *atomic units*, $\hbar = 1 = e^2/(4\pi\epsilon_0)$, and consequently $c = 4\pi\epsilon_0\hbar c/e^2 \approx 137.036$, the reciprocal of the fine-structure constant. Furthermore, it is often useful to retain factors of c in order to treat electromagnetic waves

[1] Set by the Bureau International des Poids et Mesures in 1983. See E. R. Cohen and B. N. Taylor, *Rev. Mod. Phys.*, **59**, 1121 (1987).

passing through media with different indices of refraction (and hence different speeds of light). In this text, we attempt to keep all factors of c and ϵ_0.

The constant $4\pi\epsilon_0$ relates the fields to the charges that generate them. In SI units it is determined by

$$(4\pi\epsilon_0 c)^{-1} = \mu_0 c/(4\pi) = \dot{3}0\,\mathrm{Ohms}\,, \tag{1.1}$$

where $\dot{3}$ is short hand for

$$\dot{3} \equiv 2.99\,792\,458 = \frac{c}{10^8\,\mathrm{m/s}}\,. \tag{1.2}$$

However, in Gaussian units, $4\pi\epsilon_0$ is a dimensionless quantity with the value 1, whereas in natural Heaviside-Lorentz units, $\epsilon_0 = 1 = c$ so that electromagnetic equations in such units are easily obtained from those in SI by dropping all factors of ϵ_0 and c. The Gaussian and Heaviside-Lorentz systems may both be viewed as special cases of SI in which ϵ_0 is dimensionless, and therefore in which electromagnetic and mechanical units are related, so that as we will show below, there are mechanical equivalents of electromagnetic quantities such as the statcoulomb. It is important to note, however, that since $4\pi \neq 1$, the Gaussian convention is incompatible with the Heaviside-Lorentz one. In unconstrained SI, on the other hand, $(4\pi\epsilon_0 c)^{-1}$ has the dimensions of ohms, which is the ratio of a mechanical unit of power (watts) to an electromagnetic unit of current (amperes) squared, and the mechanical and electrical units can be viewed as independent of each other.

In all common unit systems, the electric field $\mathbf{E}$ is defined by the force $q\mathbf{E}$ it exerts on a charge q :

$$\mathbf{f} = q\mathbf{E}, \tag{1.3}$$

and the electric field $\mathbf{E}$ of a point charge Q at a position $\mathbf{r}$ relative to the charge can be written

$$\mathbf{E} = \frac{1}{4\pi\epsilon_0}\frac{Q}{r^2}\hat{\mathbf{r}}\,. \tag{1.4}$$

These relations are easily combined to give the Coulomb force law

$$\mathbf{f} = \frac{1}{4\pi\epsilon_0}\frac{qQ}{r^2}\hat{\mathbf{r}}\,, \tag{1.5}$$

which can be used to relate different units. Thus, when $Q = q$,

$$q^2 = 4\pi\epsilon_0 r^2 f\,, \tag{1.6}$$

where f is the magnitude of the force $\mathbf{f}$. In Gaussian units, where $4\pi\epsilon_0 = 1$, a statcoulomb is the magnitude of two charges that repel one another by one dyne when separated by one centimeter: 1 statcoul $= 1$ (dyne)$^{1/2}$ cm. This establishes a relationship between electromagnetic units (the statcoulomb) and mechanical units (dynes and centimeters) in the Gaussian system of units.

No analogous relation exists in the SI system. The charge in SI units equivalent to 1 statcoulomb must produce the same force for the same separation. Since 1 dyne = 10^{-5} N and 1 cm = 10^{-2} m, this charge is

$$q = (4\pi\epsilon_0 f)^{1/2}\, r = \left(\frac{10^{-5}\,\mathrm{N}}{\dot{3}0\, c\ \mathrm{W/A^2}}\right)^{1/2} 10^{-2}\,\mathrm{m} = \frac{1\ \mathrm{C}}{\dot{3}\times 10^9}\,. \tag{1.7}$$

We noted that 1 W = 1 N m/s and 1 A = 1 C/s. Therefore,

$$1\ \mathrm{C} \leftrightarrow \dot{3}\times 10^9\ \text{statcoul}, \tag{1.8}$$

although it is just as correct, and perhaps simpler, to remember that

$$1\mathrm{A\ m} \leftrightarrow 10\, c\ \text{statcoul}, \tag{1.9}$$

and it helps to remind us that whereas the statcoulomb is defined in terms of the force on static charges, the SI unit is based on the definition of current.

The coulomb is a very large unit for an electrostatic charge. Note, for example, that the force between two charges each of 1 coulomb, separated by 1 km, is about one ton (see problem 2). The magnitude of charges that repel one another with a force of 1 newton when separated by 1 m is

$$q = \frac{\sqrt{10}}{\dot{3}} \times 10^{-5}\,\mathrm{C}, \tag{1.10}$$

or roughly 10 microcoulomb.

The double-headed arrow can be read here as "corresponds to," and it cannot generally be replaced by an equality because there are additional constraints on the units in the Gaussian system that do not exist in SI. (An equality may be justified within the Gaussian system but not within SI, because the mechanical equivalents in the Gaussian system are generally not valid in SI.) Other units can be related by their metric definitions. Thus the definition of a volt as the potential difference that changes the energy of a charge of 1 C by 1 joule and that of the statvolt as being the potential difference that changes the energy of a statcoul by 1 erg is enough to show (see problem 3b)

$$1\ \text{statvolt} \leftrightarrow \dot{3}00\ \text{volt}. \tag{1.11}$$

It is not really correct to speak of *conversion factors* between units with different dimensions. Unfortunately, the same symbols are used in different systems of electromagnetic units for objects with different dimensions. This is especially a problem with the macroscopic (constitutive) equations of electromagnetic theory. In Gaussian units, for example, the electric field **E** and the magnetic field[2] **B** have the same dimensions whereas in SI units, **B** has the dimensions of **E** divided by a velocity. Thus in Gaussian units, 1 statV/cm = 1 gauss, but in SI units, 1 V/m = 1 tesla m/s, where *tesla* is the modern designation for weber/m^2. Problems with extra factors of m/s are largely avoided by the expediency of relating time and distance through the speed of light: $c = 1$, but as mentioned above, this is not always convenient. In any case, caution is still needed when dealing with the macroscopic fields **D** and **H**, which have the same dimensions as the microscopic fields **E** and **B** in Gaussian units, but which are expressed in terms of the sources instead of the forces in SI units.

Exercise 1.1. Starting with 1 V/m = 1 tesla m/s, show that 1 tesla = 1 N/(A m).

Heaviside-Lorentz units can be viewed as a special case of SI units when the mechanical and electromagnetic units are related by

$$\epsilon_0 = \frac{10^{-9}}{\dot{3}^2 4\pi} \frac{\mathrm{C}^2}{\mathrm{m}^2\mathrm{N}} = 1\,, \tag{1.12}$$

and thus by

$$1\,\mathrm{C} = \dot{3}\sqrt{\frac{4\pi}{10}} \times 10^5\,\mathrm{N}^{1/2}\mathrm{m} \approx 3.363 \times 10^5\,\mathrm{N}^{1/2}\mathrm{m}\,. \tag{1.13}$$

Since 1 Nm^2 = 10^9 dyne cm^2, the mechanical equivalent of statcoulombs would apparently imply 1 C = $\sqrt{4\pi} \times \dot{3} \times 10^9$ statcoul, which clearly disagrees with the physical equivalence (1.8) by a factor of $\sqrt{4\pi}$. This illustrates the incompatibility of the conventions mentioned above. The mechanical equivalents that can be established within the Gaussian or Heaviside-Lorentz unit systems are not valid in SI units, which provides the connecting link relating Gaussian and Heaviside-Lorentz conventions.

In summary, in the text we use SI units, complete with factors of c and ϵ_0. The main advantages of these units are their consistency with material in elementary

[2]As we will discuss in more detail below, **B** is more fundamental than **H** and will therefore be referred to as the magnetic *field* rather than by its older name as the "magnetic induction."

texts, their flexibility in allowing different relations between electromagnetic and mechanical units, and the straightforward computations they permit in common SI units. The principal disadvantage is that these factors clutter equations with relatively unimportant factors of c and ϵ_0. The equations are expressed in natural Heaviside-Lorentz units by setting $c = 1 = \epsilon_0$, and the microscopic equations can be rendered in Gaussian units with $c = 1$ by putting $c = 1 = 4\pi\epsilon_0$. While both Heaviside-Lorentz and Gaussian units systems may be viewed as special cases of SI units, they are not compatible with each other.

1.2 Electromagnetic Theory and Vector Algebra

Maxwell's celebrated work, *A Treatise on Electricity and Magnetism* (1873) contains his basic theory of electromagnetism as twenty equations in component form with electromagnetic potentials. The vector form of Maxwell's equations common today was only worked out years later by the English telegraph engineer Oliver Heaviside (1850-1925) and the German physicist Heinrich Hertz (1857-1894).[3] No physicist today would question the advantages of expressing the equations in vector form. By viewing the electric and magnetic fields and the current density as vectors rather than separate components in a given coordinate system, we gain an efficiency not only of notation but of thought.

The modern vector concept was introduced by the Irish mathematical physicist William Rowan Hamilton (1805-1865) to denote the spatial part of his *quaternions*.[4] Quaternions were the final product of Hamilton's fifteen-year search for an extension of complex numbers to three dimensions. It was well known that complex numbers could be thought of as points on a plane and could be used to express the ratio or rotation of vectors in a plane. Quaternions were Hamilton's *hypercomplex* numbers: they extended the power of complex analysis to three dimensions. Instead of a single unit imaginary, Hamilton used three, $\mathbf{i}, \mathbf{j}, \mathbf{k}$, with

$$\mathbf{i}^2 = \mathbf{j}^2 = \mathbf{k}^2 = -1 . \tag{1.14}$$

They are interrelated by

$$\mathbf{ij} = -\mathbf{ji} = \mathbf{k} \tag{1.15}$$

[3] An excellent history of the development of Maxwell theory can be found in B. J. Hunt (1991).

[4] W. R. Hamilton (1866). See also M. Klein (1972).

and cyclic permutations. A general quaternion Q is the sum of a scalar S and what Hamilton called a *vector* $\mathbf{V}$:

$$Q = S + \mathbf{V} \tag{1.16}$$
$$\mathbf{V} = V_1\mathbf{i} + V_2\mathbf{j} + V_3\mathbf{k}\,. \tag{1.17}$$

By virtue of the product rules for $\mathbf{i}, \mathbf{j}, \mathbf{k}$, the product of two of Hamilton's vectors, say $\mathbf{V}$ and $\mathbf{W}$, gives both the dot and cross products:

$$\begin{aligned}\mathbf{VW} &= (V_1\mathbf{i} + V_2\mathbf{j} + V_3\mathbf{k})\,(W_1\mathbf{i} + W_2\mathbf{j} + W_3\mathbf{k}) \\ &= (V_1W_1 + V_2W_2 + V_3W_3)\,(-1) + \\ &\quad (V_2W_3 - V_3W_2)\,\mathbf{i} + (V_3W_1 - V_1W_3)\,\mathbf{j} + (V_1W_2 - V_2W_1)\,\mathbf{k} \\ &= -\mathbf{V}\cdot\mathbf{W} + \mathbf{V}\times\mathbf{W}\,. \end{aligned} \tag{1.18}$$

Unfortunately, quaternions possess features that made them unattractive to physicists: they are the sum of incompatible elements, their product is not commutative, and the square of a real nonvanishing vector is less than zero. That was too much to swallow, at least all at once!

The vector concept itself was new and still unfamiliar in Maxwell's time, and in fact Maxwell's work contributed significantly to its development. Maxwell gave his electromagnetic equations in quaternion form as well as in component form. Although little further analytical use was made of these quaternionic relations, they well illustrated the conceptual and notational efficiency of vectors.

Josiah Willard Gibbs (1839-1903) in America and Oliver Heaviside in Britain separated the scalar and "vector" parts of the quaternion and built their simpler vector analysis on just part of the quaternion algebra. To manipulate vectors of three-dimensional space, they used no more than the vector space with inner product plus the vector part of the product of pure quaternions, which they called the vector (cross) product. It is defined as a (pseudo)vector perpendicular to the plane of the vectors being multiplied. While Gibbs and Heaviside still employed a sum operation among nonparallel vectors that was quite different from the familiar sum of real numbers, the physics community had been softened for such radical notions by its grudging acceptance of complex numbers. The hardest part for students to grasp today is undoubtedly the cross product, but it is required for treatments, for example, of angular momentum and torque. The common unit vectors $\mathbf{i}, \mathbf{j}, \mathbf{k}$ come from Hamilton's quaternions, but their relations in the quaternion algebra are usually ignored.

While the Gibbs-Heaviside vector analysis has served physics well, it is not useful in more than three dimensions, where there is no unique direction perpendicular to a plane. Although one sometimes refers to the "vector algebra" of Gibbs and Heaviside, the formalism does not constitute a formal algebra. To count as a vector algebra, a vector space with inner product must be supplemented by a product of vectors that obeys most of the familiar rules of multiplication of numbers. The product does not need to be commutative, but it should be associative and distributive over addition. The Gibbs-Heaviside vector cross product is not associative, since if $\mathbf{a}, \mathbf{b}, \mathbf{c}$ are arbitrary vectors in a three-dimensional Euclidean space, then generally

$$\mathbf{a}\times(\mathbf{b}\times\mathbf{c}) \neq (\mathbf{a}\times\mathbf{b})\times\mathbf{c}\,. \tag{1.19}$$

The product is also not commutative ($\mathbf{a}\times\mathbf{b} \neq \mathbf{b}\times\mathbf{a}$), but this is not a serious problem.

Here we introduce a simple alternative to the Gibbs-Heaviside approach: the *geometric algebra* of British mathematician William K. Clifford. It gives a transparent mathematical representation of rotations, reflections, and other geometrical operations, and it extends the power of complex analysis to spaces of any number of dimensions. Furthermore, an important subspaces of the vector space of the full geometric algebra of n-dimensional Euclidean space is seen to be an $(n+1)$-dimensional Minkowski space useful in relativity.

1.3 An Introduction to Clifford Algebras

1.3.1 Associative Vector Product

Clifford investigated the consequences of assuming the existence of an associative vector product. Building on the previous work of Hamilton and the German high-school teacher Hermann Günther Grassmann (1809-1877), he showed that in an n-dimensional Euclidean space $\mathbb{E}^n$, one could define a suitable vector product constrained by the following **fundamental axiom**:

Axiom 1.1 *For any real vector* $\mathbf{v}$ *in the space,*

$$\mathbf{v}\mathbf{v} = \mathbf{v}^2 = \mathbf{v}\cdot\mathbf{v} \tag{1.20}$$

Here, the dot indicates the usual dot (or scalar or inner) product of the space. It is surprising that this one axiom is sufficient to determine the structure of the resulting geometric (Clifford) algebra $\mathcal{C}\ell_n$, but that is the case.[5] Since $\mathbf{v}$ is any real vector, we can put $\mathbf{v} = \mathbf{u} + \mathbf{w}$. The axiom then expands to

$$\mathbf{u}^2 + \mathbf{uw} + \mathbf{wu} + \mathbf{w}^2 = \mathbf{u} \cdot \mathbf{u} + 2\mathbf{u} \cdot \mathbf{w} + \mathbf{w} \cdot \mathbf{w} . \quad (1.21)$$

Subtracting $\mathbf{u}^2 = \mathbf{u} \cdot \mathbf{u}$ and $\mathbf{w}^2 = \mathbf{w} \cdot \mathbf{w}$, we obtain the alternative form of the axiom:

Corollary 1.1 *Let* $\mathbf{u}$ *and* $\mathbf{w}$ *be any real vectors in the space. Then*

$$\mathbf{uw} + \mathbf{wu} = 2\mathbf{u} \cdot \mathbf{w} . \quad (1.22)$$

In particular, if $\{\mathbf{e}_1, \mathbf{e}_2, \dots\}$ is a basis of the vector space,

$$\mathbf{e}_j\mathbf{e}_k + \mathbf{e}_k\mathbf{e}_j = 2\eta_{jk} , \quad (1.23)$$

where $\eta_{jk} = \mathbf{e}_j \cdot \mathbf{e}_k$ is the *metric tensor* of the space in the given basis.

More generally, if $\mathbf{u}$ and $\mathbf{w}$ are perpendicular, the RHS of (1.22) vanishes and the vectors anticommute: if $\mathbf{u} \cdot \mathbf{w} = 0$, then $\mathbf{uw} = -\mathbf{wu}$. However, even if $\mathbf{u} \cdot \mathbf{w} = 0$, the product $\mathbf{uw}$ of nonvanishing vectors $\mathbf{u}$ and $\mathbf{w}$ cannot be zero because, by virtue of the associativity of the vector product, multiplication by $\mathbf{u}$ from the left and $\mathbf{w}$ from the right gives a positive scalar $\mathbf{u}^2\mathbf{w}^2 > 0$.

Thus, vectors generally do not commute with each other in the vector algebra. However, it is assumed that they all commute with scalars. It follows that collinear vectors $\mathbf{u}$ and $\lambda\mathbf{u}$, where λ is any scalar factor, commute. Higher-order products of vectors can be formed, and all such products and linear combinations of such products are elements of the vector algebra. These products, although not common in elementary physics instruction, play important geometric roles in physics. In advanced courses they are often represented by tensors, but the tensor notation tends to disguise their geometrical significance.

[5] We consider here Clifford algebras of Euclidean spaces. In particular, we ignore the exceptional case of a space of characteristic 2 (see P. Lounesto, 1997). Clifford himself considered only spaces of definite metric. Much of his work on Hamilton's quaternions and Grassmann algebras treated spaces in which the unit vectors square to -1, and this is the work often quoted and sometimes used to define Clifford algebras by mathematicians. In his 1876 abstract, Clifford proposed taking the square of unit vectors to be $+1$ and thus equal to the dot product of the vector with itself. This is the form usually adopted by physicists and the one followed here.

1.3.2 Overview

The essential feature of geometric algebras is an associative product of vectors, distributive over addition $[\mathbf{u}(\mathbf{a}+\mathbf{b}) = \mathbf{ua}+\mathbf{ub}]$, that allows one to construct inverses, square roots, and other functions of vectors much as you would with complex numbers. The main differences between such real associative algebras of vectors and fields of numbers are (1) that the product of noncollinear vectors is not abelian (*i.e.*, it does not commute): $\mathbf{uw} \neq \mathbf{wu}$ and (2) there exist zero-divisors,[6] that is, elements that are not zero themselves, but which when multiplied together equal zero: $\mathbf{uv} = 0,\ \mathbf{u} \neq 0,\ \mathbf{v} \neq 0$. An example is $\mathbf{u} = 1 + \mathbf{e}_1$, $\mathbf{v} = 1 - \mathbf{e}_1$. This feature endows the algebras of vectors with a richer structure than a field and enables such powerful mathematical tools as projectors (see problem 11).

The principal advantage of geometric algebra is that it provides economy and clarity of expression for geometrical relationships on which physics is intimately based. At the same time, the algebra avoids both the cumbersome component notation of tensors found in traditional relativistic treatments and the dependence on an assumed set of coordinates implied by such notation.

The chapters that follow provide many examples, but before proceeding, we first address the basic question of the existence of vector algebras as constrained above.

1.3.3 Matrix Representations

The concept of vector products other than the dot and cross products is unfamiliar to many. The sceptical scientist may well question the very existence of algebras built on such a product. After all, it is always possible to propose mathematical structures that don't exist or are internally inconsistent. The easiest way to see that a consistent Clifford algebra exists is to show that there are matrices with all the required properties of elements of the algebra.

The simplest nontrivial geometric algebra is $C\ell_2$, where, for example, we can represent the basis vectors $\mathbf{e}_1$ and $\mathbf{e}_2$ of $\mathbb{E}^2$ by the matrices

$$\underline{e}_1 = \begin{pmatrix} 0 & 1 \\ 1 & 0 \end{pmatrix}, \ \underline{e}_2 = \begin{pmatrix} 1 & 0 \\ 0 & -1 \end{pmatrix}. \tag{1.24}$$

[6] As shown by F. Georg Frobenius (1849–1917) in *J. für Math.* **84**, 1–63 (1878), the only exceptions are the three geometric algebras identified with the real and complex fields and the quaternions; they are *division* algebras, in which every nonzero element is invertible.

We use underlines here to indicate matrices and to distinguish *algebraic elements* from their *matrix representations*.[7] The unit scalar of the algebra $\mathcal{C}\ell_2$ is represented by the 2×2 identity matrix $\underline{1}$. One readily verifies the relation (1.22) for $\underline{\mathbf{e}}_1$ and $\underline{\mathbf{e}}_2$:

$$\begin{aligned} (\underline{\mathbf{e}}_1)^2 &= \underline{1} = (\underline{\mathbf{e}}_2)^2 \\ \underline{\mathbf{e}}_1\underline{\mathbf{e}}_2 &= -\underline{\mathbf{e}}_2\underline{\mathbf{e}}_1 \, . \end{aligned}$$

While the matrices $\underline{\mathbf{e}}_1$ and $\underline{\mathbf{e}}_2$ are both taken to be hermitian and square to $+\underline{1}$, their product is antihermitian:

$$\underline{\mathbf{e}}_1\underline{\mathbf{e}}_2 = \begin{pmatrix} 0 & -1 \\ 1 & 0 \end{pmatrix}, \tag{1.25}$$

and squares to $-\underline{1}$:

$$(\underline{\mathbf{e}}_1\underline{\mathbf{e}}_2)^2 = -\underline{1} \, . \tag{1.26}$$

Furthermore, $\underline{\mathbf{e}}_1\underline{\mathbf{e}}_2$ cannot be expressed as a linear combination of $\underline{1}$, $\underline{\mathbf{e}}_1$ and $\underline{\mathbf{e}}_2$.

Exercise 1.2. Prove the last assertion by showing that the matrix trace of $(\underline{\mathbf{e}}_1\underline{\mathbf{e}}_2)^2$ is -2, whereas the trace of $\underline{\mathbf{e}}_1\underline{\mathbf{e}}_2$ times any linear combination of $\underline{1}$, $\underline{\mathbf{e}}_1$ and $\underline{\mathbf{e}}_2$ vanishes.

The four matrices $\underline{1}$, $\underline{\mathbf{e}}_1$, $\underline{\mathbf{e}}_2$, and $\underline{\mathbf{e}}_1\underline{\mathbf{e}}_2$ are linearly independent and form a complete set of 2×2 matrices. Any product or real linear combination of $\underline{1}$, $\underline{\mathbf{e}}_1$, $\underline{\mathbf{e}}_2$, and $\underline{\mathbf{e}}_1\underline{\mathbf{e}}_2$ is a real 2×2 matrix and represents an element of $\mathcal{C}\ell_2$, and any real 2×2 matrix can be expanded in the basis $\{\underline{1}, \underline{\mathbf{e}}_1, \underline{\mathbf{e}}_2, \underline{\mathbf{e}}_1\underline{\mathbf{e}}_2\}$ using real coefficients. Thus, the elements of the geometric algebra generated by the basis vectors $\underline{\mathbf{e}}_1, \underline{\mathbf{e}}_2$ are elements in the larger four-dimensional linear space spanned by $\{\underline{1}, \underline{\mathbf{e}}_1, \underline{\mathbf{e}}_2, \underline{\mathbf{e}}_1\underline{\mathbf{e}}_2\}$. Evidently at least some Clifford algebras exist.

Physicists often use the matrix representation of the larger "Pauli algebra" $\mathcal{C}\ell_3$ in which the unit vectors $\mathbf{e}_k$, $k = 1, 2, 3$, are replaced by the Pauli spin matrices $\underline{\sigma}_k$, defined by

$$\underline{\sigma}_1 = \begin{pmatrix} 0 & 1 \\ 1 & 0 \end{pmatrix}, \underline{\sigma}_2 = \begin{pmatrix} 0 & -i \\ i & 0 \end{pmatrix}, \underline{\sigma}_3 = \begin{pmatrix} 1 & 0 \\ 0 & -1 \end{pmatrix}. \tag{1.27}$$

[7] Since the relations among the algebraic elements are also obeyed by matrices of its matrix representation, it is usually not necessary to distinguish the two. In future chapters, we will generally not bother to underline matrices.

The resulting matrix representation of a vector $\mathbf{v}$ of $\mathbb{E}^3$ is then $\sum_k v^k \underline{\sigma}_k$, which is commonly written as $\mathbf{v} \cdot \boldsymbol{\sigma}$ in misleading notation that makes it look more like a scalar than a vector. However, this matrix representation is not unique. In fact, any similarity transformation of all the basis elements is also a faithful matrix representation. Indeed, many possible matrix representations can be found for any Clifford algebra. The only property that all matrix representations share is the algebra itself. It is important to understand that matrix representations do exist and form internally consistent algebraic structures, but the actual matrix representation is not important; only its algebra matters. We can concentrate on the algebra and forget explicit representations.[8]

1.4 Geometric Interpretations

For applications in physics we need geometrical interpretations of the vector products. Any product $\mathbf{ab}$ of vectors $\mathbf{a}$ and $\mathbf{b}$ can be split into a symmetric (commuting) part and an antisymmetric (anticommuting) part:

$$\mathbf{ab} = \frac{1}{2}(\mathbf{ab}+\mathbf{ba}) + \frac{1}{2}(\mathbf{ab}-\mathbf{ba}) \tag{1.28}$$

$$\equiv \mathbf{a}\cdot\mathbf{b} + \mathbf{a}\wedge\mathbf{b}\,. \tag{1.29}$$

According to the corollary (1.22), the symmetric part is a scalar, namely the dot (inner) product of the two vectors. The second line of this relation defines the antisymmetric part to be the *wedge* (*exterior*) product of the vectors:

$$\mathbf{a}\wedge\mathbf{b} = \frac{1}{2}(\mathbf{ab}-\mathbf{ba})\,. \tag{1.30}$$

Although we make little explicit use of the wedge product in this text, it is useful to note that it is not a scalar: scalars always commute whereas $\mathbf{a}\wedge\mathbf{b}$ anticommutes with $\mathbf{a}$ and $\mathbf{b}$, and hence with any linear combination of $\mathbf{a}$ and $\mathbf{b}$, that is, with any vector in the plane defined by $\mathbf{a}$ and $\mathbf{b}$. For example,

$$\mathbf{a}(\mathbf{a}\wedge\mathbf{b}) = \frac{1}{2}\left(\mathbf{a}^2\mathbf{b}-\mathbf{aba}\right) = -(\mathbf{a}\wedge\mathbf{b})\,\mathbf{a}\,. \tag{1.31}$$

We have thus shown:

---

[8] However, for students new to Clifford algebras, it can be a useful crutch to think of the elements as matrices: the rules for taking products and linear combinations are the same.

Theorem 1.2 *The product* $\mathbf{ab}$ *of vectors* $\mathbf{a}$ *and* $\mathbf{b}$ *is a scalar if and only if* $\mathbf{a}$ *and* $\mathbf{b}$ *commute.*

It is also easy to prove another theorem:

Theorem 1.3 *Vectors* $\mathbf{a}$ *and* $\mathbf{b}$ *commute if and only if they are proportional and hence collinear.*

Proof: We saw above that collinear vectors commute. Assume now that $\mathbf{a}$ and $\mathbf{b}$ commute. Multiplying $\mathbf{ab} = \mathbf{ba}$ from one side by $\mathbf{a}$ and using the associativity of vector multiplication, we obtain

$$(\mathbf{aa})\,\mathbf{b} = \mathbf{a}\,(\mathbf{ab}) = \mathbf{a}\,(\mathbf{a}\cdot\mathbf{b})\ , \tag{1.32}$$

which states that $\mathbf{a}$ and $\mathbf{b}$ are collinear. ■

1.4.1 Bivectors

If the vectors $\mathbf{a}$ and $\mathbf{b}$ are not collinear, then in addition to the scalar part $\mathbf{a}\cdot\mathbf{b}$, the product $\mathbf{ab}$ has a new part: the antisymmetric *bivector*

$$\mathbf{ab} - \mathbf{a}\cdot\mathbf{b} = \mathbf{a}\wedge\mathbf{b} = -\mathbf{b}\wedge\mathbf{a},$$

which replaces the cross product in three dimensions and generally represents a patch of the *plane* containing **a** and **b**.

Just as any vector of $\mathbb{E}^n$ can be expanded in an orthonormal vector basis $\{\mathbf{e}_1, \mathbf{e}_2, \mathbf{e}_3, \ldots, \mathbf{e}_n\}$, so, too, any bivector in the algebra can be expanded as a linear combination of the unit *basis bivectors* $\mathbf{e}_1\mathbf{e}_2$, $\mathbf{e}_2\mathbf{e}_3$, ... , which represent patches of unit areas on linearly independent planes in $\mathbb{E}^n$. The Clifford algebra $\mathcal{C}\ell_n$ thus contains an $n\,(n-1)\,/2$-dimensional linear subspace of bivectors that is distinct from the n-dimensional linear space of vectors from which the algebra is generated. Since the basis vectors $\mathbf{e}_j$ are orthonormal,

$$\begin{aligned} \mathbf{e}_j\mathbf{e}_k &= \mathbf{e}_j\wedge\mathbf{e}_k = -\mathbf{e}_k\mathbf{e}_j \\ (\mathbf{e}_j\mathbf{e}_k)^2 &= \mathbf{e}_j\mathbf{e}_k\mathbf{e}_j\mathbf{e}_k = -\mathbf{e}_j^2\mathbf{e}_k^2 = -1\,. \end{aligned} \tag{1.33}$$

In $\mathbb{E}^3$, the bivector basis has three elements, and these can be associated with the unit quaternions. For example,

$$\mathbf{i} = \mathbf{e}_3\mathbf{e}_2 \tag{1.34}$$

$$\mathbf{j} = \mathbf{e}_1\mathbf{e}_3 \tag{1.35}$$

$$\mathbf{k} = \mathbf{e}_2\mathbf{e}_1\,. \tag{1.36}$$

These satisfy the rules (1.14) and (1.15), for example,

$$\mathbf{ij} = \mathbf{e}_3\mathbf{e}_2\mathbf{e}_1\mathbf{e}_3 = \mathbf{e}_2\mathbf{e}_1 = \mathbf{k}\,. \tag{1.37}$$

Because the bivector space in $\mathbb{E}^3$ happens to have the same dimensionality as the vector space, Hamilton could use the bivector symbols to denote vectors as well, but as Clifford himself pointed out, Hamilton's quaternions are really not vectors but operators (bivectors) on vectors.[9]

1.4.2 Higher-Grade Multivectors

Similarly, by expanding vectors in the basis $\{\mathbf{e}_j\}$, multiplying them together, and eliminating squares of basis vectors, one sees that any product of three vectors is the sum of a vector and a *trivector*, and that all trivectors, which represent three-dimensional volume elements, are linear combinations of the $n(n-1)(n-2)/3!$ basis trivectors $\mathbf{e}_i\mathbf{e}_j\mathbf{e}_k$ with $i < j < k$. Continuing in this way with higher-order products of arbitrary n-dimensional vectors, one finds that such products can be expressed in a finite *mulitvector* basis with

$$\binom{n}{m} = \frac{n!}{m!\,(n-m)!} \tag{1.38}$$

vectors of grade m, or more succinctly, m-vectors, and $m = 0, 1, 2, \cdots, n$. The m-vectors form a vector subspace of the algebra. The geometric algebra acts in the linear space formed from the direct sum of the m-vector subspaces with $0 \le m \le n$. Recalling the binomial expansion of $(1+x)^n$, we see that the dimension of the full linear space of the algebra is

$$1 + n + \frac{1}{2}n(n-1) + \cdots = \sum_{m=0}^{n}\binom{n}{m} = (1+1)^n = 2^n. \tag{1.39}$$

The linear space is more than an n-dimensional vector space since it contains not only vectors but also scalars as integral elements.

Exercise 1.3. Why can the trivector basis not contain both $\mathrm{e}_1\mathrm{e}_2\mathrm{e}_3$ and $\mathrm{e}_2\mathrm{e}_1\mathrm{e}_3$? [*Hint:* the basis elements of a linear space must be linearly independent.]

[9] Clifford called $\mathbf{i}, \mathbf{j}$, and $\mathbf{k}$ *versors* (a term coined by Hamilton) because, as explained below, they can rotate vectors by a right angle. See W. K. Clifford (1882).

Exercise 1.4. Find the size of the bivector and trivector spaces in $\mathcal{C}\ell_3$ and $\mathcal{C}\ell_4$ and give explicit bases for these spaces in terms of products of the basis vectors $\mathbf{e}_k$.

Exercise 1.5. Consider a two-dimensional pseudo-Euclidean space with a metric tensor η_{jk}, whose basis elements $\mathbf{e}_1, \mathbf{e}_2$ obey

$$\mathbf{e}_j\mathbf{e}_k + \mathbf{e}_k\mathbf{e}_j = 2\eta_{jk} \tag{1.40}$$

where $\eta_{11} = -\eta_{22} = 1$ and $\eta_{12} = \eta_{21} = 0$. Find the dimensionality of the bivector space and find the square $(\mathbf{e}_j\mathbf{e}_k)^2$ of any unit bivector $(j \neq k)$.

Since the binomial coefficients $\binom{n}{m}$ and $\binom{n}{n-m}$ are equal, the dimensionality of the m-vector subspace is the same as that for the $(n-m)$-vector subspace. The two subspaces are said to be *dual* to one another. In the Euclidean case with $n = 3$, all elements in the geometric algebra $\mathcal{C}\ell_3$ can be expressed as real linear combinations of a scalar, a vector, a bivector, and a trivector, which can also be referred to as elements of grade 0, 1, 2, and 3, respectively. A full basis for the real algebra of three-dimensional space thus comprises eight elements:

$$\{1, \mathbf{e}_j, \mathbf{e}_j\mathbf{e}_k, \mathbf{e}_1\mathbf{e}_2\mathbf{e}_3\}, \quad 1 \leq j < k \leq 3 . \tag{1.41}$$

The scalar and trivector subspaces are dual to each other; both are one-dimensional. The vector subspace is dual to the bivector subspace and *vice versa*; each has three dimensions.

1.4.3 Bivectors as Generators of Rotation

One of the most basic geometric concepts in an n-dimensional vector space that is missing from the traditional vector-space treatment is the *bivector*. It was interpreted above simply as a patch of plane. Its use as a generator of rotations is more important.

Consider the effect on basis vectors $\mathbf{e}_1$ and $\mathbf{e}_2$ of the unit bivector $\mathbf{e}_1\mathbf{e}_2$, multiplied from the right:

$$\mathbf{e}_1(\mathbf{e}_1\mathbf{e}_2) = \mathbf{e}_2 \tag{1.42}$$

$$\mathbf{e}_2(\mathbf{e}_1\mathbf{e}_2) = -\mathbf{e}_1 . \tag{1.43}$$

Both basis vectors are rotated in the same sense by a right angle in the $\mathbf{e}_1\mathbf{e}_2$ plane. Any vector in the $\mathbf{e}_1\mathbf{e}_2$ plane is a linear combination of $\mathbf{e}_1$ and $\mathbf{e}_2$, and it will be

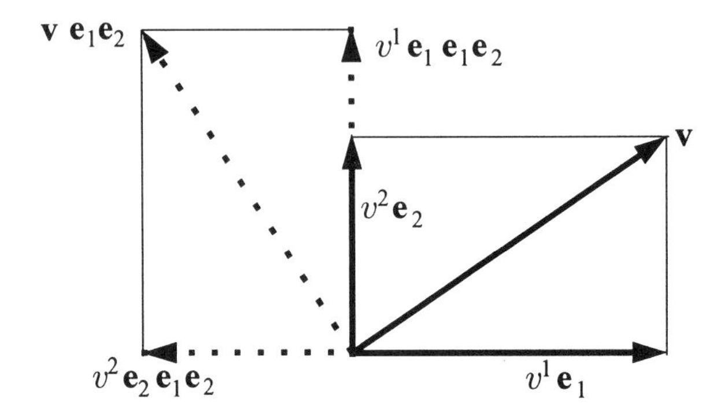

Fig. 1.1. The unit bivector $\mathbf{e}_1\mathbf{e}_2$ rotates any vector $\mathbf{v} = v^1\mathbf{e}_1 + v^2\mathbf{e}_2$ in the $\mathbf{e}_1\mathbf{e}_2$ plane by a right angle. The dashed vector and its components show the result of the rotation.

similarly rotated (see Fig. 1). Thus, for $\mathbf{v} = v^1\mathbf{e}_1 + v^2\mathbf{e}_2$,

$$\mathbf{v}\,(\mathbf{e}_1\mathbf{e}_2) = v^1\mathbf{e}_2 - v^2\mathbf{e}_1\,. \tag{1.44}$$

Exercise 1.6. Show that multiplication by the bivector from the left produces a rotation by 90 degrees in the opposite direction.

Exercise 1.7. Demonstrate that two-sided multiplication by the bivector $\mathbf{e}_1\mathbf{e}_2$, that is, the transformation

$$\mathbf{r} \rightarrow \mathbf{e}_1\mathbf{e}_2\mathbf{r}\mathbf{e}_1\mathbf{e}_2\,, \tag{1.45}$$

leaves $\mathbf{r} = \mathbf{e}_1$, $\mathbf{e}_2$, and linear combinations thereof, invariant, but if $\mathbf{r}$ is any vector perpendicular to both $\mathbf{e}_1$ and $\mathbf{e}_2$, the transformation changes its sign.

The last exercise shows how to reflect any vector $\mathbf{r}$ in the $\mathbf{e}_1\mathbf{e}_2$ plane:

$$\mathbf{r} \rightarrow (\mathbf{e}_1\mathbf{e}_2)\,\mathbf{r}\,(\mathbf{e}_1\mathbf{e}_2) = 2\left(r^1\mathbf{e}_1 + r^2\mathbf{e}_2\right) - \mathbf{r}\,. \tag{1.46}$$

All components of $\mathbf{r}$ perpendicular to the $\mathbf{e}_1\mathbf{e}_2$ plane reverse sign. A change of sign of the reflection operation changes it into a 180° rotation in the $\mathbf{e}_1\mathbf{e}_2$ plane:

$$\mathbf{r} \rightarrow (\mathbf{e}_2\mathbf{e}_1)\,\mathbf{r}\,(\mathbf{e}_1\mathbf{e}_2) = \mathbf{r} - 2\left(r^1\mathbf{e}_1 + r^2\mathbf{e}_2\right)\,. \tag{1.47}$$

Now, only components in the plane reverse sign. Rotations of vectors in the $\mathbf{e}_1\mathbf{e}_2$ plane by angles other than multiples of 90° can be effected by linear combinations of rotations by 0 and by 90°:

$$\mathbf{v} \rightarrow \mathbf{v}\left[\cos\theta + (\mathbf{e}_1\mathbf{e}_2)\sin\theta\right] . \tag{1.48}$$

By power series expansions, remembering that $(\mathbf{e}_1\mathbf{e}_2)^2 = -1$ [see (1.33)], one sees that the rotation operator can be represented by a single exponential function of the unit bivector $\mathbf{e}_1\mathbf{e}_2$:

$$\exp(\mathbf{e}_1\mathbf{e}_2\theta) = \cos\theta + (\mathbf{e}_1\mathbf{e}_2)\sin\theta . \tag{1.49}$$

This should look familiar. It is just the Euler relation for complex numbers: $\exp(i\theta) = \cos\theta + i\sin\theta$ with i replaced by the unit bivector $\mathbf{e}_1\mathbf{e}_2$. The proof of the two relations is the same: only the property $(\mathbf{e}_1\mathbf{e}_2)^2 = -1 = i^2$ plays a role.

Every vector in the $\mathbf{e}_1\mathbf{e}_2$ plane can be represented as the rotation $\mathbf{v} = r\mathbf{e}_1\exp(\mathbf{e}_1\mathbf{e}_2\theta)$ of a vector of the same length r aligned with $\mathbf{e}_1$. One can map every such vector into the scalar r times the rotation by itself:

$$\mathbf{v} \rightarrow \mathbf{e}_1\mathbf{v} = r\exp(\mathbf{e}_1\mathbf{e}_2\theta) . \tag{1.50}$$

These elements all commute and the bivector $\mathbf{e}_1\mathbf{e}_2$ in the space of such elements can be represented by i. In this way, vectors in the plane are represented by complex numbers. The Clifford-algebra approach thus helps explain why complex numbers are useful for handling vectors in two dimensions, but it does more: it also gives an extension to n dimensions. Complex analysis is a powerful tool for studying the geometry of two dimensions; Clifford analysis, using geometric algebra, is the extension of this tool to n dimensions.

Exercise 1.8. Prove that the unit bivector $\mathbf{e}_1\mathbf{e}_2$ is identical to the bivector formed from the product of the two rotated unit vectors $\mathbf{e}_1\exp(\mathbf{e}_1\mathbf{e}_2\theta)$ and $\mathbf{e}_2\exp(\mathbf{e}_1\mathbf{e}_2\theta)$. In other words, prove that bivectors are independent of rotations in the plane they specify.

The unit vector obtained from $\mathbf{e}_1$ by a rotation by θ in the $\mathbf{e}_1\mathbf{e}_2$ plane is

$$\begin{aligned} \mathbf{e}_\theta &\equiv \mathbf{e}_1\cos\theta + \mathbf{e}_2\sin\theta \\ &= \mathbf{e}_1(\cos\theta + \mathbf{e}_1\mathbf{e}_2\sin\theta) \\ &= \mathbf{e}_1\exp(\mathbf{e}_1\mathbf{e}_2\theta) , \end{aligned} \tag{1.51}$$

and this can be rearranged to give the rotation operator as a product of unit vectors

$$\mathbf{e}_1\mathbf{e}_\theta = \exp(\mathbf{e}_1\mathbf{e}_2\theta) = \cos\theta + (\mathbf{e}_1\mathbf{e}_2)\sin\theta\,, \tag{1.52}$$

or equivalently, as a product of unit bivectors:

$$\mathbf{e}_1\mathbf{e}_\theta = (\mathbf{e}_1\mathbf{e}_2)\,(\mathbf{e}_2\mathbf{e}_\theta)\ . \tag{1.53}$$

Note that $\mathbf{e}_1\mathbf{e}_\theta$ is not a pure bivector because $\mathbf{e}_1$ and $\mathbf{e}_\theta$ are not perpendicular.

1.4.4 Rotations in n Dimensions

The rotation of any vector $\mathbf{v} = v^1\mathbf{e}_1 + v^2\mathbf{e}_2$ by an angle θ in the $\mathbf{e}_1\mathbf{e}_2$ plane can be expressed in several ways:

$$\mathbf{v} \;\longrightarrow\; \mathbf{v}\exp(\mathbf{e}_1\mathbf{e}_2\theta) = \exp(\mathbf{e}_2\mathbf{e}_1\theta)\,\mathbf{v} \tag{1.54}$$

$$= \exp(\mathbf{e}_2\mathbf{e}_1\theta/2)\,\mathbf{v}\exp(\mathbf{e}_1\mathbf{e}_2\theta/2)\ . \tag{1.55}$$

The last form is particularly useful, since it also correctly describes the rotation of any element of $\mathcal{C}\ell_n$. To see this, note that it has the form of a similarity transformation

$$\mathbf{v} \to \mathbf{v}' = R\mathbf{v}R^{-1}, \tag{1.56}$$

where $R = \exp(\mathbf{e}_2\mathbf{e}_1\theta/2)$ and $R^{-1} = \exp(\mathbf{e}_1\mathbf{e}_2\theta/2)$. If x is any element such as a scalar or vector perpendicular to the plane that commutes with $\mathbf{e}_1\mathbf{e}_2$, it also commutes with R and is therefore invariant under the transformation

$$x \to RxR^{-1} = xRR^{-1} = x\,. \tag{1.57}$$

The only components of an arbitrary vector $\mathbf{v}$ of $\mathcal{C}\ell_n$ that are changed by the transformation are those in the plane of rotation, and that is precisely what we want. Now if two elements $\mathbf{v}$ and $\mathbf{w}$ of $\mathcal{C}\ell_n$ transform according to the similarity transformation (1.56), then so does their product

$$\mathbf{vw} \to \mathbf{v}'\mathbf{w}' = R\mathbf{v}R^{-1}R\mathbf{w}R^{-1} = R\mathbf{vw}R^{-1}, \tag{1.58}$$

and since the transformation is linear, so does any linear combination of such products. Thus, every element of $\mathcal{C}\ell_n$ is rotated by the same rule (1.56).

Exercise 1.9. Show that if the vector $\mathbf{v}$ has components perpendicular to the plane $\mathbf{e}_1\mathbf{e}_2$, then the expressions $\mathbf{v}\exp(\mathbf{e}_1\mathbf{e}_2\theta)$ and $\exp(\mathbf{e}_2\mathbf{e}_1\theta)\,\mathbf{v}$ are not pure vectors.

Rotations can also be expressed in terms of reflections. Using (1.53) but with the half angle, we can write

$$\begin{aligned} \mathbf{v}' &= \exp(\mathbf{e}_2\mathbf{e}_1\theta/2)\,\mathbf{v}\exp(\mathbf{e}_1\mathbf{e}_2\theta/2) \\ &= \left(\mathbf{e}_{\theta/2}\mathbf{e}_1\right)\mathbf{v}\left(\mathbf{e}_1\mathbf{e}_{\theta/2}\right) \\ &= \left(\mathbf{e}_{\theta/2}\mathbf{e}_3\right)\left(\mathbf{e}_3\mathbf{e}_1\right)\mathbf{v}\left(\mathbf{e}_1\mathbf{e}_3\right)\left(\mathbf{e}_3\mathbf{e}_{\theta/2}\right) \\ &= \left(\mathbf{e}_{\theta/2}\mathbf{e}_3\right)\left(\mathbf{e}_1\mathbf{e}_3\right)\mathbf{v}\left(\mathbf{e}_1\mathbf{e}_3\right)\left(\mathbf{e}_{\theta/2}\mathbf{e}_3\right) . \end{aligned} \tag{1.59}$$

The third line is obtained by inserting $(\mathbf{e}_3)^2 = 1$, and the last line demonstrates the equivalence of a rotation to a pair of successive reflections in intersecting planes. The angle of rotation is seen to be twice the angular opening between the planes. The bivector $\mathbf{e}_1\mathbf{e}_2$ is said to be the *generator* of the rotation.

Exercise 1.10. Verify by explicit multiplication that $\cos\theta/2 + \mathbf{e}_1\mathbf{e}_2\sin\theta/2$ and $\cos\theta/2 + \mathbf{e}_2\mathbf{e}_1\sin\theta/2$ are inverses of each other.

Exercise 1.11. Use the transformation (1.56) to show that bivectors are invariant under rotations in the plane of the bivector.

Exercise 1.12. Verify that the equations for reflections and rotations given in this and the previous subsections also work if the metric is negative definite: $(\mathbf{e}_j)^2 = -1,\ j = 1, 2, 3, \ldots, n$. However, check the sense of the rotation.

Exercise 1.13. Discuss how two full-length mirrors might be arranged so that someone looking in the mirrors will see herself rotated by 180° rather than simply reflected. What if she wants to see herself from the side, that is, rotated by 270°?

Exercise 1.14. Suppose one end of a periscope is rotated by 180° about the vertical axis so that someone looking into the lower port sees the scene behind. Show that the image in such a periscope is upside down.

In the expression $R = \exp(\mathbf{e}_2\mathbf{e}_1\theta/2)$, the rotation angle θ is a continuous scalar parameter, and the bivector $\mathbf{e}_2\mathbf{e}_1$ is called the *generator* of the rotation. The scalar factor multiplying the generator is actually the *half angle*, $\theta/2$. At least it is half the angle measured by the *length* of the angular arc traced by the end of a unit radius. However, it is equal to the full magnitude of the *area* swept out by a unit radius, and since it multiplies a unit bivector, it seems appropriate that it represent the measure of the angular area.

1.4.5 Other Clifford Algebras

Many generalizations have been made of the work by Clifford. A most important one is to vector spaces of nondefinite metric. The Clifford algebra based on a vector space of metric signature (p,q) is denoted $\mathcal{C}\ell_{p,q}$ and has p basis vectors that square to $+1$ and q that square to -1. Its structure is again determined by the axiom (1.20) and its corollary (1.22), which for basis vectors $\mathbf{e}_j$ takes the form

$$\mathbf{e}_j\mathbf{e}_k+\mathbf{e}_k\mathbf{e}_j = 2\eta_{jk}\,, \tag{1.60}$$

where η_{jk} are elements of the metric tensor of the space.

The algebra of the complex field is $\mathcal{C}\ell_{0,1}$ and Hamilton's quaternion algebra is $\mathcal{C}\ell_{0,2}$. The algebra of the 4×4 Dirac matrices, developed by P. A. M. Dirac (1928)[10] for his relativistic quantum theory of the electron, is $\mathcal{C}\ell_{1,3}$ or $\mathcal{C}\ell_{3,1}$ depending on the metric signature chosen. The usual Dirac theory uses matrix representations of $\mathcal{C}\ell_{1,3}$ (or $\mathcal{C}\ell_{3,1}$) over the complex field. Hestenes[11]. Greider[12], and others have shown how this can be reduced to $\mathcal{C}\ell_{1,3}$ (or $\mathcal{C}\ell_{3,1}$) over the reals. Many authors have used the "Dirac algebra" $\mathcal{C}\ell_{1,3}$ to describe relativistic phenomena. One must be warned, however, that in spaces of indefinite metric, some basis vectors, like Hamilton's unit quaternions, square to -1, and not all basis vectors can be made hermitian. To avoid problems of indefinite metrics, it may be useful to embed the algebra of an indefinite metric space in the algebra of a larger Euclidean space. For example, as we discuss further below, the complex field $\mathcal{C}\ell_{0,1}$ is the center (*i.e.*, commuting part) of $\mathcal{C}\ell_3$, and the quaternions $\mathcal{C}\ell_{0,2}$ are also just part (the *even* part, see below) of $\mathcal{C}\ell_3$. It may also be possible to establish an isomorphism to the algebra of a Euclidean space of the same size, just as $\mathcal{C}\ell_{1,3}$ can be shown to be isomorphic to $\mathcal{C}\ell_4$. We discuss below an alternative approach to Minkowski spacetime, one based on what are called paravectors of the smaller algebra $\mathcal{C}\ell_3$.

Mathematical models of physical phenomena are hardly ever unique. Isomorphisms (one-to-one correspondences) among mathematical models ensure mathematical equivalency, but not an equal ease of application. One's choice among

[10] P. A. M. Dirac, *Proc. Roy. Soc.* (London) **A117**, 610–624 (1928); **A118**, 351–361 (1928).

[11] D. Hestenes (1966) and *J. Math. Phys.* **16**, 556 (1975).

[12] K. R. Greider, *Phys. Rev. Letters* **44**, 1718 (1980); *Foundations of Phys.* **14**, 467 (1984).

different models may be based on personal preference. Simplicity and beauty are valuable guides, but they also largely reflect subjective judgements. Hamilton's quaternions possess the necessary mathematical structure to handle three-dimensional vectors and their rotations and reflections, but they do not extend directly to spaces of more than three dimensions, and as discussed above, the fact that Hamilton's vectors have negative squares has been viewed by many as ugly.

1.4.6 Volume Element and Hodge Duals

The product

$$\mathbf{e}_T \equiv \mathbf{e}_1 \mathbf{e}_2 \cdots \mathbf{e}_n \tag{1.61}$$

of all the orthonormal basis vectors in the Euclidean vector space $\mathbb{E}^n$ on which the geometric algebra is grounded is sometimes called the *canonical element* or *volume element* of the algebra. It is the basis element of the one-dimensional subspace of n-vectors, to which all of the highest-grade objects in the algebra are proportional. It either commutes or anticommutes with all vectors in the algebra according to whether n is odd or even, and for a Euclidean space, it squares to either $+$ or -1 according to whether $n\,(n-1)\,/2$ is even or odd (see problem 7), or equivalently according to whether or not[13] $n \bmod 4 < 2$.

The inverse of the volume element is

$$\mathbf{e}_T^{-1} = \mathbf{e}_n \mathbf{e}_{n-1} \cdots \mathbf{e}_1 \,. \tag{1.62}$$

If any m-vector V of the algebra $\mathcal{C}\ell_n$ is multiplied by $\mathbf{e}_T^{-1}$, then each m-vector basis element in V eliminates m of the n $\mathbf{e}_k$ factors in $\mathbf{e}_T^{-1}$, leaving an $(n-m)$-vector. As pointed out above, the $(n-m)$-vector subspace is dual to the m-vector subspace, and both subspaces have the same dimensionality. We define[14]

$$^*V \equiv V \mathbf{e}_T^{-1} \tag{1.63}$$

to be the *Hodge dual* of V. In $\mathcal{C}\ell_3$, the Hodge dual of a bivector is a vector. In particular, it is the vector cross product. For example,

$$^*\left(\mathbf{e}_1\mathbf{e}_2\right) = \left(\mathbf{e}_1\mathbf{e}_2\right)\left(\mathbf{e}_3\mathbf{e}_2\mathbf{e}_1\right) = \mathbf{e}_3 = \mathbf{e}_1 \times \mathbf{e}_2 \,. \tag{1.64}$$

[13] The expression $n \bmod 4$ means the remainder between 0 and 3 after n is divided by 4.

[14] This definition is equivalent within an overall sign to conventional definitions of Hodge duals that use fully antisymmetric Levi-Civita symbols. (The conventional definitions frequently differ from each other by the sign.) Our definition, which is the reverse of what Lounesto calls the *Clifford dual* (see Lounesto 1997) is not only easier to apply to m-vectors, it also works for *any element* of the algebra including a mixture of various grades.

It is this relation that allowed Gibbs and Heaviside to represent bivectors by cross products, and the same relation also let Hamilton use bivectors to represent vectors. Neither relation holds in spaces of dimension $n \neq 3$.

In the geometric algebra $\mathcal{C}\ell_3$ based on the physical space of three Euclidean dimensions, the canonical element $\mathbf{e}_1\mathbf{e}_2\mathbf{e}_3$ squares to -1 and commutes with all other elements. Its function is therefore just that of the imaginary i (or $-i$). We choose to set

$$\mathbf{e}_1\mathbf{e}_2\mathbf{e}_3 = i \tag{1.65}$$

for a right-handed coordinate system in $\mathbb{E}^3$: $\mathbf{e}_1 \times \mathbf{e}_2 = \mathbf{e}_3$. Geometrically, the element i in the algebra represents a *handedness*. For example, a twist from $\mathbf{e}_1$ to $\mathbf{e}_2$ is coupled to a thrust along $\mathbf{e}_3$. Under inversion of the basis vectors $\mathbf{e}_j \to -\mathbf{e}_j$, the coordinate system becomes left-handed and the product $\mathbf{e}_1\mathbf{e}_2\mathbf{e}_3 \to -i$.

1.4.7 Paravectors

The size of the $\mathcal{C}\ell_3$ basis can be cut in half if its natural complex structure is recognized. With the basis trivector associated with i, all trivectors of the algebra become imaginary scalars. The real scalars and the trivectors together form what is called the *center* of the algebra, that is, the subalgebra that commutes with all elements; it may be identified with the field of complex numbers, and as a result, the Pauli algebra is said to possess a natural complex structure. Furthermore, every bivector of $\mathcal{C}\ell_3$ can be written as an imaginary vector, for example

$$\mathbf{e}_1\mathbf{e}_2 = \mathbf{e}_1\mathbf{e}_2\mathbf{e}_3/\mathbf{e}_3 = i/\mathbf{e}_3 = i\mathbf{e}_3\,, \tag{1.66}$$

and more generally,

$$\mathbf{a} \wedge \mathbf{b} = i\mathbf{a} \times \mathbf{b}\,. \tag{1.67}$$

Having recognized the natural complex structure of $\mathcal{C}\ell_3$, we can now express every element as a complex linear combination of just four basis elements: $1, \mathbf{e}_1, \mathbf{e}_2, \mathbf{e}_3$. The geometric algebra generated by a real three-dimensional Euclidean space thus operates in a complex four-dimensional vector space that contains both the complex field and the original three-dimensional space as subspaces. Every element $p \in \mathcal{C}\ell_3$ is the sum of a scalar p_0 and a vector $\mathbf{p}$, both of which may be complex:

$$p = p_0 + \mathbf{p}\,. \tag{1.68}$$

It is called a *paravector*.[15]

More generally, in $C\ell_n$, the sum of a scalar and a vector is called a paravector, and the $(n+1)$-dimensional linear subspace of scalars plus vectors is the *paravector space*. We will see below that the paravector space of $C\ell_3$ provides a natural framework for relativity.

1.4.8 Involutions of $C\ell_n$

Complex conjugation is an important transformation in complex numbers. Similar conjugations, known as *involutions*, appear in Clifford algebras. Because the algebra has a richer structure than complex numbers, it has more than one conjugation. An involution of $C\ell_n$ is an invertible transformation within $C\ell_n$, that is a one-to-one mapping $p \to \bar{p} \in C\ell_n$ and whose square is unity: $\bar{p} \to \bar{\bar{p}} = p$.

The involution of *Clifford conjugation* or *spatial reversal* reverses the sign on the (complex) vector part of any paravector element:[16]

$$p = p_0 + \mathbf{p} \to \bar{p} = p_0 - \mathbf{p} \,. \tag{1.69}$$

Spatial reversal is extended to arbitrary elements of $C\ell_n$ by the rule that the spatial reversal of a product pq of elements reverses the order of multiplication[17]: $\overline{pq} = \bar{q}\bar{p}$. With this involution, any paravector element can be split into scalar and vector

[15] The name *paravector* for the sum of a scalar and a vector, is common in all Clifford algebras. It is due to J. G. Maks, Doctoral Dissertation, Technische Universiteit Delft (the Netherlands), 1989, p. 22. A general element of a Clifford algebra is called a *clifford number* or a *cliffor* (Jancewicz 1989).

[16] Mathematicians often refer to the 2×2 matrix $\underline{\bar{p}}$, representing the Clifford conjugate of an element p of $C\ell_3$ as the (matrix) *adjoint* of $\underline{p}$. This is not to be confused with the transpose matrix of complex-conjugated elements, which is also called the adjoint and which is equivalent to the hermitian conjugate for finite matrices. Rather, it is the element whose product with $\underline{p}$ is the *determinant* of $\underline{p}$:

$$\underline{p\bar{p}} = \det p \,.$$

See the following section.

[17] An invertible mapping of elements into elements that preserves the order of multiplication is called an *automorphism*. If the order is reversed, the mapping is said to be an *antiautomorphism*. Some mathematicians define involutions to be antiautomorphic, but the term is also often applied to automorphisms (that are their own inverse).

parts:

$$\begin{aligned} p &= \frac{1}{2}(p+\bar{p}) + \frac{1}{2}(p-\bar{p}) \\ &\equiv \langle p\rangle_S + \langle p\rangle_V , \end{aligned} \tag{1.70}$$

and by extension, $\langle p\rangle_S$ and $\langle p\rangle_V$ are respectively the (possibly complex) "scalar" and "vector" parts of an arbitrary element p. The scalar and vector parts of a product of elements may be thought of as generalization of the scalar (dot) and exterior (wedge) products of vectors to general elements of $\mathcal{C}\ell_n$:

$$\langle pq\rangle_S \equiv \frac{1}{2}(pq+\overline{pq}) , \tag{1.71}$$

$$\langle pq\rangle_V \equiv \frac{1}{2}(pq-\overline{pq}) . \tag{1.72}$$

The "scalar" part of an element in $\mathcal{C}\ell_n$ contains just the contributions of grades 0, 3, 4, 7, 8, 11, 12, ..., that is grades g with $g \bmod 4 = 0,3$, whereas the "vector part" contains grades 1, 2, 5, 6, 9, 10, ..., that is grades g with $g \bmod 4 = 1,2$.

Reversion or *hermitian conjugation,* $p \rightarrow p^\dagger$, is another important involution. It reverses the order of all vector products. In any matrix representation of $\mathcal{C}\ell_n$ in which the basis vectors of the $\mathbb{E}^n$ are represented by hermitian matrices, it corresponds to hermitian conjugation of the matrices. In $\mathcal{C}\ell_3$, it is like complex conjugation except that it reverses the order of products: $(pq)^\dagger \rightarrow q^\dagger p^\dagger$. It thus changes every i to $-i$ and reverses the order of multiplication. If an element $p \in \mathcal{C}\ell_3$ is expanded in the basis $\{\mathbf{e}_0 \equiv 1, \mathbf{e}_1, \mathbf{e}_2, \mathbf{e}_3\}$, then all basis elements can be taken to be hermitian and hermitian conjugation is effected by taking the complex conjugate (indicated here by an asterisk) of each coefficient:

$$p = p^\mu \mathbf{e}_\mu \rightarrow p^\dagger = p^{\mu *}\mathbf{e}_\mu , \tag{1.73}$$

where repeated indices in any term are summed over their allowed values; in (1.73), the sums are over $\mu = 0,1,2,3$. hermitian conjugation is used to split elements into *real* and *imaginary* parts:

$$\begin{aligned} p &= \frac{1}{2}(p+p^\dagger) + \frac{1}{2}(p-p^\dagger) \\ &\equiv \langle p\rangle_\Re + \langle p\rangle_\Im . \end{aligned} \tag{1.74}$$

In $\mathcal{C}\ell_n$, elements of grade g with $g \bmod 4 = 0,1$ are real, whereas those with $g \bmod 4 = 2,3$ are imaginary. By using both involutions of spatial reversal and hermitian conjugation, all four grades of a general element (real scalar, imaginary scalar, real vector, imaginary vector) can be isolated. For example, the real scalar part of p is

$$\langle p \rangle_{\Re S} = \frac{1}{4}\left(p + p^{\dagger} + \bar{p} + \bar{p}^{\dagger}\right) \tag{1.75}$$

The combination of spatial reversal and hermitian conjugation changes the signs of all vector factors without reversing their order of multiplication. It thus changes the sign of all *odd* elements, that is those constructed from products of an odd number of vectors. The *even* elements are unchanged. Any element $p \in C\ell_n$ can thus be split into even and odd parts:

$$\begin{aligned} p &= \frac{1}{2}\left(p + \bar{p}^{\dagger}\right) + \frac{1}{2}\left(p - \bar{p}^{\dagger}\right) \\ &\equiv \langle p \rangle_{+} + \langle p \rangle_{-} \end{aligned} \tag{1.76}$$

and the transformation $p \rightarrow \bar{p}^{\dagger}$ is called the *grade automorphism*. Physically it represents a *spatial inversion*.

The same subscript notation that we have applied here to *elements* can be used to specify parts of the *algebra*. For example, $\langle C\ell_n \rangle_{\Im S}$ is the part containing only imaginary scalars. Of the possible pieces of the Pauli algebra $C\ell_3$, $\langle C\ell_3 \rangle_{\Re S} \equiv \mathbb{R}$ (the real numbers), $\langle C\ell_3 \rangle_S \equiv \mathbb{C}$ (the complex numbers), and $\langle C\ell_3 \rangle_{+} \equiv \mathbb{H}$ (the real *quaternion algebra,* formed from the even elements of $C\ell_3$) are important subalgebras of $C\ell_3$: they are themselves algebras.

1.4.9 Inverses and the Metric of Paravector Space

A paravector $p \in C\ell_n$ has an inverse if and only if there exists another element q of $C\ell_n$ whose product with p is a nonzero scalar. The inverse is then $q/(pq)$ which is both a left and a right inverse, since $pq = \langle pq \rangle_S = \langle qp \rangle_S = qp$. If an involution exists that takes both $p \rightarrow q$ and $q \rightarrow p$, then the scalar pq may be interpreted as a square "length" of p and can be used to define the metric of the space. The procedure should be familiar from the field of complex numbers where to every number c another number c^* exists such that cc^* is a real scalar and the inverse of c is $c^{-1} = c^*/(cc^*)$. If p is a vector: $p = \mathbf{p}$, then q can also be taken equal to $\mathbf{p}$ and its inverse is $\mathbf{p}^{-1} = \mathbf{p}/(\mathbf{pp})$.

The product of a paravector $p \in C\ell_n$ with itself is generally not a scalar, but $p\bar{p}$ is, since $p\bar{p} = \overline{p\bar{p}}$ and is thus a scalar: $p\bar{p} = \langle p\bar{p} \rangle_S$. As long as $p\bar{p} \neq 0$, the inverse of p exists and is given by

$$p^{-1} = \bar{p} / (p\bar{p}) . \tag{1.77}$$

The metric determined by the "square length" ("square modulus")

$$p\bar{p} = p_0^2 - \mathbf{p}^2 \tag{1.78}$$

is just the *Minkowski spacetime metric*. It is given more explicitly by the scalar product of two distinct paravectors. If in the quadratic form $p\bar{p}$ we replace p by $p = q + r$, where q and r are paravectors of $C\ell_n$, the scalar

$$\frac{1}{2}(p\bar{p} - q\bar{q} - r\bar{r}) = \frac{1}{2}(q\bar{r} + r\bar{q}) = \langle q\bar{r}\rangle_S \tag{1.79}$$

is taken to be the scalar product of the paravectors q and r. In component form,

$$\langle q\bar{r}\rangle_S = q^\mu r^\nu \langle \mathbf{e}_\mu \bar{\mathbf{e}}_\nu \rangle_S = q^\mu r^\nu \eta_{\mu\nu} , \tag{1.80}$$

where

$$\langle \mathbf{e}_\mu \bar{\mathbf{e}}_\nu \rangle_S = \frac{1}{2}(\mathbf{e}_\mu \bar{\mathbf{e}}_\nu + \mathbf{e}_\nu \bar{\mathbf{e}}_\mu) = \eta_{\mu\nu} \tag{1.81}$$

are elements of the diagonal $(n+1) \times (n+1)$ paravector metric tensor η:

$$\eta_{\mu\nu} = \begin{cases} 1, & \mu = \nu = 0 \\ -1, & \mu = \nu > 0 \\ 0, & \mu \neq \nu \end{cases} . \tag{1.82}$$

An element p is said to be *unimodular* if $p\bar{p} = 1$, and two paravectors p, q are said to be (spacetime) *orthogonal* if and only if $\langle p\bar{q}\rangle_S = 0$.

The paravector metric may be compared to the metric in the underlying n-dimensional Euclidean space, which can be defined without the Clifford conjugate. Its matrix elements are the Kronecker delta

$$\langle \mathbf{e}_j \mathbf{e}_k \rangle_S = \delta_{jk} . \tag{1.83}$$

It is important to note that the "square length" of paravectors is not necessarily positive.

Exercise 1.15. Show that the spatial reversal $\bar{p}$ of any paravector $p \in C\ell_n$ is related to p by

$$p = -\frac{1}{n-1}\mathbf{e}_\mu \bar{p} \mathbf{e}^\mu ,$$

where the *reciprocal basis paravectors* $\mathbf{e}^\mu$ are defined by $\mathbf{e}^\mu = \eta^{\mu\nu}\mathbf{e}_\nu$ and $\eta^{\mu\nu}\eta_{\nu\sigma} = \delta^\mu{}_\sigma$. (Note: for an orthonormal basis, the elements $\eta_{\mu\nu}$ and $\eta^{\mu\nu}$ are equal, $\eta^{-1} = \eta$, and $\mathbf{e}^\mu = \bar{\mathbf{e}}_\mu$.)

The fact that the paravector metric of $C\ell_3$ is just the Minkowski metric of spacetime suggests that vectors in relativity can be represented by real elements (*real paravectors*) of $C\ell_3$. In the following chapter, we will use the paravector subspace of $C\ell_3$ to lay an efficient mathematical framework for the special theory of relativity. In this framework, spacetime vectors are modeled by paravectors of $C\ell_3$.[18] Examples include

- the proper velocity $u = \gamma + \mathbf{u} = \gamma\,(1 + \mathbf{v}/c)$, with $u\bar{u} = 1$,
- the momentum $p = E/c + \mathbf{p}$, with $p\bar{p} = m^2c^2$,
- the paravector potential $A = \phi/c + \mathbf{A}$,
- the current density $j = \rho c + \mathbf{j}$.

Spacetime vectors can be classified according to their "square lengths." There are three possibilities:

$$\begin{array}{ll} p\bar{p} > 0 & p \text{ is } \textit{timelike} \\ p\bar{p} = 0 & p \text{ is } \textit{lightlike} \text{ or } \textit{null} \\ p\bar{p} < 0 & p \text{ is } \textit{spacelike}. \end{array} \tag{1.84}$$

We will look at the significance of these labels in the next chapter.

Exercise 1.16. Show that every paravector orthogonal to a timelike paravector is itself spacelike, and that every lightlike paravector is orthogonal to itself.

Exercise 1.17. Consider the standard matrix representation of $C\ell_3$, in which $\mathbf{e}_0$ is replaced by the unit 2×2 matrix and $\mathbf{e}_k$ is replaced by the Pauli-spin matrix $\underline{\sigma}_k$ (see subsection 1.3.3). Find explicit matrix representations of the general element $p = p^\mu \mathbf{e}_\mu$, its Clifford conjugate $\bar{p}$, and the product $p\bar{p}$. Verify that $p\bar{p} = \det \underline{p}$.

---

[18]The term *spacetime vector* refers to the physical nature and transformation properties of the element, whereas *paravector* refers to the mathematical role of the object in the Pauli algebra, where it is the sum of a scalar and a spatial vector.

1.5 Problems

1. **Galilean transformations** relate frames moving at constant velocity with respect to one another when it is assumed that there exists a universal time valid for all frames. Under Galilean transformations between a frame A and another frame B in which A is moving with constant velocity $\mathbf{V}$, a velocity $\mathbf{v}_A$ in frame A is seen as

$$\mathbf{v}_B = \mathbf{v}_A + \mathbf{V}$$

in frame B.

   (a) Show, assuming the validity of Newton's second law and the Galilean transformation above of velocities, that forces are the same in frames A and B.

   (b) Use the result of part (a) together with the Lorentz-force law

$$\mathbf{f} = q\mathbf{E} + q\mathbf{v} \times \mathbf{B}$$

   to derive a Galilean transformation law for electric and magnetic fields.

   (c) Consider a very long line of charge at rest in frame A and use Gauss's law to find the fields. Let λ be the charge per unit length along the line. Now consider the source as seen in frame B when the relative velocity $\mathbf{V}$ lies along the line of charge. Find the fields as seen in frame B. Are they consistent with the transformation law found in (b)?

2. Show that the force between two charges, each of one coulomb and separated by one kilometer, is equivalent to the weight of about one ton.

3. A joule is 1 kg m^2 s^{-2} and an erg is 1 g cm^2 s^{-2}.

   (a) How many ergs are there in one joule?

   (b) Since an erg is 1 statvolt statcoul and a joule is 1 volt C, use part (a) to find the number of volts in a statvolt.

4. Prove that any vector in the $\mathbf{e}_1\mathbf{e}_2$ plane anticommutes with the unit bivector $\mathbf{e}_1\mathbf{e}_2$, whereas any vector perpendicular to the plane commutes with its bivector.

5. A corner cube is constructed by placing reflecting coatings on the interior surfaces of the three planes that meet at the corner of a cube. (The other half of the cube is removed.) Use the Pauli algebra to show that any ray of light incident on the corner cube and bouncing off all three surfaces will be inverted, that is, it emerges in the precise direction from which it came.

6. Show that the involutions defined by (1.69) and (1.73), when applied to products, do require a reversal in the order of multiplication.

7. Prove explicitly that the canonical element $\mathbf{e}_1\mathbf{e}_2\mathbf{e}_3$ in the Pauli algebra squares to -1 and that it commutes with all elements of the algebra. Extend the result to n-dimensional Euclidean space by calculating the square of the canonical element and determining its commutator with vectors in this more general case.

8. Use equations (1.28) and (1.67) to show that the product of two unit spatial vectors is a rotation element:

$$\hat{\mathbf{a}}\hat{\mathbf{b}} = \exp(i\boldsymbol{\theta}) ,$$

where $i\boldsymbol{\theta}$ is a bivector representing the plane containing $\hat{\mathbf{a}}$ and $\hat{\mathbf{b}}$, and θ is the angle that rotates $\hat{\mathbf{a}}$ into $\hat{\mathbf{b}}$.

9. Noncommuting rotations. Consider the rotation elements $R_1 = \exp(\mathbf{e}_2\mathbf{e}_3\pi/2)$ and $R_2 = \exp(\mathbf{e}_3\mathbf{e}_1\pi/2)$, which can be used in transformations of the form (1.56) to rotate vectors by 180° in the $\mathbf{e}_2\mathbf{e}_3$ and $\mathbf{e}_3\mathbf{e}_1$ planes, respectively.

   (a) Prove explicitly that $R_1R_2 = -R_2R_1$ and that this product represents a 180° rotation in the $\mathbf{e}_2\mathbf{e}_1$ plane.

   (b) Show that the element $\exp(\mathbf{e}_2\mathbf{e}_3\pi/2 + \mathbf{e}_3\mathbf{e}_1\pi/2)$ also represents a rotation, but that both the magnitude and the plane of the rotation are distinct from those of R_1R_2 and R_2R_1.

10. Show by induction that the hermitian conjugate, defined in $\mathcal{C}\ell_3$ by (1.73), when applied to any product of real spatial vectors, is the product in reverse order, for example

$$(\mathbf{abc})^\dagger = \mathbf{cba}$$

for any $\mathbf{a}, \mathbf{b}, \mathbf{c} \in \langle \mathcal{C}\ell_3 \rangle_{\Re V}$.

11. Projectors. Let $P = \frac{1}{2}(1 + \mathbf{n})$, where $\mathbf{n}$ is a real unit vector. The element P is called a projector. Prove the properties

$$\begin{aligned} P^2 = P &= P^\dagger \quad \text{(idempotent, real)} \\ P\bar{P} = \bar{P}P &= 0 \quad \text{(null, selforthogonal)} \\ P + \bar{P} &= 1 \quad \text{(complete).} \end{aligned} \tag{1.85}$$

Which of these properties are incompatible with a division algebra, in which every nonzero element is invertible? (Projectors are used extensively in Chapter 6.)

12. Spectral decomposition. Let $x = x^0 + \mathbf{x}$ be a real paravector.

    (a) Find a projector P of the form given in the last problem such that

    $$x = \lambda_+ P + \lambda_- \bar{P}\,, \tag{1.86}$$

    where $\lambda_\pm$ are scalars.

    (b) Find the "eigenvalues" $\lambda_\pm$.

    (c) From the properties (1.85), next show that any integer power x^n of x can be written

    $$x^n = \lambda_+^n P + \lambda_-^n \bar{P}\,. \tag{1.87}$$

    (d) Let $f(x)$ be an analytic function of its argument, with a series expansion in powers x^n . Prove the relation

    $$f(x) = f(\lambda_+)\,P + f(\lambda_-)\,\bar{P}\,, \tag{1.88}$$

    which is known as the spectral decomposition of $f(x)$. (This relation can be used to define even nonanalytic functions of paravectors. Further discussion can be found in Chapter 6.)

13. Rotating frame. Let r be a time-dependent position in a frame rotating at angular velocity $\boldsymbol{\omega}$ about an axis through the origin $r = 0$. The same position in the stationary frame is given by

    $$\mathbf{r}' = R\mathbf{r}R^{-1}, \tag{1.89}$$

    where $R = \exp(-i\boldsymbol{\omega}t/2)$. [Note that $\boldsymbol{\omega}$ is usually treated as a vector pointing along the axis of rotation. This is satisfactory in three-dimensional space, where there exists a unique axis for any rotation, but in spaces of higher dimensions, one should instead represent the angular velocity by a bivector in the plane of rotation. In three dimensions, the bivector is $i\boldsymbol{\omega}$.] Take derivatives with respect to the time t to relate the velocity and acceleration in the stationary frame to the position, velocity, and acceleration in the rotating frame. In particular, prove

    $$\ddot{\mathbf{r}}' = R\left[\ddot{\mathbf{r}} + 2\boldsymbol{\omega}\times\dot{\mathbf{r}} + \boldsymbol{\omega}\times(\boldsymbol{\omega}\times\mathbf{r})\right]R^{-1}. \tag{1.90}$$

14. Antisymmetric products. It is often convenient to consider antisymmetric ("outer") products of the paravector basis elements $\mathbf{e}_\mu,\ \mu = 0,1,2,3$ of $\mathcal{C}\ell_3$ and their spatial reversals. Such products, indicated by $\wedge\langle\ldots\rangle$, are defined to change sign under the

interchange of any two indices, thus $\wedge\langle...\mathbf{e}_\mu\bar{\mathbf{e}}_\nu...\rangle = -\wedge\langle...\mathbf{e}_\nu\bar{\mathbf{e}}_\mu...\rangle$. All permutations of the indices can be generated by successive applications of such interchanges. Prove

$$\begin{aligned}\wedge\langle\mathbf{e}_\mu\bar{\mathbf{e}}_\nu\rangle &= \langle\mathbf{e}_\mu\bar{\mathbf{e}}_\nu\rangle_V \\ \wedge\langle\mathbf{e}_\lambda\bar{\mathbf{e}}_\mu\mathbf{e}_\nu\rangle &= \langle\mathbf{e}_\lambda\bar{\mathbf{e}}_\mu\mathbf{e}_\nu\rangle_\Im\end{aligned}$$

Show that if the basis elements are all spatial vectors $i\langle\mathbf{e}_j\bar{\mathbf{e}}_k\mathbf{e}_l\rangle_- = \epsilon_{jkl}$, the fully antisymmetric (Levi-Civita) symbol in three dimensions, defined by

$$\epsilon_{123} = \epsilon_{231} = \epsilon_{312} = -\epsilon_{213} = -\epsilon_{132} = -\epsilon_{321} = 1.$$

Also note, but do not prove (too tedious!) the relations

$$\begin{aligned}\wedge\langle\mathbf{e}_\mu\bar{\mathbf{e}}_\nu\mathbf{e}_\rho\bar{\mathbf{e}}_\sigma\rangle &= \langle\mathbf{e}_\mu\bar{\mathbf{e}}_\nu\mathbf{e}_\rho\bar{\mathbf{e}}_\sigma\rangle_{\Im S} = -i\epsilon_{\mu\nu\rho\sigma} \\ \wedge\langle\mathbf{e}_\mu\bar{\mathbf{e}}_\nu\mathbf{e}_\rho\bar{\mathbf{e}}_\sigma\mathbf{e}_\lambda\rangle &= 0\end{aligned}$$

and $\epsilon_{\mu\nu\rho\sigma}$ is the fully antisymmetric (Levi-Civita) symbol in four dimensions with $\epsilon_{1230} = 1$.

15. Golden ratio and five-pointed stars. Let R be an invertible rotation operator of the form (1.49), with θ chosen such that

$$R^5 - 1 = 0\,. \tag{1.91}$$

(a) Factor (1.91) to show that unless R is the identity, it must satisfy

$$R^{-2} + R^{-1} + 1 + R + R^2 = 0\,, \tag{1.92}$$

and solve this to obtain exact numerical expressions for $\cos\theta$. There should be four distinct values of θ in the range $-\pi < \theta < \pi$ that satisfy (1.92).

(b) Let v be any vector in the e_1e_2 plane and interpret the expression

$$\mathbf{v}\left(R^{-2} + R^{-1} + 1 + R + R^2\right) = 0$$

as stating the existence of a closed five-sided polygon. Sketch the polygon, showing the directions of the vectors that make up the sides, for each of the four allowed angles θ.

(c) Prove that the ratio of the distance between nonadjacent vertices of the polygon to the length of a side is either the golden ratio g or its inverse, namely $g^{\pm1} = \frac{1}{2}\left(\sqrt{5}\pm1\right)$.

16. Fibonacci sequence. Define a paravector γ by

$$\gamma := gP - g^{-1}\bar{P}\,, \tag{1.93}$$

where P is the projector $P = \frac{1}{2}(1 + \mathbf{n})$ (see problem 11 above) with n a real unit vector, and g is the golden ratio (see previous problem).

(a) Prove that γ is antiunimodular, i.e. that $\gamma\bar{\gamma} = -1$ and that γn is unimodular.

(b) Find γ^2 and use it to show that $\gamma^{n+1} = \gamma^n + \gamma^{n-1}$.

(c) Demonstrate that any integer power γ^n of γ can be expressed in the form $\gamma^n = a_n\gamma + a_{n-1}$ where a_n is the nth member of the Fibonacci sequence, defined iteratively by

$$a_0 = 0,\ a_1 = 1,\ a_{n+1} = a_n + a_{n-1}\,. \tag{1.94}$$

(d) From the definition of γ, prove that

$$\gamma^n = g^n P + \left(-\frac{1}{g}\right)^n \bar{P}\,, \tag{1.95}$$

and use the limit of this expression for large n to show that

$$\lim_{n\to\infty} \frac{a_{n+1}}{a_n} = g\,. \tag{1.96}$$

Chapter 2

Special Relativity

This chapter uses the paravectors introduced in the previous chapter as an efficient mathematical tool for studying relativistic phenomena. Recall that Pauli algebra $\mathcal{C}\ell_3$ is the geometric algebra of real three-dimensional Euclidean space, but that all its elements can be viewed as paravectors in a complex four-dimensional space whose metric is that of Minkowski spacetime. In this chapter, we will see how the algebra provides a *covariant* (that is, frame-independent) description for special relativity while permitting any observer to use the familiar three-dimensional vectors appropriate to the chosen frame. We will also see how the paravector formalism in the algebra beautifully unites spatial rotations and velocity transformations in a single, simple formalism and provides a solid framework for a relativistic treatment of electrodynamics.

2.1 The Spacetime Manifold

Points in spacetime represent instantaneous *events,* such as the absorption or emission of a photon. They are located by a set of four coordinates. An observer, measuring points in the laboratory frame, will use the time $t \equiv x^0/c$ on the laboratory clock together with the three coordinates x^1, x^2, x^3 of the spatial position. The collection of the scalar $x^0 = ct$ plus the three spatial values labels the spacetime point (or "event"), and can be represented in many ways, for example as

$$x = \begin{pmatrix} x^0 \\ x^1 \\ x^2 \\ x^3 \end{pmatrix} = x^\mu \mathbf{e}_\mu, \quad \mathbf{e}_0 := \begin{pmatrix} 1 \\ 0 \\ 0 \\ 0 \end{pmatrix}, \mathbf{e}_1 := \begin{pmatrix} 0 \\ 1 \\ 0 \\ 0 \end{pmatrix}, \text{ etc.} \qquad (2.1)$$

Other possible representations of points include row vectors and square matrices. In special relativity, a point x is often thought of as a vector in four-dimensional spacetime with components x^μ relative to an origin of coordinates. However, if the point x is accelerating, then x does not transform as a spacetime vector. More generally, spacetime in the presence of material bodies is not flat, and the vector space appropriate at one point is generally distinct from that at another. Even though the concentration in this text is on electrodynamics in flat spacetime, we should avoid unnecessary concepts that conflict with extensions to curved spacetime. It is therefore preferable to view x simply as a *point* in spacetime, not as a vector.

Regardless of their representations, the basis vectors $\mathbf{e}_\mu$ at a point x are given by the partial derivative of the point x with respect to the coordinates:

$$\mathbf{e}_\mu = \frac{\partial x}{\partial x^\mu} \equiv \partial_\mu x \,. \tag{2.2}$$

This relation states in mathematical form the important defining property that the μth basis vector $\mathbf{e}_\mu$ points in the direction of the change in the point when only one coordinate value, namely x^μ is increased (this is, of course, just the meaning of the partial derivative $\partial_\mu = \partial/\partial x^\mu$). It is *tangent* to the curve formed in spacetime when all coordinates except x^μ are held fixed. It is a *tangent vector*, which generally exists not in the spacetime manifold itself, but in the *tangent space,* that is the space spanned by the tangent vectors (2.2), at the point x.

An observer would never confuse space and time coordinates. Displacements of positions have vector properties; they can be rotated or reflected in different planes; they can also be added, but they add as three-dimensional vectors, not as scalars. Time, on the other hand, is one dimensional. Time, unlike spatial position, is well ordered: given any two times, one must be greater than, equal to, or less than the other; there is no "sideways" direction to time. The concept of space and time united in a four-dimensional spacetime continuum becomes important when we try to relate observations that might be made in different frames, in particular, in frames in relative motion with respect to each other.

In the last chapter, we saw that the paravector space of $C\ell_3$ has the metric of Minkowski spacetime. We can, in fact, represent any spacetime vector $p = p^\mu \mathbf{e}_\mu$ by a paravector in which $\mathbf{e}_1, \mathbf{e}_2, \mathbf{e}_3$ are the spatial unit vectors in three-dimensional physical space and $\mathbf{e}_0$ is the unit scalar: $\mathbf{e}_0 = 1$. For example, a spacetime displacement

$$dx = cdt + d\mathbf{x} \tag{2.3}$$

is the sum of a scalar time displacement times c and a spatial displacement $d\mathbf{x}$. Any particle at rest in the laboratory will have a fixed spatial position $\mathbf{x}$ and change only its time coordinate. For such a particle $dx = cdt$, and therefore by integration the spacetime position of the particle is

$$x(t) = ct + \mathbf{x}(0) = ct\mathbf{e}_0 + \mathbf{x}(0) \; . \tag{2.4}$$

More generally, a moving particle in the laboratory follows a *path* x in spacetime with displacements in both time and space coordinates. The spatial displacement of a particle moving with constant velocity $\mathbf{v}$ is $d\mathbf{x} = \mathbf{v}dt$, and consequently its spacetime path x has the form

$$x(t) = (c + \mathbf{v})\, t + \mathbf{x}(0) \; . \tag{2.5}$$

The path (2.5) has been parametrized by the laboratory time t. This is the usual parameter choice in Newtonian physics, where time is universal, that is, the same in all frames. In relativity, however, time is just one coordinate $x^0 = ct$ of the spacetime position x and depends on the motion of the observed frame with respect to the observer. The transformation behavior of x is simpler if we choose an invariant scalar parameter, one that all observers can agree upon.

2.1.1 Proper Time and Velocity

The path of a particle moving less than the speed of light can be parametrized by the *proper time* τ, that is the time as observed in the rest frame of the particle (see Fig. 2.1).[1] Although $d\tau$ is just a time interval as seen in a particular frame, all observers can agree on the frame in which the particle is instantaneously at rest. Consequently, $d\tau$ can serve not only as one component of a spacetime vector $c^{-1}dx$ in a given frame, but also as the differential of an invariant scalar parameter τ. The tangent paravector to the path $x(\tau)$ then gives the proper-time rate of motion along the path and is called the *proper velocity* of the particle:

$$\begin{aligned} uc &= \frac{dx}{d\tau} = (\partial_\mu x)\frac{dx^\mu}{d\tau} = \mathbf{e}_\mu \frac{dx^\mu}{d\tau} && (2.6) \\ &= \frac{dt}{d\tau}\frac{dx}{dt} \equiv \gamma \frac{dx}{dt} \,, && (2.7) \end{aligned}$$

[1]The meaning of "proper" here is not "correct" but "own" or "self": the proper time of the particle is the particle's own time, that is, the time as would be measured by an observer moving with the particle.

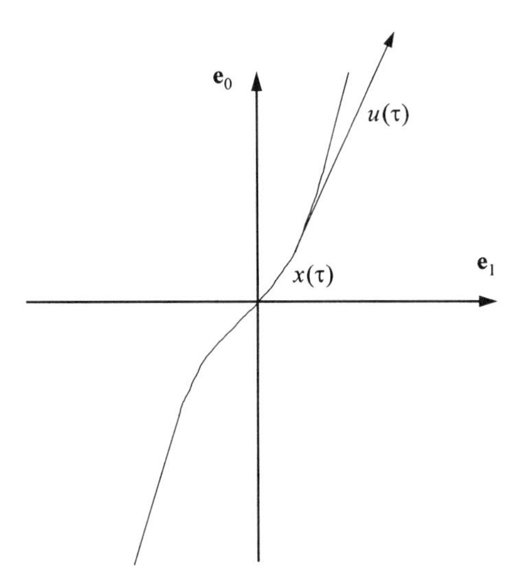

Fig. 2.1. The path or world line $x(\tau)$ of a massive particle moving in spacetime can be parametrized by the proper time τ. The tangent vector $u(\tau) = dx/d\tau$ is the proper velocity.

where $\gamma = u^0 = dt/d\tau$. When the particle is at rest, u is the scalar $u = \mathbf{e}_0 = 1$; in terms of the four-dimensional spacetime, it is a unit displacement along the time axis $\mathbf{e}_0$. Since its "square length" is $u\bar{u} = 1$, it is unimodular. More generally, the proper velocity uc of a particle moving in the lab lies in the tangent space of the spacetime manifold at x with components $u^\mu c = dx^\mu/d\tau$. The value of u for the moving particle is related to the rest-frame value 1 by a transformation that leaves $u\bar{u}$ invariant.[2]

Exercise 2.1. The coordinate velocity of a particle moving on the path (2.5) is the constant paravector $dx/dt = c + \mathbf{v}$. Show that the constant γ that normalizes $uc = \gamma dx/dt = \gamma(c + \mathbf{v})$ so that u is unimodular is given by

$$\gamma = \left(1 - v^2/c^2\right)^{-1/2}. \tag{2.8}$$

[2]Note that both the dimensionless spacetime vector u and its product uc with c can be referred to as the proper velocity of the particle. Both measure the same physical property of the particle; they differ only in the units in which this property is expressed. Physicists often choose units in which $c = 1$, but as pointed out in chapter 1, this choice conflicts with the conventional units used in some disciplines. Consequently, we retain all factors of c in this text.

The spacetime momentum of a particle of mass m and proper velocity uc is represented by the paravector

$$\begin{aligned} p &= mcu = E/c + \mathbf{p} \\ &= \frac{E}{c^2}\frac{dx}{dt}, \end{aligned} \tag{2.9}$$

where $E = \gamma mc^2$ is the particle energy and $\mathbf{p}$ is its spatial momentum. The "square length" $p\bar{p}$ gives the particle mass

$$p\bar{p} = E^2/c^2 - \mathbf{p}^2 = m^2c^2, \tag{2.10}$$

and it is the same as for the particle at rest.[3] Particles of zero mass move at the speed of light in every frame. A massless particle of velocity $\mathbf{v} = c\hat{\mathbf{v}}$ has a displacement $dx = (1 + \hat{\mathbf{v}})\, cdt$. We cannot parametrize its spacetime position by its proper time since it has no rest frame. As we will see below, as a particle approaches the speed of light, its proper time slows to a stop and its proper velocity is undefined, but its spacetime momentum is still expressed by the paravector

$$p = \frac{E}{c^2}\frac{dx}{dt}.$$

When $m = 0$, the spacetime momentum is the lightlike paravector

$$p = \frac{E}{c}(1 + \hat{\mathbf{v}}). \tag{2.11}$$

Example 2.2. **Compton effect.** The decrease in energy (and hence frequency) of electromagnetic radiation as it is scattered from slowly moving electrons is known as the Compton effect. It is a demonstration of the photon concept, since the observed frequency shift is predicted simply by the conservation of energy and momentum in the collision of an electron with a light quantum (see Fig. 2.2). Let $\hbar k$ and $\hbar k'$ be the spacetime photon momenta before and after collision, where $h = 2\pi\hbar$ is Planck's constant; they are the lightlike paravectors

$$\begin{aligned} \hbar k &= \frac{\hbar\omega}{c}\left(1 + \hat{\mathbf{k}}\right) \\ \hbar k' &= \frac{\hbar\omega'}{c}\left(1 + \hat{\mathbf{k}}'\right), \end{aligned} \tag{2.12}$$

[3] Older treatments of relativity often call γm the "relativistic mass" and refer to m as the "rest mass." For us, "mass" will always mean the Lorentz-invariant quantity m.

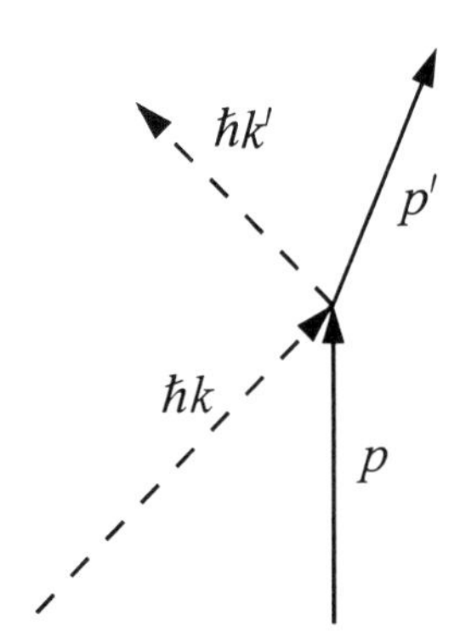

Fig. 2.2. Spacetime diagram for Compton scattering of a photon (with spacetime momentum $\hbar k$) and an electron (with spacetime momentum p).

where $\hat{\mathbf{k}}$, $\hat{\mathbf{k}}'$ are their respective propagation directions and ω, ω' are their angular frequencies. If p and p' are the electron momenta before and after collision, conservation of energy-momentum is expressed by

$$\hbar\left(k - k'\right) = p' - p\,. \tag{2.13}$$

In particular, the scalar part (times c) gives energy conservation

$$\hbar\left(\omega - \omega'\right) = E' - E\,. \tag{2.14}$$

We can also equate the square "lengths" of the paravectors (2.13). Since $k\bar{k} = 0 = k'\bar{k}'$ and $p\bar{p} = m^2c^2 = p'\bar{p}'$,

$$-\hbar^2\left(k'\bar{k} + k\bar{k}'\right) = 2m^2c^2 - p'\bar{p} - p\bar{p}', \tag{2.15}$$

and therefore

$$\begin{aligned} -\hbar^2\left\langle k'\bar{k}\right\rangle_S &= m^2c^2 - \left\langle p'\bar{p}\right\rangle_S \\ -\hbar^2\omega\omega'\left(1 - \hat{\mathbf{k}}\cdot\hat{\mathbf{k}}'\right) &= m^2c^4 - mc^3\left\langle p'\bar{u}\right\rangle_S\,. \end{aligned} \tag{2.16}$$

If the electron is initially at rest, $\bar{u} = 1$ and $E = mc^2$. Relations (2.14) and (2.16) then reduce to

$$\begin{aligned} E' - mc^2 &= \hbar\left(\omega - \omega'\right) \\ mc^2\left(E' - mc^2\right) &= \hbar^2\omega\omega'\left(1 - \hat{\mathbf{k}}\cdot\hat{\mathbf{k}}'\right). \end{aligned} \tag{2.17}$$

Eliminating E' and solving for ω', we find the frequency of the scattered radiation

$$\omega' = \frac{\omega}{1 + (\hbar\omega/mc^2)\left(1 - \hat{\mathbf{k}} \cdot \hat{\mathbf{k}}'\right)}. \tag{2.18}$$

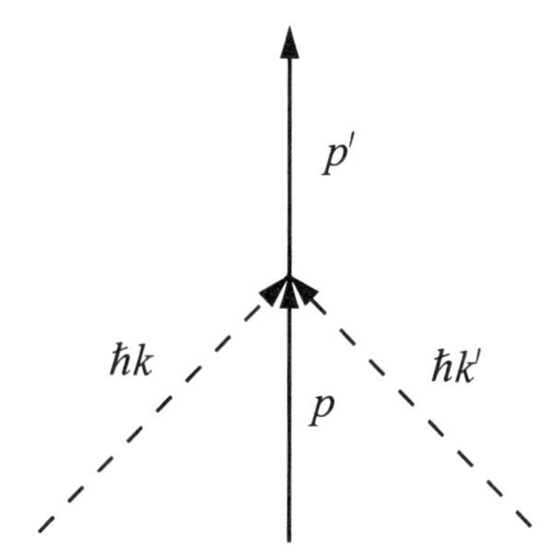

Fig. 2.3. Diagram for excitation of an atom or nucleus by photons. Energy conservation predicts that the atom or nucleus (spacetime momentum p) gains mass as a result of the excitation.

Example 2.3. **Internal energy and mass.** Let an atom or nucleus at rest undergo a two-photon excitation, where the two photons absorbed have equal energies ε and opposite spatial momenta. Since the net spatial momentum transferred to the atom or nucleus is zero, it remains at rest after the absorption. However, by energy conservation, it will have more internal energy and more mass after the absorption than before. If the mass before absorption is M, afterwards it must be larger by

$$M' - M = 2\varepsilon/c^2.$$

Example 2.4. **Threshold energies for particle creation.** High-energy physicists often search for new, short-lived particles in collisions of relativistic electrons and positrons moving with opposite spatial momenta. A frame in which the total spatial momentum vanishes, $\langle p_1 + p_2 \rangle_V = 0$, is called the *center-of-momentum frame.* The *threshold energy* is the minimum kinetic energy required to create the particle. At threshold in the center-of-momentum frame, all particles will be at rest after the collision. If the particle to be created has mass M, energy conservation then gives

$$p_1 + p_2 = (E_1 + E_2)/c = (m_1 + m_2 + M)\,c\,, \tag{2.19}$$

and the total kinetic energy in the two colliding beams must be at least

$$E_{\text{th}} = E_1 + E_2 - (m_1 + m_2)\, c^2 = Mc^2. \tag{2.20}$$

To create a 90 GeV particle, incident electrons and positrons, with equal rest energies of 0.511 MeV, must have initial kinetic energies of at least 45 GeV each. The threshold energy will generally be much higher in a frame where one of the incident particles is at rest (see next section).

With the help of partial derivatives ∂_μ, we can introduce a paravector gradient

$$\partial = \partial^\mu \mathbf{e}_\mu = c^{-1}\partial_t - \nabla. \tag{2.21}$$

Note the sign of ∇, which reflects the fact that the components[4] $\partial_j = \partial/\partial x^j$ of

$$\nabla = \mathbf{e}_1 \frac{\partial}{\partial x^1} + \mathbf{e}_2 \frac{\partial}{\partial x^2} + \mathbf{e}_3 \frac{\partial}{\partial x^3} \tag{2.22}$$

transform distinctly from vector components like ∂^j . The metric tensor η (see end of last chapter) relates spacetime vector components with upper ("contravariant") and lower ("covariant") indices

$$V_\mu = \eta_{\mu\nu} V^\nu. \tag{2.23}$$

This means that $\partial_j = -\partial^j$ and ensures, for example, that $\partial \left\langle k\bar{x}\right\rangle_S = k$, where k is a constant paravector.

Exercise 2.5. Find the paravector $\partial\,(xct + yz)$, where $ct = x^0$, $x = x^1$, $y = x^2$, and $z = x^3$. [Answer: $x\mathbf{e}_0 - ct\mathbf{e}_1 - z\mathbf{e}_2 - y\mathbf{e}_3$, with $\mathbf{e}_0 = 1$.]

Exercise 2.6. Express ∂ and $\left\langle k\bar{x}\right\rangle_S$ in terms of paravector components to verify that if k is a constant paravector, the gradient $\partial \left\langle k\bar{x}\right\rangle_S$ is the paravector

$$\partial \left\langle k\bar{x}\right\rangle_S = k\,. \tag{2.24}$$

2.2 Lorentz Transformations

Linear transformations that preserve the reality of spacetime vectors such as p and leave their "square lengths" $p\bar{p}$ invariant are called *Lorentz transformations*. In

[4] We follow the common convention that lower-case Latin letters ($j, k, \ldots$) range over integer values 1, 2, and 3, whereas lower-case Greek letters ($\mu, \nu, \ldots$) can take on values 0, 1, 2, and 3.

analogy with spatial rotations, which leave the square spatial length $\mathbf{p}^2$ unchanged, Lorentz transformations are often referred to as *spacetime rotations*. They include spatial rotations, boosts (velocity transformations), and any combination of boosts and rotations.

Since for any spacetime vector p, the product $p\bar{p}$ is invariant, the paravector properties *timelike, spacelike, null,* and *unimodular* (see chapter 1) are Lorentz-invariant concepts. Furthermore, if $p = q + r$, where q and r are also spacetime vectors, then since

$$p\bar{p} = q\bar{q} + r\bar{r} + 2\left\langle q\bar{r}\right\rangle_S , \tag{2.25}$$

then the scalar product

$$\left\langle q\bar{r}\right\rangle_S = \left\langle r\bar{q}\right\rangle_S = q^\mu r^\nu \eta_{\mu\nu} = q^\mu r_\mu = \left\langle \bar{q}r\right\rangle_S \tag{2.26}$$

is also Lorentz invariant. In particular, spacetime vectors that are orthogonal in one inertial frame are orthogonal in every such frame. Many problems in relativity can be solved without explicit Lorentz transformations, just by using Lorentz invariants.

Example 2.7. The threshold energy for particle creation is much higher in the frame of a target where one of the incident particles is initially at rest than in the center-of-momentum frame. To relate the two frames, we can use the Lorentz invariant $\left\langle p_1\bar{p}_2\right\rangle_S$

$$\left\langle p_1\bar{p}_2\right\rangle_S = E_1E_2/c^2 - \mathbf{p}_1\cdot\mathbf{p}_2 , \tag{2.27}$$

where p_1, p_2 are the spacetime momenta of the colliding particles. In the center-of-momentum frame, $\mathbf{p}_2 = -\mathbf{p}_1$ and since $p_1\bar{p}_1 = E_1^2/c^2 - \mathbf{p}_1^2 = m_1^2c^2$,

$$\begin{aligned} \left\langle p_1\bar{p}_2\right\rangle_S &= E_1E_2/c^2 + \mathbf{p}_1^2 \\ &= E_1\left(E_1+E_2\right)/c^2 - m_1^2c^2, \end{aligned} \tag{2.28}$$

whereas in the target frame with $\mathbf{p}_2 = 0$,

$$\left\langle p_1'\bar{p}_2'\right\rangle_S = E_1'm_2 . \tag{2.29}$$

Since the values of $\left\langle p_1\bar{p}_2\right\rangle_S$ are the same in every inertial frame, the threshold energy of the other particle (the projectile) is

$$E_1' - m_1c^2 = \left[E_1\left(E_1+E_2\right) - m_1\left(m_1+m_2\right)c^4\right]/\left(m_2c^2\right) \tag{2.30}$$

To create the 90 GeV particle by collisions of electrons with positrons, for which the masses and center-of-momentum energies are equal, $m_1 = m_2$ and $E_1 = E_2$, the threshold energy in the target frame is thus

$$\begin{aligned} E_1' - m_1c^2 &= 2\left[\left(\frac{E_1}{m_1c^2}\right)^2 - 1\right] m_1c^2 \\ &= 2\left[\left(\frac{M}{2m_1} + 1\right)^2 - 1\right] m_1c^2 \\ &= \left[\left(\frac{M}{2m_1} + 2\right)\right] Mc^2 \\ &\approx 8 \times 10^6 GeV\,, \end{aligned} \tag{2.31}$$

roughly $(M/2m_1) \approx$ 90,000 times larger than the 90GeV required for both particles together in the center-of-momentum frame.

2.2.1 Active and Passive Transformations

Lorentz transformations in special relativity relate observations made in different inertial frames. The observations themselves generally involve at least two inertial frames, at least implicitly: the inertial frame of the observer and that of the object (or system) observed. The transformations can be *active* and—for a given observer—change the inertial frame of the system being observed, or they can be *passive* and—for a given object—change the frame of the observer, or they can change both the observer and the observed system. Physically, it is only the *relative velocity and orientation* of the observed system (object frame) relative to the observer (the "lab" frame) that is significant, and an appropriate mathematical apparatus should reflect this dependence.

Any physical Lorentz transformation of a spacetime vector p (of an observed system relative to the lab frame of a given observer) into the spacetime vector p' can be written in the linear form (also called the *spinorial* form)

$$p \to p' = LpL^\dagger\,. \tag{2.32}$$

For example, if p is a property such as the spacetime momentum of an object when the system frame is at rest in the lab, then L gives the motion and orientation of the system frame with respect to the lab frame after the transformation. The condition $p\bar{p} = p'\bar{p}'$ implies that $\left|L\bar{L}\right| = 1$, but since the transformation (2.32) is unchanged if L is multiplied by a complex phase factor $e^{i\beta}$, we can choose $L\bar{L}$ to be real and positive:

$$L\bar{L} = 1. \tag{2.33}$$

Then L is *unimodular.* The transformation (2.32) is said to be a *restricted Lorentz transformation.* It is restricted to being physically realizable. As we will see below, restricted Lorentz transformations include all boosts, rotations, and their products. The larger group of unrestricted Lorentz transformations also include *improper* transformations such as reflections and spatial inversion, as well as *non-orthochronous* transformations which entail time reversal.[5] The full unrestricted homogenous group[6] can be represented by combining the physical transformations (2.32) with $p \to \bar{p}$ and $p \to -p$. We can therefore concentrate here on the restricted transformations, which are all *proper* and *orthochronous.* Since every L is *unimodular,* $\bar{L} = L^{-1}$ and the *inverse transformation* to (2.32) is

$$p' \to p = \bar{L}p'\bar{L}^{\dagger}\,. \tag{2.34}$$

In $\mathcal{C}\ell_3$, every unimodular element can be interpreted as a proper Lorentz transformation.[7] Such elements can be written in the exponential form

$$L = \exp[\mathbf{W}/2]\ ,\ \mathbf{W} = \mathbf{w} - i\boldsymbol{\theta}\,. \tag{2.35}$$

For an active transformation, if $\mathbf{w} = 0$, then L describes a pure rotation of the observed system in the plane $i\boldsymbol{\theta}$ (or equivalently, about the axis $\widehat{\boldsymbol{\theta}}$) by the angle θ, as discussed in chapter 1. On the other hand, if $\theta = 0$, then L is a boost, in other words a velocity transformation, with the *rapidity* (or *boost parameter*) $\mathbf{w}$. In an active transformation, L boosts the observed object(s), but in a passive transformation, the transformation L gives the motion and orientation of the initial observer relative to the final observer (in the new lab frame).

If L is real ($L = L^{\dagger}$), it represents a boost,

$$B = \exp(\mathbf{w}/2) = \cosh(w/2) + \hat{\mathbf{w}}\sinh(w/2) \tag{2.36}$$

---

[5] See also the discussion in the subsection on reflections in spacetime later in this chapter.

[6] A further extension to inhomogenous transformations such as $p \to LpL^{\dagger} + q$ allows one to include translations, as well. The group of inhomogenous Lorentz transformations is also known as the Poincaré group.

[7] It is often convenient to refer to the element L itself as "the transformation." Such terminology is not precise, since the transformation is not an element but the action (2.32) in which the element is a factor, but it is useful and hardly ever a source of confusion.

whereas if it is even ($L = \bar{L}^\dagger$), it represents a rotation:

$$R = \exp(-i\boldsymbol{\theta}/2) = \cos\theta/2 - i\widehat{\boldsymbol{\theta}}\sin\theta/2. \tag{2.37}$$

The exponential and other functions are related by power series expansions as shown in chapter 1. Every restricted Lorentz transformation L can be written as the product

$$L = BR \tag{2.38}$$

of a boost $B = \left(LL^\dagger\right)^{1/2}$ and a rotation $R = \langle L\rangle_+ / \langle B\rangle_S$.

Exercise 2.8. An element whose inverse is its spatial reversal is said to be *unimodular*: $L^{-1} = \bar{L}$, and an element whose inverse is its hermitian conjugate is said to be *unitary*: $R^{-1} = R^\dagger$. Show that any even, unimodular element is also unitary.

Exercise 2.9. Take the even part of $L = BR$ to show that since R is even and B is real, $\langle L\rangle_+ = \langle B\rangle_S R$. Then use this result to show that the rotation part R is determined entirely by the even part of L:

$$R = \frac{\langle L\rangle_+}{\sqrt{\langle L\rangle_+ \langle L\rangle_+^\dagger}}. \tag{2.39}$$

2.2.2 Linearity

The Lorentz transformation (2.32) is seen to be *linear*: for any spacetime vectors p, q and any scalar λ,

$$\begin{aligned} L\,(p+q)\,L^\dagger &= LpL^\dagger + LqL^\dagger \\ L\lambda pL^\dagger &= \lambda LpL^\dagger \end{aligned} \tag{2.40}$$

The linearity lets us simplify the calculation of any boost or rotation by breaking the paravector p to be transformed into parts that commute $\left(p^0 + \mathbf{p}_\parallel\right)$ and anticommute $\left(\mathbf{p}_\perp \equiv \mathbf{p} - \mathbf{p}_\parallel\right)$ with the direction $\hat{\mathbf{w}}$ of the boost B or the axis $\widehat{\boldsymbol{\theta}}$ of rotation R. Here $\mathbf{p}_\parallel$ is the part of $\mathbf{p}$ parallel to $\hat{\mathbf{w}}$ or $\widehat{\boldsymbol{\theta}}$. Thus,

$$\begin{aligned} BpB &= B^2\left(p^0 + \mathbf{p}_\parallel\right) + \mathbf{p}_\perp \\ RpR^\dagger &= \left(p^0 + \mathbf{p}_\parallel\right) + R^2\mathbf{p}_\perp \end{aligned} \tag{2.41}$$

and, for example,

$$\begin{aligned} R^2 \mathbf{p}_\perp &= e^{-i\boldsymbol{\theta}} \mathbf{p}_\perp = \left(\cos\theta - i\widehat{\boldsymbol{\theta}}\sin\theta\right)\mathbf{p}_\perp \\ &= \mathbf{p}_\perp \cos\theta + \widehat{\boldsymbol{\theta}} \times \mathbf{p}_\perp \sin\theta\,. \end{aligned} \tag{2.42}$$

While the vector component parallel to the axis of rotation is invariant under the rotation, it is the vector part of the paravector that is perpendicular to the boost direction that is unchanged by the boost.

By applying the involutions to the basic Lorentz transformation of spacetime vectors (2.32) , the equivalent transformations

$$\begin{aligned} \bar{p} &\to \bar{L}^\dagger \bar{p} \bar{L} \\ p^\dagger &\to L p^\dagger L^\dagger \\ \bar{p}^\dagger &\to \bar{L}^\dagger \bar{p}^\dagger \bar{L} \end{aligned} \tag{2.43}$$

can be derived.

Exercise 2.10. Note that the transformation of a paravector is the same as that of its hermitian conjugate, and prove that this is sufficient to guarantee that a transformed real paravector is real and a transformed imaginary paravector is imaginary. In other words, prove that the reality of a paravector (*i.e.*, whether it is real or not) is preserved under Lorentz transformations.

Exercise 2.11. Extend the results of the previous exercise to show that for any two elements x and L,

$$\begin{aligned} L\langle x\rangle_{\Re} L^\dagger &= \left\langle L x L^\dagger \right\rangle_{\Re} \\ L\langle x\rangle_{\Im} L^\dagger &= \left\langle L x L^\dagger \right\rangle_{\Im}\,. \end{aligned} \tag{2.44}$$

2.3 Spacetime Planes

Any two linearly independent real paravectors p and q span a two-dimensional subspace or *plane* of paravector space. Every real linear combination of p and q lies in the same plane. We show here how to represent that plane by products of p and q.

Transformations of products of paravector elements are determined by those of the paravectors themselves. The transformations are particularly simple for products in which paravectors and Clifford conjugates of paravectors alternate. The simplest nontrivial example is $p\bar{q}$ where p and q are real paravectors. The product transforms as

$$p\bar{q} \rightarrow LpL^{\dagger}\overline{LqL^{\dagger}} = LpL^{\dagger}\bar{L}^{\dagger}\bar{q}\bar{L} = Lp\bar{q}\bar{L}\,. \tag{2.45}$$

Higher-order alternating products $p\bar{q}r\cdots$ of paravectors transform either like paravectors $L\left(\cdots\right)L^{\dagger}$ or like biparavectors $L\left(\cdots\right)\bar{L}$. If L is a rotation, $L = R$, it is even: $\bar{R} = R^{\dagger}$, and the paravector and biparavector transformations are equivalent.

2.3.1 Biparavector Basis

The scalar part $\langle p\bar{q}\rangle_S$ of the paravector product $p\bar{q}$ should look familiar. It is the Lorentz-invariant scalar product of spacetime vectors. The vector part $\langle p\bar{q}\rangle_V$, called a *biparavector*, represents the plane in spacetime formed from p and q. In $C\ell_n$, the biparavectors are elements of a linear space spanned by the $n(n+1)/2$ distinct biparavector basis vectors $\langle \mathbf{e}_\mu \bar{\mathbf{e}}_\nu\rangle_V$

$$\langle p\bar{q}\rangle_V = p^{\mu}q^{\nu}\left\langle \mathbf{e}_\mu \bar{\mathbf{e}}_\nu\right\rangle_V \,. \tag{2.46}$$

There are up to $n(n+1)$ nonvanishing terms in the implicit sum on the RHS, with $\mu \neq \nu = 0, 1, \ldots, n$, but since

$$\left\langle \mathbf{e}_\mu \bar{\mathbf{e}}_\nu\right\rangle_V = \frac{1}{2}\left(\mathbf{e}_\mu \bar{\mathbf{e}}_\nu - \mathbf{e}_\nu \bar{\mathbf{e}}_\mu\right) = -\left\langle \mathbf{e}_\nu \bar{\mathbf{e}}_\mu\right\rangle_V \,, \tag{2.47}$$

only half of the basis biparavectors in the relation (2.46) are distinct. One can also write the biparavector in forms

$$\begin{aligned} \langle p\bar{q}\rangle_V &= \frac{1}{2}\left(p^{\mu}q^{\nu} - p^{\nu}q^{\mu}\right)\left\langle \mathbf{e}_\mu \bar{\mathbf{e}}_\nu\right\rangle_V \\ &= \sum_{\mu<\nu}\left(p^{\mu}q^{\nu} - p^{\nu}q^{\mu}\right)\mathbf{e}_\mu \bar{\mathbf{e}}_\nu \,, \end{aligned} \tag{2.48}$$

that emphasize the antisymmetric nature of the biparavector coefficient under an interchange of the indices. The second line is an expansion in terms of the $n(n+1)/2$ independent basis biparavectors, and we have noted that $\left\langle \mathbf{e}_\mu \bar{\mathbf{e}}_\nu\right\rangle_V = \mathbf{e}_\mu \bar{\mathbf{e}}_\nu$ if the paravector basis $\{\mathbf{e}_\mu\}$ is orthonormal and $\mu \neq \nu$. Any linear superposition of biparavectors is a biparavector. The biparavector space is identical to the direct sum of the vector and bivector spaces of $C\ell_n$. The total dimension of the biparavector space is the sum of the dimensions of the vector and bivector spaces

$$n + n(n-1)/2 = n(n+1)/2. \quad (2.49)$$

In $C\ell_3$, each biparavector has three real plus three imaginary vector components and can be written in the form $\mathbf{a} + i\mathbf{b}$, where $\mathbf{a}$ and $\mathbf{b}$ are real vectors.

Theorem 2.1 *If the biparavector* $\langle p\bar{q}\rangle_V = 0$, *the paravectors* p *and* q *are parallel.*

Proof: We expand $\langle p\bar{q}\rangle_V$ in basis biparavectors and use the independence of the basis elements to obtain

$$p^\mu q^\nu = p^\nu q^\mu \quad (2.50)$$

and hence

$$\frac{p^\mu}{p^\nu} = \frac{q^\mu}{q^\nu}. \quad (2.51)$$

This is the condition for p and q to be parallel. ■

The basis biparavectors $\langle \mathbf{e}_\mu \bar{\mathbf{e}}_\nu\rangle_V$ are orthonormal in the following sense:

$$\left\langle \langle \mathbf{e}_\alpha \bar{\mathbf{e}}_\beta\rangle_V \langle \mathbf{e}_\mu \bar{\mathbf{e}}_\nu\rangle_V \right\rangle_{\Re S} = \eta_{\alpha\nu}\eta_{\beta\mu} - \eta_{\alpha\mu}\eta_{\beta\nu}. \quad (2.52)$$

This is a special case of the following theorem:

Theorem 2.2 *Let* a, b, c, d *be real paravectors. Then*

$$\left\langle \langle a\bar{b}\rangle_V \langle c\bar{d}\rangle_V \right\rangle_{\Re S} = \langle \bar{b}c\rangle_S \langle \bar{a}d\rangle_S - \langle \bar{a}c\rangle_S \langle \bar{b}d\rangle_S. \quad (2.53)$$

Proof: Since $\left\langle \langle a\bar{b}\rangle_V \langle c\bar{d}\rangle_S \right\rangle_{\Re S} = 0$,

$$\begin{aligned}
\left\langle \langle a\bar{b}\rangle_V \langle c\bar{d}\rangle_V \right\rangle_{\Re S} &= \left\langle \langle a\bar{b}\rangle_V c\bar{d} \right\rangle_{\Re S} \\
&= \frac{1}{2}\left\langle a\bar{b}c\bar{d} - b\bar{a}c\bar{d} \right\rangle_{\Re S} \\
&= \frac{1}{2}\left\langle \bar{b}c\bar{d}a + \bar{b}c\bar{a}d - \bar{b}c\bar{a}d - b\bar{a}c\bar{d} \right\rangle_{\Re S} \\
&= \langle \bar{b}c\rangle_S \langle \bar{a}d\rangle_S - \frac{1}{2}\left\langle b\bar{c}a\bar{d} + b\bar{a}c\bar{d} \right\rangle_{\Re S} \\
&= \langle \bar{b}c\rangle_S \langle \bar{a}d\rangle_S - \langle \bar{a}c\rangle_S \langle \bar{b}d\rangle_S.
\end{aligned}$$

In the third line, the same term $\bar{b}c\bar{a}d$ was added and subtracted. In the fourth line, the bar-dagger conjugate (the grade automorphism) was taken of the third term (the real scalar part is unchanged by the conjugation). We also noted that the scalar product $\langle \bar{a}c \rangle_S = \langle c\bar{a} \rangle_S$ of real paravectors is real. ■

Exercise 2.12. Show that the result (2.53) is a generalization of the vector result

$$(\mathbf{a} \times \mathbf{b}) \cdot (\mathbf{c} \times \mathbf{d}) = (\mathbf{a} \cdot \mathbf{c})(\mathbf{b} \cdot \mathbf{d}) - (\mathbf{a} \cdot \mathbf{d})(\mathbf{b} \cdot \mathbf{c}) \ ,$$

which is obtained when all the scalar parts of a, b, c, d vanish.

Exercise 2.13. Show that Lorentz transformations generally mix the real and imaginary parts of products $p\bar{q}$ of paravectors p and q, but that the scalar and vector parts transform independently. In other words, the vector part $\langle p\bar{q} \rangle_V$ transforms to a spatial vector, and the scalar part $\langle p\bar{q} \rangle_S$ transforms to a scalar.

Exercise 2.14. Show generally for any two elements x and L of the algebra that

$$\begin{aligned} L \langle x \rangle_V \bar{L} &= \langle L x \bar{L} \rangle_V \\ L \langle x \rangle_S \bar{L} &= \langle L x \bar{L} \rangle_S \ . \end{aligned} \tag{2.54}$$

Let $\mathbf{F}$ be a biparavector

$$\mathbf{F} = \frac{1}{2} F^{\mu\nu} \langle \mathbf{e}_\mu \bar{\mathbf{e}}_\nu \rangle_V \ . \tag{2.55}$$

Since only that part of $F^{\mu\nu}$contributes that is antisymmetric in the indices μ, ν, we can take $F^{\mu\nu}$ itself to be antisymmetric:

$$F^{\mu\nu} = -F^{\nu\mu}. \tag{2.56}$$

By the orthonormality condition (2.52), the relation can be inverted to give the coefficients

$$F^{\mu\nu} = \eta^{\mu\beta} \eta^{\nu\alpha} \langle \mathbf{F} \langle \mathbf{e}_\alpha \bar{\mathbf{e}}_\beta \rangle_V \rangle_{\Re S} \ . \tag{2.57}$$

The $F^{\mu\nu}$ are elements of a *second-rank antisymmetric tensor*. Such tensor elements play prominent roles in traditional formulations of relativity, but we usually find it more efficient to work directly with the biparavectors $\mathbf{F}$.

2.3.2 Simple and Compound Biparavectors

It is not always possible to associate $\mathbf{F}$ with a single spacetime plane. Biparavectors of the form $\langle p\bar{q}\rangle_V$, where p and q are real paravectors, do represent single paravector planes, namely the plane containing all real linear combinations of p and q. They are called *simple biparavectors* and square to real scalars. Examples are $\mathbf{e}_1\bar{\mathbf{e}}_0$, which squares to $+1$, and $\mathbf{e}_1\bar{\mathbf{e}}_2$, which squares to -1. To show that $\langle p\bar{q}\rangle_V$ generally squares to a real scalar, consider first the case where at least one of the paravectors, say p, is not null. Let $q^{\perp}$ be the part of q that is orthogonal to p in paravector space:

$$q^{\perp} = q - \frac{\langle q\bar{p}\rangle_S}{\langle p\bar{p}\rangle_S} p . \tag{2.58}$$

Since it is easily verified that $\langle q^{\perp}\bar{p}\rangle_S = 0$, then

$$\langle q\bar{p}\rangle_V = q^{\perp}\bar{p} = -p\bar{q}^{\perp} = -\langle p\bar{q}\rangle_V ,$$

and consequently its square is real:

$$\langle q\bar{p}\rangle_V^2 = \langle p\bar{q}\rangle_V^2 = -q^{\perp}\bar{p}p\bar{q}^{\perp} = -\left(q^{\perp}\bar{q}^{\perp}\right)(p\bar{p}) .$$

In the case where both paravectors are null, $p\bar{p} = 0 = q\bar{q}$,

$$\begin{aligned} \langle q\bar{p}\rangle_V^2 &= \frac{1}{4}(q\bar{p} - p\bar{q})(q\bar{p} - p\bar{q}) \\ &= \frac{1}{4}(q\bar{p}q\bar{p} + p\bar{q}p\bar{q}) \\ &= \frac{1}{4}[q\bar{p}(q\bar{p} + p\bar{q}) + p\bar{q}(p\bar{q} + q\bar{p})] \\ &= \frac{1}{2}[q\bar{p}\langle q\bar{p}\rangle_S + p\bar{q}\langle p\bar{q}\rangle_S] \\ &= \langle q\bar{p}\rangle_S^2 , \end{aligned} \tag{2.59}$$

which is a nonnegative real scalar.

However, the square of a biparavector such as $\mathbf{F} = 2\mathbf{e}_0\bar{\mathbf{e}}_1 + \mathbf{e}_2\bar{\mathbf{e}}_3$ is not necessarily real:

$$\begin{aligned} \mathbf{F}^2 &= (2\mathbf{e}_0\bar{\mathbf{e}}_1 + \mathbf{e}_2\bar{\mathbf{e}}_3)^2 \\ &= 4 + 4\mathbf{e}_0\bar{\mathbf{e}}_1\mathbf{e}_2\bar{\mathbf{e}}_3 - 1 \\ &= 3 + 4\mathbf{e}_0\bar{\mathbf{e}}_1\mathbf{e}_2\bar{\mathbf{e}}_3 . \end{aligned}$$

The result is a scalar plus a multiparavector of grade 4, which in the Pauli algebra $C\ell_3$ is an imaginary scalar, namely $4i$. Biparavectors that are not simple are *compound.* They represent a superposition of commuting dual paravector planes with no common line of intersection. Their distinguishing feature is $\langle \mathbf{F}^2 \rangle_{\Im} \neq 0$. The concept of planes that do not intersect is not familiar from three-dimensional space, where any two distinct (nonparallel) planes intersect in a line and where the sum of any two planes is another plane, but in four or more dimensions, two planes (such as the paravector planes $\mathbf{e}_0\bar{\mathbf{e}}_1$ and $\mathbf{e}_2\bar{\mathbf{e}}_3$ considered above) can be orthogonal and not intersect. Their sum is not a single plane. We can also distinguish compound biparavectors according to the sign of $-i\langle \mathbf{F}^2 \rangle_{\Im}$. This appears related to a handedness, and we refer to a biparavector as *right handed* if $-i\langle \mathbf{F}^2 \rangle_{\Im} > 0$ and as *left handed* if $-i\langle \mathbf{F}^2 \rangle_{\Im} < 0$.

Exercise 2.15. Determine whether the following biparavectors are simple or compound. Express each simple paravector in the form $\langle p\bar{q}\rangle_V$, where p and q are real paravectors, and write each compound biparavector as a simple biparavector times a complex number:

$$\begin{gathered} \mathbf{e}_1\bar{\mathbf{e}}_2+\mathbf{e}_2\bar{\mathbf{e}}_3 \\ \mathbf{e}_1\bar{\mathbf{e}}_0+\mathbf{e}_2\bar{\mathbf{e}}_0 \\ \mathbf{e}_1\bar{\mathbf{e}}_2 + 2\mathbf{e}_2\bar{\mathbf{e}}_0 \\ \mathbf{e}_3\bar{\mathbf{e}}_0+2\mathbf{e}_1\bar{\mathbf{e}}_2 \\ \mathbf{e}_1\bar{\mathbf{e}}_0+\mathbf{e}_0\bar{\mathbf{e}}_3 - \mathbf{e}_1\bar{\mathbf{e}}_2 \,. \end{gathered}$$

The square of a biparavector transforms like a biparavector, and since it is a (complex) scalar, it is Lorentz invariant. One can therefore classify biparavectors in a Lorentz-invariant manner not only in terms of its imaginary part, which determines whether they are simple or compound,

$$\begin{aligned} \left\langle \mathbf{F}^2 \right\rangle_{\Im} &= 0, \quad \mathbf{F} \text{ is simple} \\ -i\left\langle \mathbf{F}^2 \right\rangle_{\Im} &> 0, \quad \mathbf{F} \text{ is compound and right handed} \\ -i\left\langle \mathbf{F}^2 \right\rangle_{\Im} &< 0, \quad \mathbf{F} \text{ is compound and left handed,} \end{aligned} \tag{2.60}$$

but also according to the sign of their real parts, similar to the classification of paravectors at the end of chapter 1, as timelike, spacelike, or lightlike

$$\begin{aligned} \left\langle \mathbf{F}^2 \right\rangle_{\Re} &> 0, \quad \mathbf{F} \text{ is timelike (or hyperbolic)} \\ \left\langle \mathbf{F}^2 \right\rangle_{\Re} &= 0, \quad \mathbf{F} \text{ is lightlike (or null)} \\ \left\langle \mathbf{F}^2 \right\rangle_{\Re} &< 0, \quad \mathbf{F} \text{ is spacelike (or elliptic).} \end{aligned} \tag{2.61}$$

The significance of these classification names will become clearer in the following sections.

Exercise 2.16. Let $k = k^0 \left(1 + \hat{\mathbf{k}}\right)$ be a null paravector and let $\mathbf{a}$ be a real vector perpendicular to the unit vector $\hat{\mathbf{k}}$. Show that $k\mathbf{a}$ is a simple null biparavector and that k is spacetime orthogonal to every paravector in the plane spanned by k and $\mathbf{a}$.

Exercise 2.17. Let p and q be real orthogonal paravectors: $\langle p\bar{q}\rangle_S = 0$. Show that the biparavector $\langle p\bar{q}\rangle_V$ is spacelike if and only if p and q are both spacelike. (Note that if two real paravectors are timelike, they cannot be orthogonal.)

It is also useful to classify relations between spacetime planes. Let $\mathbf{F}$ and $\mathbf{G}$ be simple biparavectors. If the product $\mathbf{FG}$ is a real scalar, then $\mathbf{F}$ and $\mathbf{G}$ are coplanar. On the other hand, if $\mathbf{FG}$ is an imaginary scalar, then $\mathbf{F}$ and $\mathbf{G}$ are *orthogonal spacetime planes.* In both cases, $\mathbf{F}$ and $\mathbf{G}$ commute.

2.4 Spacetime Rotations

An important biparavector is the exponent of the Lorentz transformation (2.35). We can expand

$$\mathbf{W} = \frac{1}{2} W^{\mu\nu} \left\langle \mathbf{e}_\mu \bar{\mathbf{e}}_\nu \right\rangle_V , \tag{2.62}$$

where (see 2.55) $W^{\mu\nu} = -W^{\nu\mu}$ and

$$\langle \mathbf{W} \rangle_{\Re} = \mathbf{w} \tag{2.63}$$
$$\langle \mathbf{W} \rangle_{\Im} = -i\boldsymbol{\theta} . \tag{2.64}$$

If $\mathbf{W}$ is a simple biparavector, the Lorentz transformation $L = \exp(\mathbf{W}/2)$ is also called simple and induces a *spacetime rotation* in the plane of $\mathbf{W}$. If the simple transformation is a boost, $\mathbf{W}$ is timelike; if it is a spatial rotation, $\mathbf{W}$ is spacelike. Although it is convenient to refer to both as spacetime rotations, the nature of boosts and spatial rotations are different in several respects. Whereas the spatial rotations form a compact subgroup of Lorentz transformations, boosts form a group only if restricted to one spatial direction (*i.e.*, to a single timelike plane in spacetime). The reason is that the product of two boosts in different directions is not a pure boost, but rather a boost together with a rotation.[8] Furthermore, even

[8]This is discussed more fully in the section on Thomas Precession in Chapt. 4.

the group of boosts in one direction is not compact: the boost parameter $|\mathbf{w}|$ can always be made larger and is not cyclic like the rotation parameter $|\boldsymbol{\theta}|$. As we saw in chapter 1, the basic action of a spatial bivector such as $\mathbf{e}_1\bar{\mathbf{e}}_2$ on a vector $\mathbf{v} = v^1\mathbf{e}_1 + v^2\mathbf{e}_2$ in the same plane is to rotate it by a right angle:

$$\mathbf{e}_1\bar{\mathbf{e}}_2\mathbf{v} = v^1\mathbf{e}_2 - v^2\mathbf{e}_1 , \tag{2.65}$$

but a timelike biparavector such as $\mathbf{e}_1\bar{\mathbf{e}}_0$ reflects a paravector $p = p^0\mathbf{e}_0 + p^1\mathbf{e}_1$ in the light cone direction $\mathbf{e}_0 + \mathbf{e}_1$:

$$\mathbf{e}_1\bar{\mathbf{e}}_0 p = p^0\mathbf{e}_1 + p^1\mathbf{e}_0 . \tag{2.66}$$

A boost $B = \exp(w\mathbf{e}_1\bar{\mathbf{e}}_0/2)$ of p therefore yields a paravector more closely aligned with the light cone

$$BpB = \exp(w\mathbf{e}_1\bar{\mathbf{e}}_0)\, p = \gamma p + \sqrt{\gamma^2 - 1}\mathbf{e}_1\bar{\mathbf{e}}_0 p , \tag{2.67}$$

where $\gamma = \cosh w$. Thus, "rotations" in timelike planes differ fundamentally from the more common spatial rotations.

2.4.1 Generators

The transformation L for a spacetime rotation in the spacetime plane of the unit biparavector $\langle \mathbf{e}_\mu\bar{\mathbf{e}}_\nu\rangle_V$ has the exponential form $\exp\left(\langle \mathbf{e}_\mu\bar{\mathbf{e}}_\nu\rangle_V W/2\right)$, where $W/2$ is a real measure of the amount of rotation. The unit biparavector $\langle \mathbf{e}_\mu\bar{\mathbf{e}}_\nu\rangle_V$ is called the (unit) *generator* of the spacetime rotation. These generators come in commuting pairs representing orthogonal spacetime planes, one timelike and one spacelike (see Fig. 2.4). The two generators of any pair are Hodge duals of each other (to within a sign) and anticommute with the other four unit biparavectors. The spacelike unit biparavectors square to -1, are purely imaginary, and generate rotations as seen in the last chapter. The timelike ones square to $+1$, are purely real, and generate boosts. All unit biparavectors are unitary and change sign under Clifford (bar) conjugation. They also obey characteristic commutation relations (see problem 20), and the symmetry group that relates them to each other is the direct product of SU(2) (spatial rotations) and U(1) (duality rotations).

Simple Lorentz transformations act in a single spacetime plane. They mix the paravector components in that plane and leave all paravectors in the orthogonal plane invariant. The biparavectors both of the plane in which the transformation acts and of the orthogonal plane are invariant.

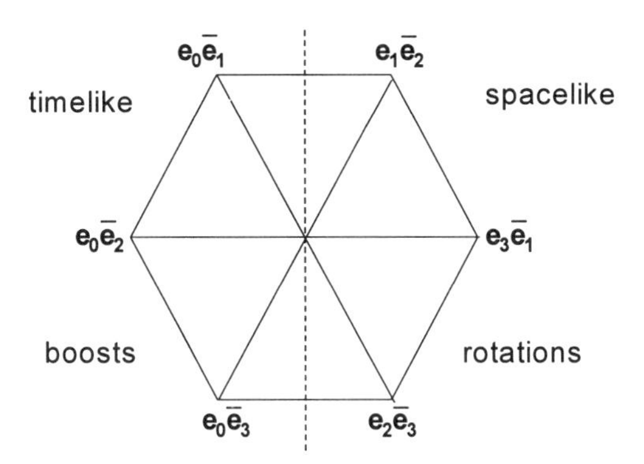

Fig. 2.4. The unit biparavectors that generate restricted Lorentz transformations come in commuting pairs, shown on opposite sides of the hexagon. The timelike ones to the left generate boosts, whereas the spacelike ones to the right generate spatial rotations.

Exercise 2.18. Show that a simple paravector plane $\langle p\bar{q}\rangle_V$ is invariant under spacetime rotations (Lorentz transformations) of p and q in the plane $\langle p\bar{q}\rangle_V$ or its dual.

The following theorem regarding paravector planes is useful in establishing the structure of the associated Lie algebra: (see problem 20).

Theorem 2.3

$$\frac{1}{2}\left[\langle p\bar{q}\rangle_V, \langle r\bar{s}\rangle_V\right] = \langle q\bar{r}\rangle_S \langle p\bar{s}\rangle_V + \langle p\bar{s}\rangle_S \langle q\bar{r}\rangle_V - \langle p\bar{r}\rangle_S \langle q\bar{s}\rangle_V - \langle q\bar{s}\rangle_S \langle p\bar{r}\rangle_V .$$

Its proof is left as a problem (see problem 19).

The ability to perform Lorentz transformations easily has great practical utility. It means we can work out a problem in the simplest frame, often the rest frame, and then transform the result to the desired frame, usually the lab frame. If the information sought can be expressed as a Lorentz invariant, the calculation is even simpler since no Lorentz transformation is needed. For example, the scalar product $\langle A\bar{u}\rangle_S$ of the vector potential A with the proper velocity is just its value in the rest frame where $u = 1$: $\langle A\bar{u}\rangle_S = \langle A_{rest}\rangle_S = \phi_{rest}/c$. Knowing the Lorentz transformations also helps us find the fundamental physical laws, since we expect such laws to look the same to all inertial observers. The first advantage allows us to determine the Lorentz-force law from the force on a charge at rest (see problem 11). The second will be used in the next chapter to determine the form of Maxwell's equations from the Coulomb law.

2.5 Proper Velocity

A simple example of a spacetime-vector transformation is the transformation of the dimensionless proper velocity from one frame to another. In the rest frame of the particle, its proper velocity is $u_{rest} = 1$. In the lab frame, it is $u = Lu_{rest}L^\dagger = LL^\dagger$. If L is written as the product of a boost with a rotation: $L = BR$, then u is independent of the rotation

$$u = LL^\dagger = B^2 = \exp(\mathbf{w})\,. \tag{2.68}$$

We can identify any boost with the timelike square root of a proper velocity:

$$B = u^{1/2}. \tag{2.69}$$

In an active transformation, B boosts an object at rest up to the proper velocity u, whereas in a passive transformation, the boost changes the observer's frame to one in which the original observer moves with proper velocity u. Since only the relative motion of the observer and the object is important, these transformations are fully equivalent. Note that boosts are commonly referred to by the proper velocity $u = B^2$ they induce in a rest frame (or by the corresponding coordinate velocity $\mathbf{v} = c\,\langle u\rangle_V / \langle u\rangle_S$, see below) rather than by the rapidity $\mathbf{w} = \ln u = \hat{\mathbf{u}}\ln(\gamma + |\mathbf{u}|)$.[9] Of course, at low velocities, $\mathbf{v}/c \approx \mathbf{u} \approx \mathbf{w}$.

Example 2.19. The boost

$$B = \frac{(9 + \mathbf{e}_1)}{4\sqrt{5}} \tag{2.70}$$

induces a proper velocity

$$\begin{aligned} u &= B^2 = \left[9^2 + 1 + 18\mathbf{e}_1\right]/80 \qquad (2.71)\\ &= (41 + 9\mathbf{e}_1)/40\,. \qquad (2.72) \end{aligned}$$

That is, B would boost a frame from rest to one moving with coordinate velocity $\mathbf{v} = (9/41)\,c\mathbf{e}_1 \approx 0.2195\,c\mathbf{e}_1$. This may be compared to $\mathbf{u} = (9/40)\,\mathbf{e}_1 = 0.225\,\mathbf{e}_1$ and to the boost parameter $\mathbf{w} = \ln u = \ln(50/40)\;\mathbf{e}_1 \approx 0.2231\,\mathbf{e}_1$.

---

[9] Functions of paravectors can be evaluated with the help of spectral decomposition, see problem 12 of Chapt. 1. Here, we can alternatively use

$$u = \gamma + \mathbf{u} = \cosh w + \hat{\mathbf{w}}\sinh w$$

to find $w = \ln(\cosh w + \sinh w)$.

Exercise 2.20. Find the proper velocity induced by the boost $B = (2 + \mathbf{e}_1)/\sqrt{3}$. Show that the corresponding coordinate velocity is $\mathbf{v} = 0.8\, c\mathbf{e}_1$.

Example 2.21. Use the unimodularity of $B = u^{1/2}$ to show that

$$B = \frac{u+1}{\sqrt{2(\gamma+1)}}. \tag{2.73}$$

Solution: Consider $B\left(B + \bar{B}\right) = 2\langle B\rangle_S B = B^2 + 1 = u + 1$. Take the scalar parts of both sides to find

$$2\langle B\rangle_S^2 = \langle u+1\rangle_S = \gamma + 1.$$

The result follows:

$$B = u^{1/2} = \frac{u+1}{2\langle B\rangle_S} = \frac{u+1}{\sqrt{2(\gamma+1)}}.$$

The Lorentz-invariant product $u\bar{u}$ is unity: $u\bar{u} = u_{rest}\bar{u}_{rest} = 1$, which also follows from (2.68) and the unimodularity (2.33) of proper Lorentz transformations. As a result, u itself is unimodular: $u^{-1} = \bar{u}$. It is said to be a *unit paravector.* Since the proper velocity u of an object is the tangent vector $u = dx/cd\tau$ of its world line $x(\tau)$, its scalar part γ gives the relative rates of coordinate (lab-frame) and rest-frame clocks

$$\langle u\rangle_S = \gamma = \frac{dt}{d\tau}, \tag{2.74}$$

sometimes called the *time-dilation factor*. For every second that clicks on the clock moving at constant proper velocity u, the lab clock advances by γ seconds. The moving clock appears to be running too slowly. The vector part of u is related to the coordinate velocity $\mathbf{v} = d\mathbf{x}/dt$ by

$$\langle u\rangle_V = \mathbf{u} = \frac{d\mathbf{x}}{cd\tau} = \frac{dt}{d\tau}\frac{d\mathbf{x}}{cdt} = \gamma\frac{\mathbf{v}}{c}, \tag{2.75}$$

which can be combined with the previous result to give

$$u = \gamma(1 + \mathbf{v}/c). \tag{2.76}$$

The unit size of u means that γ and $\mathbf{v}$ are related by

$$u\bar{u} = \gamma^2\left(1 - \mathbf{v}^2/c^2\right) = 1 \tag{2.77}$$

and thus $\gamma = (1 - \mathbf{v}^2/c^2)^{-1/2} \geq 1$. In the limit $v \to c$, γ becomes singular. Note that the invariant interval of proper time can also be expressed

$$(cd\tau)^2 = (ucd\tau)(\bar{u}cd\tau) = dx\, d\bar{x} \tag{2.78}$$

from which its Lorentz invariance is obvious. Alternatively, one can write

$$cd\tau = \langle cd\tau \rangle_S = \langle \bar{u}ucd\tau \rangle_S = \langle \bar{u}dx \rangle_S \tag{2.79}$$

which also shows that $d\tau$ is identical with the interval of time in the rest frame (see the end of the previous section).

Exercise 2.22. Expand $dx = \mathbf{e}_\mu dx^\mu$ in the scalar relation (2.78) and use $\eta_{\mu\nu} = \langle \mathbf{e}_\mu \bar{\mathbf{e}}_\nu \rangle_S$ to show that

$$(cd\tau)^2 = \eta_{\mu\nu} dx^\mu dx^\nu. \tag{2.80}$$

Consider two frames $\mathcal{S}$ and $\mathcal{S}'$ related by a pure boost $L = B$ so that any spacetime vector p in $\mathcal{S}$ is seen in $\mathcal{S}'$ to be

$$p' = LpL^\dagger = BpB\,. \tag{2.81}$$

The inverse transformation $\bar{L} = \bar{B}$ gives spacetime vectors in $\mathcal{S}'$ as seen by an observer in $\mathcal{S}$: $p = \bar{L}p'\bar{L}^\dagger = \bar{B}p'\bar{B}$. In particular, the proper velocity of $\mathcal{S}$ in $\mathcal{S}'$ is $u = B^2$ whereas that of $\mathcal{S}'$ in $\mathcal{S}$ is $\bar{\mathrm{B}}^2 = \bar{u}$.

A particle of mass m and proper velocity u has spacetime momentum

$$p = E/c + \mathbf{p} = mcu \tag{2.82}$$

where $E = \gamma mc^2$ is the rest energy plus the kinetic energy. At low speeds

$$E = mc^2\gamma = mc^2\left(1 - \mathbf{v}^2/c^2\right)^{-1/2} \approx mc^2 + \frac{1}{2}m\mathbf{v}^2 \tag{2.83}$$

and the kinetic energy takes its familiar nonrelativistic form. Note that γ may be considered the total energy of a particle in units of its rest energy mc^2, and that the velocity $\mathbf{v}$ is just the spatial momentum $\mathbf{p}$ in units of the energy divided by c^2:

$$\gamma = E/mc^2,\ \mathbf{v} = \mathbf{p}c^2/E\,. \tag{2.84}$$

Exercise 2.23. Find the next nonvanishing correction to the kinetic energy expression $\frac{1}{2}mv^2$ in the expansion (2.83).

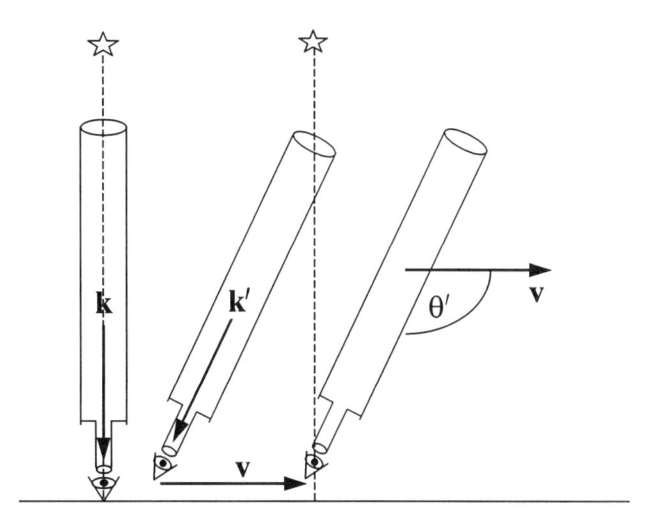

Fig. 2.5. Stellar aberration as seen from the frame of the sun. The observer at the telescope on the left is stationary in the frame of the sun and sees the star directly overhead, whereas the observer on the right is moving with the earth at velocity $\mathbf{v}$. In order for the light from the star to pass through the telescope and be seen, the moving observer on the right must tilt the telescope.

Example 2.24. **Stellar aberration and the headlight effect.** Let a photon from a distant star have the momentum paravector $\hbar k = \hbar\omega\left(1+\hat{\mathbf{k}}\right)/c$ in the inertial frame of the sun. Its momentum in the frame of the earth, which moves around the sun with velocity $\mathbf{v}$, is $\hbar$ times the wave paravector

$$\begin{aligned} k' &= LkL^{\dagger} = \bar{u}\left(k^0 + \mathbf{k}_{\parallel}\right) + \mathbf{k}_{\perp} \\ &= \gamma k^0 \left(1 - \mathbf{v}/c\right)\left(1 + \hat{\mathbf{v}}\cos\boldsymbol{\theta}\right) + \mathbf{k}_{\perp} \end{aligned} \tag{2.85}$$

where $u = \gamma\left(1 + \mathbf{v}/c\right)$ is the proper velocity of the earth in the frame of the sun, and

$$\mathbf{k}_{\parallel} = \mathbf{k} - \mathbf{k}_{\perp} = \mathbf{k}\cdot\hat{\mathbf{v}}\,\hat{\mathbf{v}} = k^0\hat{\mathbf{v}}\cos\theta\,. \tag{2.86}$$

Note that the transformation in this case is $L = B = \bar{u}^{1/2}$. It may be considered a *passive* transformation of the observer from the frame of the sun to the frame of the earth; it is equivalent to an *active* transformation of the photon from the frame of the earth to the frame of the sun, which has proper velocity $\bar{u}$ as seen from the earth. Let the angle between $\mathbf{k}'$ and $\mathbf{v}$ be θ'. Then

$$\begin{aligned} \langle k'\hat{\mathbf{v}}\rangle_S &= k'^0 \cos\theta' = \gamma k^0 (\cos\theta - v/c) \\ \langle k'\rangle_S &= k'^0 = \gamma k^0 [1 - (v/c)\cos\theta] \end{aligned} \tag{2.87}$$

and consequently

$$\cos\theta' = \frac{c\cos\theta - v}{c - v\cos\theta}. \tag{2.88}$$

For example, light that in the frame of the sun strikes the earth from directions perpendicular to its velocity $\mathbf{v}$ (*i.e.*, $\cos\theta = 0$) appears to have a component in the direction of $-\mathbf{v}$ as seen from the earth. The result might be called the **snowflake effect**, after a similar effect that is well known to anyone who has driven through a snow storm: the snow flakes seem to strike the car predominantly head-on. (The analogy is not fully obvious, because the speed of light—but not the speed of the snowflakes—is the same in every frame. Nevertheless, one can still combine the velocity of the photon in the frame of the sun and the velocity $-\mathbf{v}$ of the sun as seen from earth.) From the earth, the position of the stars is thereby shifted toward the $+\mathbf{v}$ direction by the motion of the earth. The effect is known as **stellar aberration** (see Fig. 2.5).
Perhaps less intuitive is the equivalent transformation viewed as an active transformation: recall that the transformation above is equivalent to an active boost of the light source from the earth to the frame of the sun, moving with velocity $-\mathbf{v}$. A source of light that radiates more or less isotropically in its rest frame radiates mainly in the forward direction when boosted to high velocities. The light "picks up" the velocity of the source not by changing its speed, but by changing its direction. At very high velocities, the angular spread of the beam can be shown (see problem 17) to be about γ^{-1}. This is known as the **headlight effect**[10] (see Fig. 2.6).

Exercise 2.25. In frame A, a beam of monoenergetic neutrons moves with coordinate velocity $\mathbf{v} = 0.36\,c\mathrm{e}_1 + 0.48\,c\mathrm{e}_2$. Determine the proper velocity u in A. Check your result by verifying the unimodularity of u. By about how much is the mean lifetime of a neutron in the beam going to be increased beyond its value of about 15 minutes when at rest? Now find the proper velocity and the coordinate velocity of the neutrons in a frame B that moves in frame A with velocity $\mathbf{V} = 0.8\,c\mathrm{e}_1$.

[10] See E. F. Taylor and J. A. Wheeler (1992), p. 115.

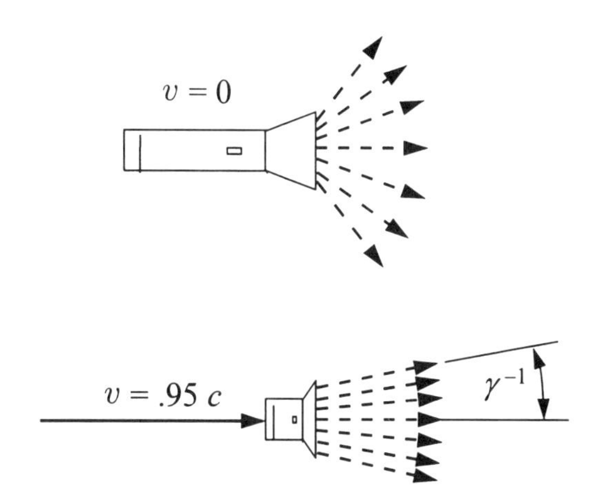

Fig. 2.6. A schematic representation of the headlight effect: the light from a rapidly moving source is thrown forward in the direction of motion. The Lorentz contraction of the flashlight is indicated.

2.6 Tetrads and Matrix Representations

Lorentz transformations, as stated above, relate inertial frames. They also relate paravector bases, $\{\mathbf{e}_\mu\}$, called *tetrads,* of different *inertial frames relative to an observer*. We will generally assume, unless otherwise noted, that the tetrads are *orthonormal*:

$$\langle \mathbf{e}_\mu \bar{\mathbf{e}}_\nu \rangle_S = \eta_{\mu\nu} = \begin{cases} 1, & \mu = \nu = 0 \\ -1, & \mu = \nu = 1,2,3 \\ 0, & \mu \neq \nu \end{cases} . \tag{2.89}$$

Here, η is the (flat-space) metric tensor introduced in the previous chapter. The orthonormality condition is seen to constrain the Lorentz-invariant scalar $\langle \mathbf{e}_\mu \bar{\mathbf{e}}_\nu \rangle_S$, and consequently, it is a Lorentz-invariant condition.

Note that the basis paravectors are spacetime vectors corresponding to the separation of a pair of *events* rather than the enduring spatial vectors familiar from Galilean mechanics. The latter sweep out *spacetime planes* in time and are represented by timelike biparavectors in Minkowski spacetime.

A Lorentz transformation of a spacetime vector p,

$$p = p^{\mu}\mathbf{e}_{\mu} \to LpL^{\dagger} = Lp^{\mu}\mathbf{e}_{\mu}L^{\dagger}\,, \tag{2.90}$$

can be simply expressed

$$p^{\mu}\mathbf{e}_{\mu} \to p^{\mu}\mathbf{u}_{\mu}\,, \tag{2.91}$$

where the basis paravectors (*tetrad elements*) before and after transformation are related by:

$$\mathbf{u}_{\mu} = L\mathbf{e}_{\mu}L^{\dagger} = \mathbf{e}_{\nu}\mathcal{L}^{\nu}_{\ \mu}\,. \tag{2.92}$$

The coefficients $\mathcal{L}^{\nu}_{\ \mu}$ are elements of a 4×4 matrix that is often used to represent the transformation. They may be found for any Lorentz transformation L by algebraically evaluating the product $L\mathbf{e}_{\mu}L^{\dagger}$. We can write a formal expression for $\mathcal{L}^{\nu}_{\ \mu}$ by introducing the reciprocal basis paravector $\bar{\mathbf{e}}^{\nu}$ defined by the Lorentz-invariant relation[11]

$$\left\langle \bar{\mathbf{e}}^{\nu}\mathbf{e}_{\mu}\right\rangle_{S} = \delta^{\nu}_{\ \mu}\,. \tag{2.93}$$

If $\{\mathbf{e}_{\mu}\}$ is a *rest tetrad*, that is, if it is a tetrad for a frame at rest with respect to the observer, then $\mathbf{e}_0 = 1$ and $\bar{\mathbf{e}}^{\nu} = \mathbf{e}_{\nu}$, but these relations do not generally hold for other frames.

Exercise 2.26. Show that the reciprocal basis paravectors can be used to determine paravector components. Thus, if $p = p^{\mu}\mathbf{e}_{\mu}$, then $p^{\mu} = \left\langle p\bar{\mathbf{e}}^{\mu}\right\rangle_{S}$.

Using (2.93), we find that the transformation elements are the scalar products of the tetrad elements before and after transformation:

$$\mathcal{L}^{\nu}_{\ \mu} = \left\langle \bar{\mathbf{e}}^{\nu}L\mathbf{e}_{\mu}L^{\dagger}\right\rangle_{S} = \left\langle \bar{\mathbf{e}}^{\nu}\mathbf{u}_{\mu}\right\rangle_{S} \tag{2.94}$$

As mentioned above, the transformation may be considered as *active* (the observer frame is unchanged, and the object frame is transformed), *passive* (the object frame is the same, but the observer has been transformed backwards), or a mixture of the two. Only the *relative* orientation and motion of the observer and object frames is important.

---

[11] It is important to remember that $\mathbf{e}_{\mu}$ and $\mathbf{e}^{\nu}$ are paravectors, not scalar coefficients.

Exercise 2.27. Show that the transformation matrix $\mathcal{L}^{\nu}_{\ \mu}$ in (2.94) can also be written as $\left\langle \overline{(\bar{L}\mathbf{e}^{\nu}\bar{L}^{\dagger})}\mathbf{e}_{\mu} \right\rangle_S$. Interpret physically. [*Hint*: note that $\langle pq \rangle_S = \langle qp \rangle_S$.]

Exercise 2.28. Use the result (2.92) to show that the metric tensor η with components $\eta_{\mu\nu} = \langle \mathbf{e}_{\mu}\bar{\mathbf{e}}_{\nu} \rangle_S = \langle \mathbf{u}_{\mu}\bar{\mathbf{u}}_{\nu} \rangle_S$ obeys

$$\widetilde{\mathcal{L}}\eta\mathcal{L} = \eta\,.$$

The real matrices $\mathcal{L}$ form the group $SO_+(1,3)$ of spacetime rotations (restricted Lorentz transformations). They also describe the relation of components, taken on the observer's tetrad $\{\mathbf{e}_{\mu}\}$ in a Lorentz transformation. Combining (2.90) and (2.92), we can write the transformation

$$p \rightarrow p^{\mu}\mathbf{u}_{\mu} = p^{\mu}\mathbf{e}_{\nu}\mathcal{L}^{\nu}_{\ \mu} =: p'^{\nu}\mathbf{e}_{\nu} = p'\,. \tag{2.95}$$

The transformed components are thus

$$p'^{\nu} = \mathcal{L}^{\nu}_{\ \mu}p^{\mu}\,. \tag{2.96}$$

Example 2.29. Consider a rotation in the $\mathbf{e}_1\bar{\mathbf{e}}_2$ plane by the angle ϕ. In this case, $L = R = \exp(\mathbf{e}_1\bar{\mathbf{e}}_2\phi/2)$ and $R^{\dagger} = R^{-1}$. Since $\mathbf{e}_0$ and $\mathbf{e}_3$ commute with $\mathbf{e}_1\bar{\mathbf{e}}_2$ whereas $\mathbf{e}_1$ and $\mathbf{e}_2$ anticommute with it, $\mathbf{u}_{\mu} = L\mathbf{e}_{\mu}L^{\dagger}$ is

$$\begin{aligned}
\mathbf{u}_0 &= \mathbf{e}_0 \\
\mathbf{u}_1 &= R\mathbf{e}_1R^{\dagger} = \exp(\mathbf{e}_1\bar{\mathbf{e}}_2\phi)\,\mathbf{e}_1 \\
&= \mathbf{e}_1\cos\phi + \mathbf{e}_2\sin\phi \\
\mathbf{u}_2 &= R\mathbf{e}_2R^{\dagger} = \exp(\mathbf{e}_1\bar{\mathbf{e}}_2\phi)\,\mathbf{e}_2 \\
&= \mathbf{e}_2\cos\phi - \mathbf{e}_1\sin\phi \\
\mathbf{u}_3 &= \mathbf{e}_3\,.
\end{aligned} \tag{2.97}$$

The transformation matrix of elements $\mathcal{L}^{\nu}_{\ \mu} = \langle \bar{\mathbf{e}}^{\nu}\mathbf{u}_{\mu} \rangle_S$ thus gives

$$(\mathbf{u}_0, \mathbf{u}_1, \mathbf{u}_2, \mathbf{u}_3) = (\mathbf{e}_0, \mathbf{e}_1, \mathbf{e}_2, \mathbf{e}_3)\begin{pmatrix} 1 & 0 & 0 & 0 \\ 0 & \cos\phi & -\sin\phi & 0 \\ 0 & \sin\phi & \cos\phi & 0 \\ 0 & 0 & 0 & 1 \end{pmatrix} \tag{2.98}$$

and

$$\begin{pmatrix} p'^0 \\ p'^1 \\ p'^2 \\ p'^3 \end{pmatrix} = \begin{pmatrix} 1 & 0 & 0 & 0 \\ 0 & \cos\phi & -\sin\phi & 0 \\ 0 & \sin\phi & \cos\phi & 0 \\ 0 & 0 & 0 & 1 \end{pmatrix}\begin{pmatrix} p^0 \\ p^1 \\ p^2 \\ p^3 \end{pmatrix}. \tag{2.99}$$

Example 2.30. Let the transformation now be a boost in the $\mathbf{e}_1$ direction to a velocity v. Then $L = B = \exp(\mathbf{e}_1\bar{\mathbf{e}}_0 w/2)$ with $LL^\dagger = \mathbf{u}_0 = \gamma(\mathbf{e}_0 + v\mathbf{e}_1/c)$. One finds [see equation (2.41)]

$$\begin{aligned} \mathbf{u}_0 &= \gamma(\mathbf{e}_0 + v\mathbf{e}_1/c) \\ \mathbf{u}_1 &= \mathbf{u}_0\mathbf{e}_1 = \gamma(\mathbf{e}_1 + v\mathbf{e}_0/c) \\ \mathbf{u}_2 &= \mathbf{e}_2 \\ \mathbf{u}_3 &= \mathbf{e}_3 , \end{aligned} \tag{2.100}$$

and consequently, from (2.94),

$$(\mathbf{u}_0, \mathbf{u}_1, \mathbf{u}_2, \mathbf{u}_3) = (\mathbf{e}_0, \mathbf{e}_1, \mathbf{e}_2, \mathbf{e}_3) \begin{pmatrix} \gamma & \gamma v/c & 0 & 0 \\ \gamma v/c & \gamma & 0 & 0 \\ 0 & 0 & 1 & 0 \\ 0 & 0 & 0 & 1 \end{pmatrix} \tag{2.101}$$

and

$$\begin{pmatrix} p'^0 \\ p'^1 \\ p'^2 \\ p'^3 \end{pmatrix} = \begin{pmatrix} \gamma & \gamma v/c & 0 & 0 \\ \gamma v/c & \gamma & 0 & 0 \\ 0 & 0 & 1 & 0 \\ 0 & 0 & 0 & 1 \end{pmatrix} \begin{pmatrix} p^0 \\ p^1 \\ p^2 \\ p^3 \end{pmatrix} . \tag{2.102}$$

Exercise 2.31. Verify that the two generators of the transformations R and B in the two examples above are orthogonal and that R and B commute. Check that the corresponding 4×4 matrices $\mathcal{L}$ also commute.

Exercise 2.32. Show that R and B above are simple transformations, but their product RB is compound. (The compound transformation $L = RB$ can be called a *rifle transformation,* because it rotates the object about the boost direction, analogous to the transformation that a rifle imparts to a bullet.)

The transformation elements $\mathcal{L}^\nu_{\ \mu} = \langle \bar{\mathbf{e}}^\nu \mathbf{u}_\mu \rangle_S$ are commonly expressed in terms of partial derivatives relating the coordinates in the lab and transformed frames. Consider the transformation of the infinitesimal displacement $dx = \mathbf{e}_\mu dx^\mu$ to

$$\begin{aligned} dx' &= \mathbf{e}_v dx'^\nu = L dx L^\dagger = L\mathbf{e}_\mu L^\dagger dx^\mu \\ &= \mathbf{u}_\mu dx^\mu . \end{aligned} \tag{2.103}$$

Evidently,

$$\begin{aligned} dx'^\nu &= \langle \bar{\mathbf{e}}^\nu \mathbf{u}_\mu \rangle_S dx^\mu \\ &= \frac{\partial x'^\nu}{\partial x^\mu} dx^\mu , \end{aligned} \tag{2.104}$$

and therefore

$$\mathcal{L}^{\nu}_{\ \mu} = \langle \bar{\mathbf{e}}^{\nu} \mathbf{u}_{\mu} \rangle_S = \frac{\partial x'^{\nu}}{\partial x^{\mu}}. \tag{2.105}$$

The change in coordinates that accompanies the Lorentz transformation also requires a transformation of the gradient operator $\partial = \mathbf{e}^{\mu}\partial_{\mu}$:

$$\partial' = \mathbf{e}^{\nu} \frac{\partial}{\partial x'^{\nu}} = L\partial L^{\dagger} = L\mathbf{e}^{\mu} L^{\dagger} \partial_{\mu} = \mathbf{u}^{\mu} \frac{\partial}{\partial x^{\mu}} . \tag{2.106}$$

By virtue of the symmetries of the scalar product and the chain rule,

$$\frac{\partial}{\partial x^{\mu}} = \langle \bar{\mathbf{u}}_{\mu} \mathbf{e}^{\nu} \rangle_S \frac{\partial}{\partial x'^{\nu}} = \langle \bar{\mathbf{e}}^{\nu} \mathbf{u}_{\mu} \rangle_S \frac{\partial}{\partial x'^{\nu}} \tag{2.107}$$

also implies relation (2.105). However, note that in equation (2.107) the primed quantity stands on the RHS whereas in (2.104) the transformed (primed) quantity in on the LHS: the partial derivatives transform *contragrediently* (inversely) to the displacements.

2.6.1 Covariant vs. Invariant

All paravectors, including basis paravectors, are relative to an observer. The most common basis paravectors for observers to use are those of their own frame, in which $\mathbf{e}_0 = 1$. A Lorentz transformation $p \to LpL^{\dagger} = p^{\mu} L\mathbf{e}_{\mu} L^{\dagger}$ can be considered either an active or a passive transformation. After the transformation, the proper velocity of the object frame with respect to the observer is $\mathbf{u}_{\mu} = L\mathbf{e}_{\mu} L^{\dagger}$.

Traditional treatments of relativity often consider spacetime vectors and bivectors to be well-defined quantities independent of any observer, and the equations of physics relating such quantities are said to be *invariant*. The only transformations considered within such a framework are passive.

In our treatment, we wish to consider active as well as passive transformations, and perhaps also transformations where both the observer and the observed object are transformed. The geometric-algebra approach makes immediate an important property of physical relations: only the *relative* orientation and motion of the observer and the observed object is significant. When the relative motion or orientation of pairs of object/observer frames differ, the equations have terms whose values differ for the different pairs, but even then the *forms* of the equations are the same. Both sides of any fundamental physical relation transform in the same way, that is *covariantly,* so that its form is the same for all object/observer pairs.

A simple example of a covariant equation is the relation between the paramomentum p of a particle of mass m and its dimensionless proper velocity u :

$$p = mu\,. \tag{2.108}$$

If the system is boosted or observer is changed, the spacetime vectors on the two sides of the equation must be transformed, but they will be transformed covariantly, that is, in the same way, and the relationship will remain valid.

The identification of reciprocal basis paravectors in the observer's rest frame, mentioned in the previous section, is an example of a noncovariant relation:

$$\bar{\mathbf{e}}^{\mu} = \mathbf{e}_{\mu}\,. \tag{2.109}$$

It is not covariant because the barred spacetime vector on the left transforms distinctly from the unbarred one on the right. The relation among moving tetrad elements and their reciprocals is generally different. To find the reciprocal element $\bar{\mathbf{u}}^{\mu}$, we must transform both sides of (2.109) in the same way:

$$\begin{aligned} \bar{\mathbf{u}}^{\mu} &= \overline{(L\mathbf{e}^{\mu}L^{\dagger})} = \bar{L}^{\dagger}\bar{\mathbf{e}}^{\mu}\bar{L} \\ &= \bar{L}^{\dagger}\mathbf{e}_{\mu}\bar{L}\,. \end{aligned} \tag{2.110}$$

Now, since $\mathbf{u}_{\mu} = L\mathbf{e}_{\mu}L^{\dagger}$, we can apply the inverse relation

$$\mathbf{e}_{\mu} = \bar{L}\mathbf{u}_{\mu}\bar{L}^{\dagger} \tag{2.111}$$

to obtain

$$\bar{\mathbf{u}}^{\mu} = \bar{L}^{\dagger}\bar{L}\mathbf{u}_{\mu}\bar{L}^{\dagger}\bar{L} = \bar{\mathbf{u}}^{0}\mathbf{u}_{\mu}\bar{\mathbf{u}}^{0} \tag{2.112}$$

in place of relation (2.109). We note here that since $\mathbf{e}_0 = 1 = \mathbf{e}^0$, the tetrad element $\mathbf{u}^0 = L\mathbf{e}^0 L^{\dagger} = LL^{\dagger}$ is the proper velocity u of the transformation. As a check, one can verify that $\bar{\mathbf{u}}^{\mu} = \bar{\mathbf{u}}^{0}\mathbf{u}_{\mu}\bar{\mathbf{u}}^{0}$ is covariant and reduces to the correct relation in the rest frame. Furthermore, the Lorentz scalar

$$\langle \mathbf{u}_{\mu}\bar{\mathbf{u}}^{\nu}\rangle_S = \left\langle L\mathbf{e}_{\mu}L^{\dagger}\bar{L}^{\dagger}\mathbf{e}_{\mu}\bar{L}\right\rangle_S = \langle \mathbf{e}_{\mu}\bar{\mathbf{e}}^{\nu}\rangle_S = \delta_{\mu}^{\nu} \tag{2.113}$$

is seen to be invariant.

Thus, frame-specific noncovariant relations are often useful. In particular, we can transform them in order to derive covariant relations. For another example, see problem (11).

2.6.2 Reflections in Spacetime

In chapter 1, we saw how two reflections produced a rotation. Here we extend the argument to spacetime. Spatial reflections are examples of improper Lorentz transformations. In such reflections, the sign of the component of the reflected vector along one spatial direction, namely the normal to the reflecting plane, is changed. The scalar or time component remains unaffected. In terms of paravector space, a spatial reflection occurs not in a plane but in a three-dimensional subspace that includes the time axis. Only the components of paravectors in the direction dual to the subspace change sign.

More generally, if p is any non-null paravector direction, the component of a paravector x along p is $\langle p\bar{x}\rangle_S \, p/\left(p\bar{p}\right)$, and consequently the reflection of x in the three-dimensional subspace of ${}^*p$ is

$$\begin{aligned} x &\rightarrow x - \frac{2\left\langle p\bar{x}\right\rangle_S p}{p\bar{p}} \\ &= \frac{p\bar{p}x - \left(p\bar{x} + x\bar{p}\right)p}{p\bar{p}} \\ &= -\frac{p\bar{x}p}{p\bar{p}}. \end{aligned} \tag{2.114}$$

This is a Lorentz transformation, since it preserves $x\bar{x}$, but it does not have the form (2.32). In fact it is an *improper* transformation.

The product of two such reflections yields a proper Lorentz transformation

$$x \rightarrow \frac{q\bar{p}x\bar{p}q}{p\bar{p}q\bar{q}} = \pm LxL^\dagger, \tag{2.115}$$

with $L = q\bar{p}\left|p\bar{p}q\bar{q}\right|^{-1/2}$, where the sign in the expression $\pm LxL^\dagger$ is the sign of $p\bar{p}q\bar{q}$. If both p and q are spacelike or both are timelike, then $p\bar{p}q\bar{q} > 0$, L is unimodular, and the transformation (2.115) has the form (2.32). Otherwise, L is *antiunimodular* ($L\bar{L} = -1$) and the transformation of x picks up a minus sign: $x \rightarrow -LxL^\dagger$. The significance of this sign is seen by applying the transformation to the time axis $\mathbf{e}_0$. Let $\mathbf{u}_0$ be the transformed paravector

$$\mathbf{e}_0 \rightarrow \mathbf{u}_0 = \pm L\mathbf{e}_0L^\dagger. \tag{2.116}$$

Its time component $\langle \mathbf{u}_0\rangle_S = \pm\left\langle L\mathbf{e}_0L^\dagger\right\rangle_S$ gives the lab time required for clocks of the transformed frame to advance by one unit. By expanding L in complex components, $L = L^{\mu}\mathbf{e}_{\mu}$ one easily sees that

$$\left\langle L\mathbf{e}_0 L^{\dagger}\right\rangle_S = \sum_{\mu} |L^{\mu}|^2 > 0\,, \tag{2.117}$$

and consequently, the sign of $\langle \mathbf{u}_0 \rangle_S$ is the same as the sign of the transformation (2.115); the direction of time on commoving clocks is preserved if and only if $p\bar{p}q\bar{q} > 0$. We therefore say that the double reflection (2.115) is *orthochronous* if both p and q are timelike or both spacelike, but it is *nonorthochronous* is one is timelike and the other spacelike. Physically realizable Lorentz transformations are orthochronous and can be equated to rotations in spacetime.

Exercise 2.33. Show that if a double reflection $L = q\bar{p}\,|p\bar{p}q\bar{q}|^{-1/2}$ with real paravectors p and q is a pure boost, the vector parts of p and q must be aligned (or one must vanish), and if it is a pure rotation by more than zero degrees, then the scalar parts of p and q must vanish.

Exercise 2.34. Demonstrate that the orthochronicity of a Lorentz transformation is determined by the sign of the matrix element $\mathcal{L}^0_{\ 0}$ [see equation (2.94)].

To reflect a biparavector, we reflect the paravector parts. Thus, if x and y are real paravectors, the reflection of $x\bar{y}$ in the real paravector direction p is given by

$$x\bar{y} \to \left(-\frac{p\bar{x}p}{p\bar{p}}\right)\overline{\left(-\frac{p\bar{y}p}{p\bar{p}}\right)} = \frac{p\bar{x}y\bar{p}}{p\bar{p}} = \frac{p\overline{(x\bar{y})}^{\dagger}\bar{p}}{p\bar{p}}\,. \tag{2.118}$$

Time reversal (T) is reflection in $p = \mathbf{e}_0$:[12]

$$\mathsf{T} : x \to -\bar{x}\,,\ x\bar{y} \to \bar{x}y = \overline{(x\bar{y})}^{\dagger}. \tag{2.119}$$

A *spatial reflection* ($\Sigma_{\hat{\mathbf{n}}}$) in a real unit vector $\hat{\mathbf{n}}$ can be written for both real paravectors and their alternating products in the form

$$\Sigma_{\hat{\mathbf{n}}} :\ x \to \hat{\mathbf{n}}\bar{x}^{\dagger}\hat{\mathbf{n}}\,,\ x\bar{y}\cdots \to \hat{\mathbf{n}}\overline{(x\bar{y}\cdots)}^{\dagger}\hat{\mathbf{n}}\,. \tag{2.120}$$

[12]Caution is required if explicit or implicit derivatives with respect to proper time are involved, since an additional factor of -1 is generally introduced for the time reversal of each such derivative.

Spatial inversion (the *parity* operation P) in $\mathbb{E}^3$ is equivalent to successive reflections in $\mathbf{e}_1, \mathbf{e}_2$, and $\mathbf{e}_3$:

$$\mathsf{P} \;:\; x \to \mathbf{e}_3\mathbf{e}_2\mathbf{e}_1\bar{x}\mathbf{e}_1\mathbf{e}_2\mathbf{e}_3 = \bar{x} = \bar{x}^\dagger \,, \tag{2.121}$$

$$x\bar{y}\cdots \;\to\; \mathbf{e}_3\mathbf{e}_2\mathbf{e}_1\overline{(x\bar{y}\cdots)}^\dagger\mathbf{e}_1\mathbf{e}_2\mathbf{e}_3 = \overline{(x\bar{y}\cdots)}^\dagger . \tag{2.122}$$

2.7 Spacetime Diagrams

A simple application of Lorentz transformations can help us construct quantitative spacetime diagrams that assist and sharpen our visual image of relativistic phenomena. Consider a frame at rest with respect to observer A, but moving with velocity $\mathbf{v} = v\mathbf{e}_3$ with respect to observer B. The frame relative to A has a tetrad $\{\mathbf{e}_\mu\}$ with $\mathbf{e}_0 = 1$, whereas the same frame relative to B is moving and has the tetrad $\{\mathbf{u}_\mu = L\mathbf{e}_\mu L^\dagger\}$ and its proper velocity is $u = \mathbf{u}_0$. The transformation L is a boost of the form $L = \exp(\mathbf{e}_3 w/2)$, which commutes with $\mathbf{e}_0, \mathbf{e}_3$ and anticommutes with $\mathbf{e}_1, \mathbf{e}_2$. Thus,

$$\begin{aligned}
\mathbf{u}_0 &= u\mathbf{e}_0 = \gamma\left(\mathbf{e}_0 + v\mathbf{e}_3/c\right) \\
\mathbf{u}_1 &= \mathbf{e}_1 \\
\mathbf{u}_2 &= \mathbf{e}_2 \\
\mathbf{u}_3 &= u\mathbf{e}_3 = \gamma\left(\mathbf{e}_3 + v\mathbf{e}_0/c\right) .
\end{aligned} \tag{2.123}$$

Consider the case $v/c = 0.6$, $\gamma = (1 - 0.36)^{-1/2} = 1.25$, $\gamma v/c = 0.75$. The spacetime diagram (Fig. 2.7) shows part of the moving tetrad $\{\mathbf{u}_0, \mathbf{u}_3\}$ in terms of $\mathbf{e}_0$ and $\mathbf{e}_3$. Note that $\mathbf{u}_0, \mathbf{u}_3$ are still unit paravectors, because their square "lengths" are given by $\mathbf{u}_0\bar{\mathbf{u}}_0 = \gamma^2\left(1 - v^2/c^2\right) = 1 = -\mathbf{u}_3\bar{\mathbf{u}}_3$. However, because the metric of the space is nondefinite, our picture of spacetime in the definite metric of a Euclidean plane requires a ruler whose scale depends on its orientation. In particular, as the ruler approaches the light cone angle, its unit distance grows without limit.

Exercise 2.35. The locus of points equidistant from the origin of a Euclidean plane form a circle centered at the origin. Show that the points equidistant from the origin in the Minkowski metric of a spacetime diagram form hyperbolae $c^2t^2 - \mathbf{r}^2 = \text{const.}$ with asymptotes on the light cone.

If the tetrad $\{\mathbf{e}_\mu\}$ is thought of as the tetrad of A as seen by A, as suggested above, then the transformation $\mathbf{e}_\mu \to \mathbf{u}_\mu$ represents a passive transformation of

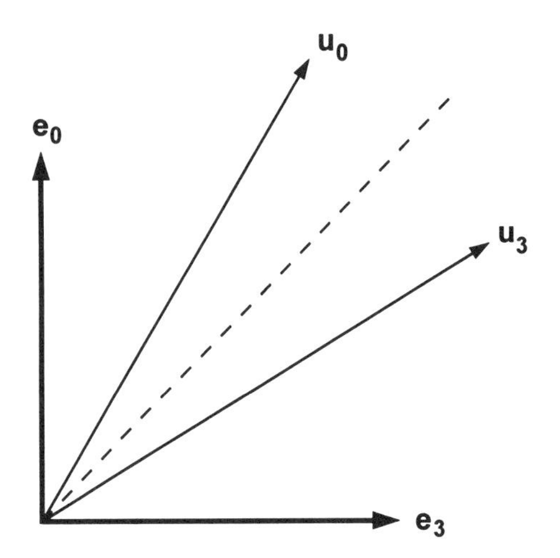

Fig. 2.7. Spacetime diagram for $v = 0.6\,c$. Here, the transformed time axis $\mathbf{u}_0$ is also the proper velocity of the moving tetrad. The dashed line is the world line of a light signal.

the same object as viewed by two observers. However, $\{\mathbf{e}_\mu\}$ is also the tetrad of B as seen by observer B, and with this interpretation, $\mathbf{e}_\mu \to \mathbf{u}_\mu$ may be thought of as the active transformation of an object (the tetrad) as it is boosted from rest in B to the velocity of frame A as seen in B. Note that $\mathbf{u}_0$ lies along the path of the object, and everywhere along the $\mathbf{u}_0$, the spatial position of the object in its rest frame is fixed. The length of $\mathbf{u}_0$ gives a unit of time for a clock fixed in B. Its component on $\mathbf{e}_0$ is greater than one unit, indicating the effect of time dilation: one unit of time on a moving clock takes a time of γ units in the lab.

Similarly, but less intuitively, $\mathbf{u}_3$ is the line of equal times in the initial frame B. However, the times along $\mathbf{u}_3$ are different in frame A. This means that synchronized clocks in one frame do not appear synchronized in another frame; it is the source of a number of surprising results in relativity. To measure the length of a moving rod, the observer determines the distance between its two ends at a given instant. The component of $\mathbf{u}_3$ on $\mathbf{e}_3$, namely γ, shows how large a rod, fixed in the lab frame A and aligned along the direction of relative motion, must be in order that it be seen as having unit length in the moving frame B.

Exercise 2.36. For the observers described above, draw the spacetime diagram of the axes $\{\mathbf{e}_0, \mathbf{e}_3\}$ of frame B relative to B and $\{\mathbf{u}_0, \mathbf{u}_3\}$ of frame B relative to observer A.

Exercise 2.37. Use (2.94) to calculate the 4×4 transformation matrix $\mathcal{L}^{\nu}{}_{\mu}$ for the explicit transformation $\mathbf{e}_\mu \to \mathbf{u}_\mu$ given above.

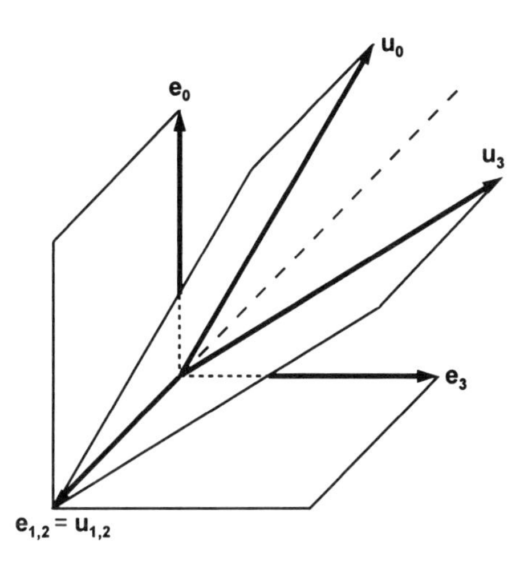

Fig. 2.8. Spacetime diagram showing the boost of spacetime planes. The unit parabivectors $\langle \mathbf{u}_\mu \bar{\mathbf{u}}_\nu \rangle_V$ result from a boost of the rest-frame planes $\langle \mathbf{e}_\mu \bar{\mathbf{e}}_\nu \rangle_V$ to the velocity $0.6\, c\mathbf{e}_3$.

By taking products of the unit paravectors before and after transformation, we can learn how the unit biparavectors transform. For the boost along $\mathbf{e}_3$ treated above, the spacetime planes $\mathbf{e}_3\bar{\mathbf{e}}_0$ and $\mathbf{e}_1\bar{\mathbf{e}}_2$ are invariant: $\mathbf{u}_3\bar{\mathbf{u}}_0 = \mathbf{e}_3\bar{\mathbf{e}}_0$, $\mathbf{u}_1\bar{\mathbf{u}}_2 = \mathbf{e}_1\bar{\mathbf{e}}_2$. However, the other four basis biparavectors, $\mathbf{e}_k\bar{\mathbf{e}}_0$ and $\mathbf{e}_k\bar{\mathbf{e}}_3$ with $k = 1, 2$, change, becoming tilted spacetime planes:

$$\begin{aligned} \mathbf{u}_k\bar{\mathbf{u}}_0 &= \gamma\left(\mathbf{e}_k\bar{\mathbf{e}}_0 + v\mathbf{e}_k\bar{\mathbf{e}}_3/c\right) \\ \mathbf{u}_3\bar{\mathbf{u}}_k &= \gamma\left(\mathbf{e}_3\bar{\mathbf{e}}_k + v\mathbf{e}_0\bar{\mathbf{e}}_k/c\right) , \end{aligned} \tag{2.124}$$

where $k = 1, 2$. This transformation, shown above in Fig. (2.8), can help us understand Lorentz transformations of biparavectors such as the electromagnetic field.

2.8 Electromagnetic Field

One of the most important biparavectors is the electromagnetic field. The electric field $\mathbf{E}$ is defined as the force on a static unit charge, as given in the previous chapter. The magnetic field may similarly be determined by the law that gives the force on a current element $Id\mathbf{l}$:

$$d\mathbf{f} = Id\mathbf{l} \times \mathbf{B}\,, \tag{2.125}$$

where I is the current and $d\mathbf{l}$ is the element of wire, oriented in the direction of current flow. (With the condition $c = 1$, this relation is valid in any consistent set of units.)

This result may be derived from the Lorentz-force equation

$$\mathbf{f} = q\mathbf{E} + q\mathbf{v} \times \mathbf{B}\,. \tag{2.126}$$

We consider an element $d\mathbf{l}$ of wire that is electrically neutral, so that its incremental positive charge dq is balanced by an equal negative charge $-dq$. Suppose, as is usually the case with metal conductors, that only the negative charge moves. The magnitude of the moving charge in $d\mathbf{l}$ is $dq = Idt$, where dt is the average time for any electron to traverse $d\mathbf{l}$. The average velocity of the electrons is $\mathbf{v} = -d\mathbf{l}/dt$. Since the total charge vanishes, the interaction with the electric field averages to zero, and the net force $d\mathbf{f}$ on the wire segment is due solely to the interaction of the moving electrons with the magnetic field:

$$d\mathbf{f} = (-dq)\,\mathbf{v} \times \mathbf{B} = (-Idt)\,(-d\mathbf{l}/dt) \times \mathbf{B} = Id\mathbf{l} \times \mathbf{B}\,. \tag{2.127}$$

Note that no matter what the orientation of the wire, the force always lies in the plane perpendicular to $\mathbf{B}$.

The electromagnetic field is one or two planes in spacetime (a *spacetime bivector*), represented by a biparavector in $C\ell_3$ whose real part is the electric field $\mathbf{E}$ and whose imaginary (spatial bivector) part is the magnetic field $ic\mathbf{B}$:

$$\mathbf{F} = \mathbf{E} + ic\mathbf{B}\,. \tag{2.128}$$

It is called the *Faraday*.[13] The electric field is a persistent spatial vector that sweeps out a timelike plane in spacetime, in which one of the directions of the plane is the time direction. On the other hand, the magnetic field is more accurately thought of as acting in a spacelike plane perpendicular to $\mathbf{B}$. The form of $\mathbf{F}$, with $\mathbf{B}$ entering as the spatial bivector $ic\mathbf{B}$ thus makes geometrical sense.

The "split" of the electromagnetic field $\mathbf{F}$ into electric- and magnetic-field parts is different for observers who are moving with respect to each other; the

[13]See C. W. Misner *et al.* (1973), p. 73.

relationship is given by the Lorentz transformation for spacetime bivectors, (2.45). In particular, an electromagnetic field $\mathbf{F}$ in the frame $\mathcal{S}$ is boosted to

$$\mathbf{F} \to \mathbf{F}' = L\mathbf{F}\bar{L} = u^{1/2}\mathbf{F}\bar{u}^{1/2} = \mathbf{F}_{\|} + u\mathbf{F}_{\perp} \tag{2.129}$$

for an observer in frame $\mathcal{S}'$, for whom frame $\mathcal{S}$ moves with proper velocity u. (In $\mathcal{S}$, frame $\mathcal{S}'$ moves with proper velocity $\bar{u}$.) Here, $\mathbf{F}_{\|} = \mathbf{F} \cdot \hat{\mathbf{u}}\,\hat{\mathbf{u}}$ is the part of the field parallel to the boost direction, and $\mathbf{F}_{\perp} = \mathbf{F} - \mathbf{F}_{\|}$ is that part perpendicular. This boost transformation may be contrasted to that for spacetime vectors (2.41):

$$p \to p' = LpL^{\dagger} = u^{1/2}pu^{1/2} = u\left(p^0 + \mathbf{p}_{\|}\right) + \mathbf{p}_{\perp}. \tag{2.130}$$

Whereas the parallel part $\mathbf{F}_{\|}$ of the spacetime bivector is unchanged, it is the perpendicular part $\mathbf{p}_{\perp}$ of the spacetime vector that is unaffected by the boost. (See also problem 3).

As noted earlier in this chapter, the square of a biparavector is Lorentz invariant. In particular, this applies to the complex scalar $\mathbf{F}^2 = (\mathbf{E}^2 - c^2\mathbf{B}^2) + 2ic\mathbf{E} \cdot \mathbf{B}$. Thus, both the real and imaginary parts of $\mathbf{F}^2$ are separately Lorentz invariant. An electromagnetic field $\mathbf{F}$ that is timelike and hence predominantly electric in any inertial frame is therefore timelike and predominantly electric in every inertial frame. Similarly, if $\mathbf{F}$ is spacelike and hence predominantly magnetic in any one frame, it is predominantly magnetic in every frame. Furthermore, if either $\mathbf{E}$ or $\mathbf{B}$ vanishes in one frame, then $\mathbf{E} \cdot \mathbf{B} = 0$ in every inertial frame, and $\mathbf{F}$ is simple and acts in a single spacetime plane. Otherwise, when $\mathbf{E} \cdot \mathbf{B} \neq 0$, $\mathbf{F}$ is compound. Then there is an inertial frame in which $\mathbf{E}$ and $\mathbf{B}$ are parallel vectors in space and, equivalently, orthogonal planes in spacetime. We return to this interesting distinction between simple and compound fields in chapter 4.

As an example, consider a static electric field $\mathbf{E} = E_0\mathbf{e}_1$ where the scalar E_0 is a real constant. In a spacetime diagram, this field sweeps out a timelike plane represented by the spacetime bivector $\mathbf{F} = E_0\mathbf{e}_1\bar{\mathbf{e}}_0$. In another frame, one in which the original frame is boosted to a proper velocity $u = \mathbf{u}_0 = \gamma\left(\mathbf{e}_0 + v\mathbf{e}_3/c\right)$, the electromagnetic field sweeps through the spatial plane $\mathbf{e}_3\mathbf{e}_1$ and thereby picks up a magnetic component in the $\mathbf{e}_2$ direction. Formally, the transformed field is

$$\begin{aligned} \mathbf{F}' &= u^{1/2}\mathbf{F}\bar{u}^{1/2} \\ &= E_0\mathbf{u}_1\bar{\mathbf{u}}_0 \\ &= \gamma E_0\left(\mathbf{e}_1\bar{\mathbf{e}}_0 + v\mathbf{e}_1\bar{\mathbf{e}}_3/c\right), \end{aligned} \tag{2.131}$$

which lies in a spacetime plane with a slope v/c with respect to the $\mathbf{e}_0$ (time) axis. This tilted plane is the sum of a vertical timelike plane $\mathbf{E}' = \gamma E_0\mathbf{e}_1\bar{\mathbf{e}}_0$ and a horizontal spacelike plane $ic\mathbf{B}' = \gamma E_0 v \mathbf{e}_1 \bar{\mathbf{e}}_3 / c = i\gamma E_0 v \mathbf{e}_2 / c$. The horizontal plane, proportional to the spatial bivector $\mathbf{e}_1 \bar{\mathbf{e}}_3$, represents the magnetic field $\mathbf{B}' = \gamma E_0 v \mathbf{e}_2 / c^2$.

Exercise 2.38. Split $\mathbf{F}$ and $\mathbf{F}'$ into real and imaginary parts in the transformation (2.129) to derive the field relations

$$\mathbf{E}' = \mathbf{E}_{\|} + \gamma \mathbf{E}_{\perp} - \gamma \mathbf{v} \times \mathbf{B} \tag{2.132}$$

$$\mathbf{B}' = \mathbf{B}_{\|} + \gamma \mathbf{B}_{\perp} + \gamma \mathbf{v} \times \mathbf{E}/c^2. \tag{2.133}$$

Exercise 2.39. Show explicitly for the above example that $\mathbf{E}' \cdot \mathbf{B}' = \mathbf{E} \cdot \mathbf{B}$ and $\mathbf{E}'^2 - c^2 \mathbf{B}'^2 = \mathbf{E}^2 - c^2 \mathbf{B}^2$.

On the other hand, if one boosts a spatial ("horizontal") plane such as a magnetic field in a coplanar direction, the boosted plane will tilt "upward," away from the boost direction, and it will pick up a temporal ("vertical") component representing an electric field. The covariant picture of the electromagnetic field is as a spacetime bivector. In place of the Newtonian view of an electric field as a vector that acts in a given direction (along a line) in space and a magnetic field as a vector that acts in the plane perpendicular to the vector, we can see the electric and magnetic fields together as a single bivector that acts in one or two planes in spacetime. Different observers will perceive their planes to have different tilts and thus different timelike (electric) and spacelike (magnetic) components. The nature of the action of the field is that of a spacetime rotation, as will be specified in detail in the chapter below on the Lorentz force.

The field lines of a static electric charge radiate isotropically from the charge. If the charge is moving rapidly at velocity $\mathbf{v}$, their directions change: the electric-field components perpendicular to $\mathbf{v}$ are increased by a factor γ more than the parallel components. Thus the $\mathbf{E}$ field lines bunch toward the plane perpendicular to $\mathbf{v}$. See Fig. (2.9) and problem 15.

2.9 Problems

1. Consider a spacetime vector $A = 3\mathbf{e}_0 + 6\mathbf{e}_1 + 5\mathbf{e}_2$ at the origin of the lab frame. Find the spacetime vector seen at the same spacetime point in a frame that travels with velocity $\mathbf{v} = 0.8c\mathbf{e}_1$ in the lab frame.

2. Find the boost that transforms the rest frame into a frame moving with proper veloc-

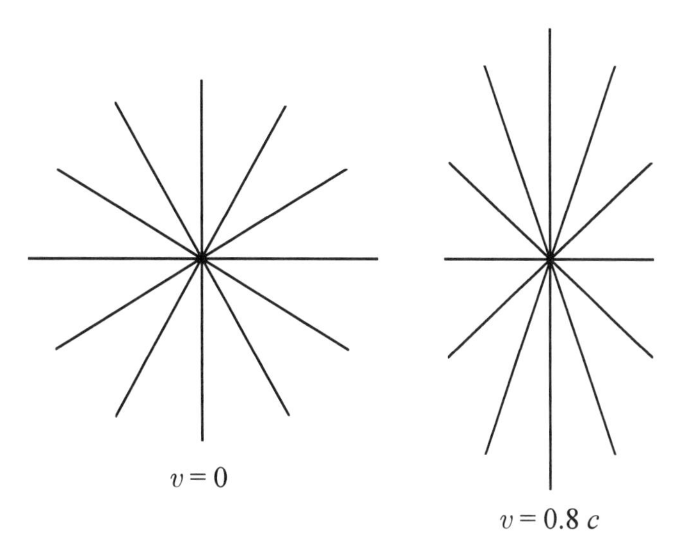

Fig. 2.9. The field lines of a rapidly moving charge are bunched toward the plane perpendicular to the velocity. The charge on the left is stationary, whereas the charge on the right is moving at a velocity of $0.8\,c$ toward the right. See problem 15 in this chapter.

ity $u = (5 + 3\,\mathbf{e}_3)\,/4$. [See relation (2.73).]

3. Recall that the boost of a spacetime vector leaves the components perpendicular to the boost direction invariant, whereas the boost of a spacetime bivector leaves those components parallel to the boost direction invariant. Apply the boost transformation (2.130) to paravectors p and q and form the biparavector $\langle p'\bar{q}'\rangle_V$ from the results. Show that $\langle p'\bar{q}'\rangle_V$ and $\langle p\bar{q}\rangle_V$ are related by the spacetime bivector transformation (2.129).

4. Show that the boost $p \to u^{1/2}pu^{1/2}$ (a proper Lorentz transformation) of the spacelike paravector $p = \mathbf{e}_0 - 2\mathbf{e}_1$ changes the sign of its time component. Does this indicate a change in the direction of time flow, or is there a better way to interpret the result?

5. Use the relation (2.41) in (2.94) to find the 4×4 matrix $\mathcal{L}\,(\mathbf{w})$ representing a boost from rest to a velocity

$$\mathbf{v} = v\,(\mathbf{e}_1 \cos\phi + \mathbf{e}_2 \sin\phi)\,.$$

Show that the matrix $\mathcal{L}(\mathbf{w})$ can be obtained from the product

$$\mathcal{L}(\mathbf{w}) = \mathcal{L}(\phi\mathbf{e}_3)\,\mathcal{L}(w\mathbf{e}_1)\,\mathcal{L}(-\phi\mathbf{e}_3)$$

of matrices for a boost by the same magnitude in the $\mathbf{e}_1$ direction with rotations about the $\mathbf{e}_3$ axis by $\pm\phi$.

6. **Time dilation.** Consider the boost transformation shown in Fig. (2.7). Sketch the world lines of clocks that are equally spaced in the moving frame along the $\mathbf{u}_3$ axis. Draw lines to show the locus of equal unit times in the moving frame, and use the result to explain why lab clocks will be seen to be running too slowly by observers in the moving frame.

7. **Lorentz contraction.** Consider a meter stick at rest and oriented along the direction of motion $\mathbf{e}_3$ as viewed by observer A. The two ends of the meter stick follow the world lines $x_a(\tau) = c\tau\mathbf{e}_0$ and $x_b(\tau) = c\tau\mathbf{e}_0 + l_0\mathbf{e}_3$ where $l_0 = 1\,\mathrm{m}$ is the length of the meter stick as observed by A. According to B, frame A moves with velocity $\mathbf{v} = 0.6c\mathbf{e}_3$ as shown in Fig. (2.7), and these two world lines are $x'_a(\tau) = c\tau\mathbf{u}_0$ and $x'_b(\tau) = c\tau\mathbf{u}_0 + l_0\mathbf{u}_3$. Express x'_a and x'_b in terms of the basis elements $\mathbf{e}_\mu$ of the tetrad at rest with respect to B. Show that a given instant τ in the rest frame of the stick (frame A) corresponds to two different times $\langle x'_a(\tau)\rangle_S \neq \langle x'_b(\tau)\rangle_S$ in B at the two ends of the stick. Demonstrate that the length of the stick as measured by observer B at any instant $\langle x'_a\rangle_S = \langle x'_b\rangle_S$ has the Lorentz-contracted value $\gamma^{-1}l_0$. Sketch a spacetime diagram showing the two world lines as seen by B and indicating how points at the same proper time τ on the two lines are related.

8. **Velocity composition.** Boost a spacetime velocity to obtain a rule for "composing velocities" in relativity. In particular, show that if C moves in frame B with proper velocity u_{BC} and B moves in frame A with proper velocity u_{AB}, and if $\mathbf{u}_{BC}$ and $\mathbf{u}_{AB}$ are collinear, then the proper velocity of C as seen by A is $u_{AC} = u_{AB}u_{BC}$ and the rapidity vectors of the transformations simply add. Use this result to find the coordinate velocity $\mathbf{v}_{AC} = c\langle u_{AC}\rangle_V / \langle u_{AC}\rangle_S$ of a galaxy C as seen by an observer in galaxy A when, according to an observer B in between, A and C are receding in opposite directions at speeds of 0.9 the speed of light.

9. Show that the proper velocity u induced by n applications of the same boost $B = (3+\mathbf{e}_1)/\sqrt{8}$ is given by

$$u = B^{2n} = \frac{1}{2}\left(2^n + 2^{-n}\right) + \frac{1}{2}\left(2^n - 2^n\right)\mathbf{e}_1\,.$$

[*Hint*: one way to solve this is to recall the form

$$B = e^{\mathbf{w}/2} = \frac{1}{2}\left(e^{w/2} + e^{-w/2}\right) + \frac{1}{2}\left(e^{w/2} - e^{-w/2}\right)\hat{\mathbf{w}}\,. \tag{2.134}$$

An alternative approach uses spectral decomposition (see chapter 1, problem 12).]

10. **Motional electric field.** An atom moves at a thermal speed of 1 km/s through a perpendicular magnetic field of 2 Tesla, which is the same as 2×10^4 gauss. Find the "motional" electric field experienced by the atom in its rest frame. Check your result by working out the answer in both Gaussian and SI units.

11. **Lorentz Force.** Lorentz-transform the equation $\mathbf{f} = \dot{p} = q\mathbf{E}$ for the force on a static charge q to derive the covariant form of the Lorentz-force equation

$$\dot{p} = q \left\langle \mathbf{F}u \right\rangle_{\Re} ,$$

where the dot indicates differentiation with respect to the proper time τ. [*Hint*: write the electric field in the rest frame in terms of a covariant quantity, that is, something you know how to transform. Try $\mathbf{E} = \frac{1}{2}\left(\mathbf{F} + \mathbf{F}^{\dagger}\right)$. Then transform the proper time derivative of p as a spacetime vector and note $LL^{\dagger} = u$.]

12. **Doppler shift**. The angular frequency ω of an electromagnetic wave in vacuum is c times the scalar part of a *propagation paravector* $k = \omega/c + \mathbf{k} = \frac{\omega}{c}\left(1 + \hat{\mathbf{k}}\right)$. Show that the frequency "seen" in the rest frame of an atom which moves with proper velocity $u = \gamma\left(1 + \mathbf{v}/c\right)$ is

$$\omega_0 = \left\langle ck\bar{u}\right\rangle_S = \gamma\omega\left(1 - \hat{\mathbf{k}}\cdot\mathbf{v}/c\right) .$$

13. **Twin "paradox".** Sketch the world line of Vito, as viewed by his twin sister Valerie at home on earth while he explores the neighborhood of the solar system. Suppose he travels for three of his years in one direction at a speed of $0.8c$ relative to the sun and then turns around relatively abruptly and returns to earth at the same speed. Also show the paths of the light signals that each sends the other on the anniversaries of the launch that sent Vito on his way. How much older is Valerie than Vito when they meet after his return to earth? (For simplicity, ignore the periods of acceleration, even though they may be important on such a "short" trip as this.)

14. **Speedy space repair.** Many apparent paradoxes in special relativity rely on the surprising result that two events separated by a spacelike interval can be ordered differently in time in different inertial frames. Consider a future speedy service garage in space that exchanges water, air, and antimatter fuel of a spaceship as it passes through at high velocity. In the frame of the garage, the spaceship, which is greatly Lorentz contracted, first enters the service bay through the front door. After the tail of the ship has passed through the front door, that door is closed, the ship is serviced (very rapidly, of course!), and the back door of the garage is then opened to let the still moving, but now serviced, spaceship continue on its way. However,

in the inertial frame of the spaceship, the garage is greatly Lorentz contracted and there is no way the spaceship could fit inside. Describe the sequence of events as viewed from the inertial frame of the spaceship. You may idealize the servicing by an event at the middle of the ship. Support your description with a spacetime diagram showing world lines of the front and back of both the garage and the spaceship.

15. **Field lines of moving charge.** Sketch the field lines of a charge at rest; show 16 lines in a spatial plane, radiating isotropically from the charge. Now consider the transformed field in a frame in which the charge moves along a straight trajectory in the plane with speed $0.8\,c$. Find and sketch the directions of the transformed electric field lines (see Fig 2.9). With the aid of a sketch, explain the magnetic field around the moving charge in terms of the spatial bivectors swept out in three-dimensional space by the moving electric field lines. Does the direction of the magnetic field make sense as the vector dual to the spatial planes?

16. **Doppler-reduced spectroscopy.** In an effort to reduce the *Doppler width* of spectral lines arising from the thermal motion of atoms, spectroscopists sometimes observe atoms in well-collimated beams of controlled velocity $\mathbf{v}$. The part of the Doppler effect proportional to the speed v of the atoms (the "first-order Doppler effect") is eliminated by observing light that is radiated perpendicular to the beam.

    (a) At some angle θ of emission to the beam direction, the radiation is entirely free of the Doppler effect (free "to all orders"). Derive the relation

    $$\cos\theta_0 = -\sqrt{\frac{\gamma-1}{\gamma+1}}$$

    for this angle, where $\gamma = \left(1 - v^2/c^2\right)^{-1/2}$. Find θ_0 for the beam speed $v = 0.8\,c$.

    (b) In any real experiment, there is a distribution of kinetic energies $(\gamma - 1)\,mc^2$ among atoms in the beam and a range of angles θ accepted by the optics. Find $\cos\theta$ at which the relative spectral spread $d\omega'/d\gamma$ of the Doppler-shifted frequencies ω' vanishes.

17. A photon of spacetime momentum $\hbar k = \hbar\omega\left(1 + \hat{\mathbf{k}}\right)/c$ is incident on an electron moving with proper velocity $u = \gamma\,(1 + \mathbf{v}/c)$ in the lab.

    (a) Derive the momentum $\hbar k' = \hbar\omega'\left(1 + \hat{\mathbf{k}}'\right)/c$ of the photon in the electron rest frame.

(b) Show that $\cos\theta' := \hat{\mathbf{k}}'\cdot\hat{\mathbf{v}}$ is given in terms of $\cos\theta := \hat{\mathbf{k}}\cdot\hat{\mathbf{v}}$ by

$$\cos\theta' = \frac{c\cos\theta - v}{c - v\cos\theta},$$

and that this relation is consistent with the inverse relation

$$\cos\theta = \frac{v + c\cos\theta'}{c + v\cos\theta'}.$$

(Hint: substitute the expression for $\cos\theta$ into the equation for $\cos\theta'$ and see if $\cos\theta' = \cos\theta'$ results.)

(c) Take $v = 0.8c$ and sketch the vectors $\mathbf{k}$ and $\mathbf{k}'$ for a set of photons that is isotropic in the lab frame (limit your sketch to one plane containing $\mathbf{v}$ and take evenly spaced angles in the electron rest frame, for example $\theta = n\pi/6,\ n = 1, 2, \ldots, 12$.)

(d) Show that the half-angle spread of incident photon directions in the rest frame of an electron is roughly γ^{-1} when $\gamma \gg 1$. [Hint: calculate $\sin\theta' \approx \theta'$ when $\theta = 0$.]

18. **Inverse Compton scattering.** In the usual Compton effect, discussed in example 2.1.1 of this chapter, a photon scatters from an electron at rest and loses a small fraction of its energy in the process. In "inverse Compton scattering" the electron has high energy and transfers some of it to the photons. The term "inverse" is misleading since the physical processes are the same, only the frames of reference differ. Use the Doppler effect and the results of the previous problem to show how such scattering provides a mechanism through which ultrarelativistic electrons can produce high-energy gamma rays from black-body photons. You may assume that in the initial rest frame of the electron, photons of energies much less than the rest energy mc^2 of the electron are scattered nearly elastically and roughly isotropically by the electron. Consider roughly 500 GeV electrons scattering an isotropic distribution of "soft" 2.7 K black-body photons, a remnant of the "big bang" with typical energies of a fraction of a milli-eV. Show that in the electron rest frame, a highly collimated beam of photons with typical energies of hundreds of eV is incident on the electron. After the usual Compton scattering in the initial rest frame of the electron, transform the distribution of scattered photons back to the lab frame, and show how a collimated beam of gamma rays, hundreds of MeV in energy, results. Sketch the photon distributions in lab and electron rest frames both before and after scattering.

19. **Theorem proof.** Prove theorem 2.3 by expanding the commutator into four pairs of terms of the form $p\bar{q}r\bar{s} - r\bar{s}p\bar{q}$ and showing that

$$p\bar{q}r\bar{s} - r\bar{s}p\bar{q} = 2\left(\langle\bar{q}r\rangle_S\, p\bar{s} - \langle p\bar{r}\rangle_S\, q\bar{s} + r\bar{p}\,\langle q\bar{s}\rangle_S - r\bar{q}\,\langle\bar{p}s\rangle_S\right). \tag{2.135}$$

20. **Structure equations for SL(2,C).** The Lorentz transformations L introduced in this chapter form the Lie group SL(2,C): they are unimodular (= Special) and Linear transformations with a faithful representation by complex 2×2 matrices. As an application of theorem 2.3, show that the generators $\langle \mathbf{e}_\mu \bar{\mathbf{e}}_\nu \rangle_V$ of SL(2,C) obey the *structure equations* of the group,

$$\begin{aligned} \frac{1}{2}\left[\langle \mathbf{e}_\alpha \bar{\mathbf{e}}_\beta \rangle_V , \langle \mathbf{e}_\mu \bar{\mathbf{e}}_\nu \rangle_V\right] &= \eta_{\beta\mu} \langle \mathbf{e}_\alpha \bar{\mathbf{e}}_\nu \rangle_V + \eta_{\alpha\nu} \langle \mathbf{e}_\beta \bar{\mathbf{e}}_\mu \rangle_V \\ &\quad -\eta_{\alpha\mu} \langle \mathbf{e}_\beta \bar{\mathbf{e}}_\nu \rangle_V - \eta_{\beta\nu} \langle \mathbf{e}_\alpha \bar{\mathbf{e}}_\mu \rangle_V , \end{aligned} \tag{2.136}$$

where the commutator bracket is defined by

$$[\mathbf{A}, \mathbf{B}] := \mathbf{AB} - \mathbf{BA}. \tag{2.137}$$

Show further that in the particular cases of timelike and spacelike planes, the structure relations reduce to

$$\begin{aligned} \frac{1}{2}\left[\langle \mathbf{e}_0 \bar{\mathbf{e}}_j \rangle_V , \langle \mathbf{e}_0 \bar{\mathbf{e}}_k \rangle_V\right] &= -\langle \mathbf{e}_j \bar{\mathbf{e}}_k \rangle_V \\ \frac{1}{2}\left[\langle \mathbf{e}_0 \bar{\mathbf{e}}_k \rangle_V , \langle \mathbf{e}_j \bar{\mathbf{e}}_k \rangle_V\right] &= \langle \mathbf{e}_0 \bar{\mathbf{e}}_j \rangle_V \\ \frac{1}{2}\sum_k \left[\langle \mathbf{e}_j \bar{\mathbf{e}}_k \rangle_V , \langle \mathbf{e}_k \bar{\mathbf{e}}_l \rangle_V\right] &= -\langle \mathbf{e}_j \bar{\mathbf{e}}_l \rangle_V , \end{aligned} \tag{2.138}$$

which with the substitutions $\frac{1}{2}\langle \mathbf{e}_0 \bar{\mathbf{e}}_j \rangle_V = -K_j$ and $\frac{1}{2}\langle \mathbf{e}_j \bar{\mathbf{e}}_k \rangle_V = -i\varepsilon_{jkl} J^l$ take the more familiar forms

$$\begin{aligned} [K_j, K_k] &= i\varepsilon_{jkl} J^l \\ \left[J^j, K_k\right] &= i\varepsilon_k^{\ lj} K_l \\ \left[J^j, J^k\right] &= i\varepsilon^{jk}_{\ \ l} J^l. \end{aligned} \tag{2.139}$$

Here, $\varepsilon_{jkl} = \varepsilon_j^{\ kl} = \varepsilon^{jk}_{\ \ l}$ is the usual Levi-Civita symbol, which is fully antisymmetric among its indices and is defined to have the value $\varepsilon_{123} = 1$.

Chapter 3

Maxwell's Equation(s)

The core relations in electromagnetic theory are Maxwell's equations, developed in the mid-nineteenth century. With these equations, Maxwell unified electricity and magnetism and predicted electromagnetic (*e.g.*, radio and light) waves.[1] Most undergraduate texts on electromagnetic theory present Maxwell's equations as four differential equations. The equations relate derivatives of the fields in a vacuum to sources of charges and currents; their physical interpretation relies to a large extent on Faraday's concept of electric and magnetic field lines. The four equations take the following *microscopic forms:*

- $\varepsilon_0 \nabla \cdot \mathbf{E} = \rho$, a form of *Gauss' law* that gives ε_0^{-1} as the number of electric field lines (in SI units) that originate (or end) on each unit charge; in regions of space where the charge density ρ vanishes, the field lines are continuous (divergence-free); since the electric field is the number of field lines per unit area perpendicular to the line, *Coulomb's law* for the field $\mathbf{E}$ of a static charge q follows directly by spherical symmetry: $\mathbf{E} = \varepsilon_0^{-1} q \hat{\mathbf{r}} \left(4\pi r^2\right)^{-1}$;

- $\nabla \cdot \mathbf{B} = 0$, which asserts that magnetic field lines exist only as continuous loops; they never start or stop, and hence there are *no magnetic monopoles*;

- $\nabla \times \mathbf{E} + \partial \mathbf{B}/\partial t = 0$, *Faraday's law of induction*, which relates the change in potential (the "electromotive force," emf $= \oint d\mathbf{l} \cdot \mathbf{E}$) around any loop to the rate of change of the magnetic flux through it; and

[1] Many of the details were cleared up and refined by Fitzgerald and others after Maxwell's death in 1879. See B. J. Hunt (1991).

- $\nabla \times \mathbf{B}/\mu_0 - \varepsilon_0 \partial \mathbf{E}/\partial t = \mathbf{j}$, where $\mu_0^{-1} = c^2 \varepsilon_0$, *Ampere's law*, giving the magnetic field generated by a current[2], with Maxwell's addition of the displacement current. For a long straight steady current, $\mathbf{j} = I\mathbf{e}_3 \delta(x)\, \delta(y)$, integration and symmetry give the magnetic field

$$\mathbf{B} = \frac{\mu_0 I}{2\pi\rho} \mathbf{e}_3 \times \hat{\boldsymbol{\rho}}, \tag{3.1}$$

where $\boldsymbol{\rho} = \mathbf{e}_1 x + \mathbf{e}_2 y$ is the position relative to the wire. For more general steady currents, one can use the *Biot-Savart law* to determine the field contribution at $\mathbf{r}$ relative to an element $Id\mathbf{l}$ of the current loop:

$$d\mathbf{B} = \frac{\mu_0}{4\pi} \frac{Id\mathbf{l} \times \hat{\mathbf{r}}}{r^2}. \tag{3.2}$$

The *macroscopic form* splits the current j into bound plus free parts and re-expresses the first and last of the equations (*i.e.*, the *inhomogeneous equations* with the currents) in terms of only the free-current contributions. If the medium is linear and isotropic, one simply replaces $\varepsilon_0 \mathbf{E}$ and $\mu_0^{-1}\mathbf{B}$ in the inhomogeneous equations by the *constitutive equations* $\mathbf{D} = \varepsilon \mathbf{E}$ and $\mathbf{H} = \mu^{-1}\mathbf{B}$, respectively. The second and third equations above, which are *homogeneous* (every term is first-order in the fields $\mathbf{E}$ and $\mathbf{B}$) since they do not involve sources, are exactly the same in their macroscopic and microscopic forms. The macroscopic equations will be discussed in more detail in Chapter 8.

Maxwell's equations are tied tightly to the relativistic transformation properties of the electromagnetic field as well as to Faraday's concept of field lines. Faraday's law, for example, describes what happens when the magnetic flux through a fixed area $\mathbf{S}$, conceptually the number of magnetic field lines passing through $\mathbf{S}$, is increased. According to the relation $\nabla \cdot \mathbf{B} = 0$, the magnetic field lines are continuous and have no start or end points. The only way to increase the flux in $\mathbf{S}$ is to bring in additional field lines through the boundary $\partial\mathbf{S}$ of $\mathbf{S}$ (see Fig. 3.1). Since the magnetic field $\mathbf{B}$ is the flux density and hence the line density, the time-rate of change of the flux through $\mathbf{S}$ is just the integral around the boundary $\partial\mathbf{S}$ of $\mathbf{B}$ times the component of the velocity of the field lines that is perpendicular to

[2]The vector $\mathbf{j}$ gives the amount of charge flowing per second through a unit area normal to the flow. It has units of amperes per square meter and is traditionally called the "current density" in electromagnetic theory. It is the vector part of the covariant paravector $j = \rho c + \mathbf{j}$. We adopt here the modern usage in theoretical physics that refers to j and its vector part simply as the "current."

both $\mathbf{B}$ and the boundary element $d\mathbf{l}$:

$$\frac{d}{dt}\int_{\mathbf{S}} d\boldsymbol{\sigma}\cdot\mathbf{B} = \oint_{\partial\mathbf{S}} d\mathbf{l}\cdot\mathbf{v}\times\mathbf{B}\,, \tag{3.3}$$

where $d\boldsymbol{\sigma}$ is an element of the surface $\mathbf{S}$. However, as we saw in the last chapter, a moving magnetic field $\mathbf{F}' = ic\mathbf{B}'$ generates an electric field $\mathbf{E} = -\gamma\mathbf{v}\times\mathbf{B}'$,

$$\begin{aligned} \mathbf{F} &= L\mathbf{F}'\bar{L} = u\mathbf{F}'_{\perp} + \mathbf{F}'_{\parallel} & (3.4)\\ &= \mathbf{E} + ic\mathbf{B} = \gamma\left(ic\mathbf{B}'_{\perp} - \mathbf{v}\times\mathbf{B}'\right) + ic\mathbf{B}'_{\parallel}\,, & (3.5) \end{aligned}$$

that is related to $\mathbf{B} = \gamma\mathbf{B}'_{\perp} + \mathbf{B}'_{\parallel}$ by

$$\mathbf{E} = -\mathbf{v}\times\mathbf{B}\,. \tag{3.6}$$

We are thus led directly to the integral form of Faraday's induction law:

$$\frac{d}{dt}\int_{\mathbf{S}} d\boldsymbol{\sigma}\cdot\mathbf{B} = -\oint_{\partial\mathbf{S}} d\mathbf{l}\cdot\mathbf{E}\,. \tag{3.7}$$

Because of the intimate relation between Maxwell's equations and relativity, we may anticipate that a relativistic (covariant) formulation of the Maxwell theory can offer both formal simplification and new insight.

3.1 Covariant Form

The four Maxwell equations given above are not in covariant form. The fields $\mathbf{E}$ and $\mathbf{B}$, the charge density ρ and vector current density $\mathbf{j}$, and the time and space coordinates t and $\mathbf{x}$, are mixed by transformations to different inertial frames.

In the paravector space of the Clifford algebra $C\ell_3$, Maxwell's equations are expressed as a single covariant equation[3]

$$\bar{\partial}\mathbf{F} = \left(\varepsilon_0 c\right)^{-1}\bar{j} \tag{3.8}$$

which displays directly the relation between the electromagnetic field $\mathbf{F}$ and the source term $\bar{j}$. The constant $\left(\varepsilon_0 c\right)^{-1} \equiv Z_0 = \mu_0 c = \sqrt{\mu_0/\varepsilon_0}$ is expressed in ohms

[3]The term *covariant* is used in two distinct ways in relativity. On the one hand, it refers to tensor elements with lower indices, such as covariant vectors (see Chapt. 2). On the other, as here, it refers to equations whose form is preserved under Lorentz transformations.

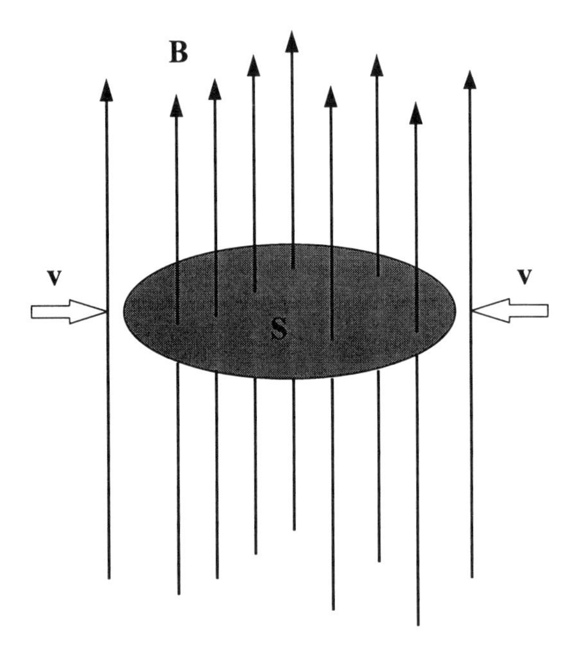

Fig. 3.1. To increase the field through the fixed surface area S field lines must move in through the boundary. A moving magnetic field in seen to have an electric field component which is sufficient to explain Faraday induction.

and is called the *characteristic impedance of the vacuum* (or of free space). Its numerical value is

$$Z_0 \equiv 4\pi K_0 = 4\pi \times \dot{3}0 \text{ ohms } \approx 376.73 \text{ ohms}. \tag{3.9}$$

Maxwell's equation (3.8) may be remembered as a free-space analog of the familiar relation $V = ZI$ between voltage V and current I in a one-dimensional circuit of impedance Z. The vacuum permittivity ε_0 has units of polarization (dipole-moment density) per unit electric field: $\mathrm{C\,m\,m^{-3}}$/(V/m). However, rather than viewing it as the volume polarizability of free space, we may be more accurate by thinking of it as relating a given vacuum polarization to the field it creates. (See the last section of this chapter.)

Expansions of $\partial = c^{-1}\partial_t - \nabla$ and $j = \rho c + \mathbf{j}$ into scalar and vector parts and of the spacetime bivector $\mathbf{F} = \mathbf{E} + ic\mathbf{B}$ into real and imaginary parts give

$$\begin{aligned}\bar{\partial}\mathbf{F} &= \left(c^{-1}\partial_t + \nabla\right)(\mathbf{E} + ic\mathbf{B}) \\ &= \nabla\cdot\mathbf{E} + \left(c^{-1}\partial_t\mathbf{E} - c\nabla\times\mathbf{B}\right) + ic\nabla\cdot\mathbf{B} + i\left(\partial_t\mathbf{B} + \nabla\times\mathbf{E}\right) \\ &= (\varepsilon_0 c)^{-1}(\rho c - \mathbf{j}) \;. \end{aligned} \tag{3.10}$$

The real scalar part of (3.10) gives Gauss' law, whereas the real vector part is Ampere's law (with Maxwell's correction). The imaginary scalar part gives the condition for the nonexistence of magnetic monopoles, and the imaginary vector part is Faraday's law of induction.

Example 3.1. Find the current density that gives rise to the magnetic field

$$\mathbf{F} = i\mathbf{e}_3 cBx/x_0 \,.$$

Solution: From (3.8), a constant current in the $-\mathbf{e}_2$ direction will do the trick:

$$\bar{j} = \varepsilon_0 c\bar{\partial}\mathbf{F} = i\mathbf{e}_1\mathbf{e}_3\varepsilon_0 c^2 B/x_0 = \frac{\mathbf{e}_2 B}{\mu_0 x_0} \,.$$

One important feature of Maxwell's equation is its *linearity*. Thus, if we double the current density j, then the field $\mathbf{F}$ is also doubled. Furthermore, we can use the *superposition principle* to solve problems: if source j_1 gives field $\mathbf{F}_1$ and source j_2 gives field $\mathbf{F}_2$, then the combined source $j_1 + j_2$ gives the total field $\mathbf{F}_1 + \mathbf{F}_2$.

3.1.1 Symmetries

Another important feature is the *covariance* of Maxwells' equation: both sides of (3.8) transform in the same way, namely as barred paravectors. In another frame, in which the original observer moves with proper velocity $LL^\dagger$, the electromagnetic field is $\mathbf{F}' = L\mathbf{F}\bar{L}$, the gradient is $\partial' = L\partial L^\dagger$, and the current density is $j' = LjL^\dagger$. Combining these relations with their conjugates, one readily transforms (3.8) to the same form in the new frame:

$$\bar{\partial}'\mathbf{F}' = (\varepsilon_0 c)^{-1}\,\bar{j}' \,. \tag{3.11}$$

The form is the same in all inertial frames. We can be sure that if $\mathbf{F}$ is a solution of Maxwell's equation for a certain current j, then $L\mathbf{F}\bar{L}$ is a solution for the current $LjL^\dagger$, where L is an arbitrary Lorentz transformation.

The transformation $\mathbf{F} \to \mathbf{F}' = L\mathbf{F}\bar{L}$ transforms the six biparavector components of the electromagnetic field $\mathbf{F}$. The transformed field $\mathbf{F}'$ should be expressed in terms of the transformed spacetime positions x', consistent with $dx' = LdxL^{\dagger}$ and the transformation $\partial' = L\partial L^{\dagger}$ of the gradient operator. The complete field transformation can thus be written

$$\mathbf{F}(x) \to \mathbf{F}'(x') = L\mathbf{F}(x)\,\bar{L} \tag{3.12}$$

in order to emphasize that not only the components of the field but also their functional dependences on position have changed.

The expansion of the paravectors into scalar plus vector parts, or of the field $\mathbf{F}$ into real and imaginary parts is, however, *not* covariant: the parts are generally different in different frames. A table of the covariant quantities is given below.

Physical Quantity	**Type**	**Symbols**	**SI Units**
speed of light	scalar	$c = \dot{3} \times 10^8$	m/s
vacuum impedance	scalar	$Z_0 = 4\pi \times \dot{3}0$	V/A = ohm
vacuum permittivity	scalar	$\varepsilon_0 = 1/\left(cZ_0\right)$	C/(V m) = F/m
vacuum permeability	scalar	$\mu_0 = Z_0/c$	N/A^2 = H/m
mass	scalar	m	kg
charge	scalar	q, Q, e	C
proper-time displ.	scalar	$d\tau = c^{-1}\left(dx\,d\bar{x}\right)^{1/2}$	s
D'Alembertian	scalar	$\square = c^{-2}\partial_t^2 - \nabla^2$	m^{-2}
current density	paravector	$j = \rho c + \mathbf{j}$	A/m^2
electromag. potential	paravector	$A = \phi/c + \mathbf{A}$	V s/m = T m
coord. displacement	paravector	$dx = cdt + d\mathbf{x}$	m
gradient	paravector	$\partial = c^{-1}\partial_t - \nabla$	m^{-1}
proper velocity	paravector	$u = \gamma\left(1 + \mathbf{v}/c\right)$	(dimensionless)
momentum	paravector	$p = E/c + \mathbf{p}$	kg m/s = N s
electromagnetic field	biparavec.	$\mathbf{F} = \mathbf{E} + ic\mathbf{B}$	N/C = V/m
Lorentz parameters	biparavec.	$\mathbf{W} = \mathbf{w} - i\boldsymbol{\theta}$	(dimensionless)

Table 3.1. Covariant electromagnetic quantities. The abbreviation $\dot{3}$ stands for 2.99792458. Note that in the column listing the covariant type, a paravector in the Pauli algebra represents a spacetime vector, and a biparavector represents a spacetime bivector.

The *Hodge dual* of the electromagnetic field $\mathbf{F} = \mathbf{E} + ic\mathbf{B}$ is

$$^{*}\mathbf{F} := -i\mathbf{F} = c\mathbf{B} - i\mathbf{E}\,. \tag{3.13}$$

It can be reached from $\mathbf{F}$ by a "duality rotation" $\mathbf{F} \to \exp(-i\beta)\,\mathbf{F}$ by the angle $\beta = \pi/2$.

The concept of relativistic covariance is a powerful tool in physics. The principle of relativity states that all inertial observers are physically equivalent: they should all experience the same physical laws. To satisfy this principle, the fundamental laws of physics should have covariant forms. The familiar Gauss-Coulomb law $\nabla \cdot \mathbf{E} = \varepsilon_0^{-1}\rho$ is not covariant, but with the help of Table 3.1, its covariant extension can be found: the electric field is the real part of the covariant biparavector representing the electromagnetic field (the Faraday) $\mathbf{F}$, the differential operator ∇ is part of the paravector gradient ∂, and ρc is the time component of the paravector current j. These must be assembled in a covariant equation that transforms in the same way on both sides and also gives the Gauss-Coulomb law. The equation (3.8) seems the only choice.

The beauty of the single Maxwell's equation (3.8) is thus more than its compactness and simplicity. It shows that all four of Maxwell's more traditional equations are the appropriate relativistic generalization of the Gauss-Coulomb law $\nabla \cdot \mathbf{E} = \varepsilon_0^{-1}\rho$.

We expect the form of Maxwell's equation also to be maintained under the improper Lorentz transformations of T, parity inversion P, and reflection $\Sigma_{\hat{\mathbf{n}}}$ in the real spatial direction $\hat{\mathbf{n}}$. For T, since

$$\begin{aligned} \mathsf{T}: \partial &\to -\bar{\partial}^\dagger = -\bar{\partial} \\ j &\to \bar{j}^\dagger = \bar{j}, \end{aligned} \tag{3.14}$$

the field must transform as

$$\mathsf{T}: \mathbf{F} \to -\bar{\mathbf{F}}^\dagger = \mathbf{F}^\dagger, \tag{3.15}$$

which means that $\mathbf{E}$ is symmetric under time reversal whereas $\mathbf{B}$ is antisymmetric. This is consistent with the invariance under T of the electric field of a static charge distribution and with the change in sign of the magnetic field of a current distribution when the flow of time and hence the direction of the current is reversed.

Under spatial inversion

$$\mathsf{P}: \partial \to \bar{\partial}^\dagger = \bar{\partial} \tag{3.16}$$

$$j \to \bar{j}^\dagger = \bar{j} \tag{3.17}$$

so that

$$\mathsf{P}: \mathbf{F} \to \bar{\mathbf{F}}^\dagger = -\mathbf{F}^\dagger. \tag{3.18}$$

Thus, $\mathbf{E}$ has *odd parity* and $\mathbf{B}$ has *even parity*. This is reasonable since $\mathbf{E}$ gives the force on a unit charge whereas $\mathbf{B}$ is normal to the spatial plane of the force. Algebraically, $\mathbf{B}$ is the Hodge dual of the plane $i\mathbf{B}$ in $\mathbb{E}^3$; since both vector directions that define the spatial plane change sign under P, the plane itself is invariant. One often distinguishes their parity by calling $\mathbf{E}$ a *vector* and $\mathbf{B}$ a *pseudovector*.

Under a reflection $\Sigma_{\hat{\mathbf{n}}}$ in the direction $\hat{\mathbf{n}}$, all alternating elements (as discussed in chapter 2) transform as

$$\Sigma_{\hat{\mathbf{n}}} : x\bar{y}\cdots \to \hat{\mathbf{n}}\overline{(x\bar{y}\cdots)}^{\dagger}\hat{\mathbf{n}} . \tag{3.19}$$

It follows that Maxwell's equation is also invariant under spatial reflection. Consequently, if we know the field $\mathbf{F}$ for a given current j, then $\hat{\mathbf{n}}\bar{\mathbf{F}}^{\dagger}\hat{\mathbf{n}}$ must be a solution of Maxwell's equation for the reflected current.

Exercise 3.2. Verify that the parity transformation is equivalent to successive reflections in the directions $\mathbf{e}_1, \mathbf{e}_2$, and $\mathbf{e}_3$.

3.1.2 Component Form

Maxwell's equation (3.8) can be expanded in products of the tetrad basis elements:

$$\bar{\partial}\mathbf{F} = \frac{1}{2}\partial^{\lambda}F^{\mu\nu}\bar{\mathbf{e}}_{\lambda}\left\langle \mathbf{e}_{\mu}\bar{\mathbf{e}}_{\nu}\right\rangle_{V} = (c\varepsilon_0)^{-1}\, j^{\rho}\bar{\mathbf{e}}_{\rho} . \tag{3.20}$$

Both sides are odd products of paravectors. The LHS contains both paravector and triparavector parts whereas the RHS has only a paravector contribution. Two equations can be isolated by taking real and imaginary parts (see Table 3.2). Thus

$$\left\langle \bar{\partial}\mathbf{F}\right\rangle_{\Re} = \frac{1}{2}\partial^{\lambda}F^{\mu\nu}\left\langle \bar{\mathbf{e}}_{\lambda}\left\langle \mathbf{e}_{\mu}\bar{\mathbf{e}}_{\nu}\right\rangle_{V}\right\rangle_{\Re} = (c\varepsilon_0)^{-1}\, j^{\rho}\bar{\mathbf{e}}_{\rho} \tag{3.21}$$

$$\left\langle \bar{\partial}\mathbf{F}\right\rangle_{\Im} = \partial^{\lambda}F^{\mu\nu}\left\langle \bar{\mathbf{e}}_{\lambda}\left\langle \mathbf{e}_{\mu}\bar{\mathbf{e}}_{\nu}\right\rangle_{V}\right\rangle_{\Im} = 0 . \tag{3.22}$$

These are known as the inhomogeneous and homogenous Maxwell's equations, respectively, since they are inhomogeneous and homogeneous in the field $\mathbf{F}$.

3.1.3 Geometrical Theorems

To appreciate the geometrical significance of these expressions, we introduce a few theorems and corollaries:

pv-grade	pv-type	no.	v-grades	basis elements
0	scalar	1	0	$1 = \mathbf{e}_0$
1	paravector	4	$0+1$	$\mathbf{e}_\mu$
2	biparavector	6	$1+2$	$\langle \mathbf{e}_\mu \bar{\mathbf{e}}_\nu \rangle_V$
3	triparavector	4	$2+3$	$\langle \mathbf{e}_\lambda \bar{\mathbf{e}}_\mu \mathbf{e}_\nu \rangle_{\Im}$ or $i\mathbf{e}_\rho$
4	pseudoscalar	1	3	$\langle \mathbf{e}_\lambda \bar{\mathbf{e}}_\mu \mathbf{e}_\nu \bar{\mathbf{e}}_\rho \rangle_{\Im S}$ or i

Table 3.2. The multiparavectors of spacetime and their relations to the multivectors of a 3-dimensional Euclidean space. Here, "pv-grade" refers to the paravector grade, "v-grades" give the corresponding spaces in Euclidean space, and "No." gives the number of linearly independent elements.

Theorem 3.1 *Let a, b, c be real paravectors. Then*

$$\left\langle \left\langle a\bar{b} \right\rangle_V c \right\rangle_{\Re} = a \left\langle \bar{b}c \right\rangle_S - b \left\langle \bar{a}c \right\rangle_S . \tag{3.23}$$

Proof: Expand the brackets

$$\left\langle \left\langle a\bar{b} \right\rangle_V c \right\rangle_{\Re} = \frac{1}{4} \left\{ a\bar{b}c - b\bar{a}c + c\bar{b}a - c\bar{a}b \right\} .$$

Then add and subtract two terms

$$0 = \frac{1}{4} \left\{ a\bar{c}b - b\bar{c}a + b\bar{c}a - a\bar{c}b \right\} ,$$

and recombine into scalar parts

$$\begin{aligned} \left\langle \left\langle a\bar{b} \right\rangle_V c \right\rangle_{\Re} &= \frac{1}{2} \left(a \left\langle \bar{b}c \right\rangle_S - b \left\langle \bar{a}c \right\rangle_S + \left\langle c\bar{b} \right\rangle_S a - \left\langle c\bar{a} \right\rangle_S b \right) \\ &= a \left\langle \bar{b}c \right\rangle_S - b \left\langle \bar{a}c \right\rangle_S , \end{aligned}$$

where we noted that $\left\langle \bar{b}c \right\rangle_S = \left\langle c\bar{b} \right\rangle_S$. ■

The bar conjugate (spatial reversal) of Theorem 3.1 gives another form of the equality:

Corollary 3.2 $\left\langle \bar{c} \left\langle b\bar{a} \right\rangle_V \right\rangle_{\Re} = \left\langle \bar{c}b \right\rangle_S \bar{a} - \left\langle \bar{c}a \right\rangle_S \bar{b}$.

Corollary 3.3 *The paravector $\left\langle \left\langle a\bar{b} \right\rangle_V c \right\rangle_{\Re}$ is orthogonal to c.*

Proof: $\left\langle \left\langle \left\langle a\bar{b}\right\rangle_V c\right\rangle_{\Re} \bar{c}\right\rangle_S = \langle a\bar{c}\rangle_S \left\langle \bar{b}c\right\rangle_S - \langle b\bar{c}\rangle_S \langle \bar{a}c\rangle_S = 0$, since $\langle a\bar{c}\rangle_S = \langle \bar{a}c\rangle_S$. ■

The theorem is an extension of the $\mathbb{E}^3$ vector-product rule

$$(\mathbf{a}\times\mathbf{b})\times\mathbf{c} = \mathbf{b}\,(\mathbf{a}\cdot\mathbf{c}) - \mathbf{a}\,(\mathbf{b}\cdot\mathbf{c}) ,$$

which in fact is just the statement of the theorem when the scalar parts vanish: $a^0 = b^0 = c^0 = 0$. More generally, the theorem states that the paravector part of the product of the biparavector $\left\langle a\bar{b}\right\rangle_V$ with the paravector c lies in the plane of a and b and is orthogonal to c. The paravector $\left\langle \left\langle a\bar{b}\right\rangle_V c\right\rangle_{\Re}$ vanishes if and only if c is orthogonal to every paravector in the plane spanned by a and b.

Theorem 3.4 *The triple product of real paravectors* a, b, c

$$\left\langle a\bar{b}c\right\rangle_{\Im} = \left\langle \left\langle a\bar{b}\right\rangle_V c\right\rangle_{\Im} = \langle\langle b\bar{c}\rangle_V a\rangle_{\Im} = \langle\langle c\bar{a}\rangle_V b\rangle_{\Im} \tag{3.24}$$

is fully antisymmetric under the exchange of any two paravectors (but keeping the middle one barred):

$$\left\langle a\bar{b}c\right\rangle_{\Im} = \frac{1}{6}\left(a\bar{b}c + b\bar{c}a + c\bar{a}b - b\bar{a}c - a\bar{c}b - c\bar{b}a\right) . \tag{3.25}$$

Proof: Expand

$$\left\langle a\bar{b}c\right\rangle_{\Im} = \frac{1}{2}\left(a\bar{b}c - c\bar{b}a\right)$$

and add and subtract a term

$$0 = -\left\langle a\bar{b}\right\rangle_S c + c\left\langle \bar{a}b\right\rangle_S$$

to obtain

$$\left\langle a\bar{b}c\right\rangle_{\Im} = \frac{1}{2}\left(c\bar{a}b - b\bar{a}c\right) .$$

Alternatively, add and subtract

$$0 = -a\left\langle \bar{b}c\right\rangle_S + \langle b\bar{c}\rangle_S\, a$$

to get

$$\left\langle a\bar{b}c\right\rangle_{\Im} = \frac{1}{2}\left(b\bar{c}a - a\bar{c}b\right) .$$

The average of all three expressions gives the fully antisymmetric result (3.25). Each of the equalities in (3.24) can be shown by similar expansions of the two parts. For example, in

$$\left\langle \left\langle a\bar{b}\right\rangle_V c\right\rangle_{\Im} = \frac{1}{2}\left(\left\langle a\bar{b}c\right\rangle_{\Im} - \left\langle b\bar{a}c\right\rangle_{\Im}\right)$$

one sees that both $\left\langle a\bar{b}c\right\rangle_{\Im}$ and $-\left\langle b\bar{a}c\right\rangle_{\Im}$ have the same expansion (3.25). ■

The antisymmetric product $\left\langle a\bar{b}c\right\rangle_{\Im}$ is a *triparavector* and represents a three-dimensional volume in paravector space. In $\mathcal{C}\ell_3$ it has imaginary vector and imaginary scalar parts (see Table 3.2), and we can establish two important corollaries:

Corollary 3.5 *Every paravector in the volume* $\left\langle a\bar{b}c\right\rangle_{\Im}$, *that is every real linear combination of* a, b, *and* c, *is orthogonal to the real paravector* $-i\left\langle a\bar{b}c\right\rangle_{\Im} = {}^*\left\langle a\bar{b}c\right\rangle_{\Im}$, *which is the dual of the triparavector* $\left\langle a\bar{b}c\right\rangle_{\Im}$.

Proof: It is the antisymmetry of the triparavector that makes this work. Consider, for example

$$\begin{aligned}\left\langle \left\langle a\bar{b}c\right\rangle_{\Im} \bar{a}\right\rangle_S &= \left\langle \frac{1}{2}\left(a\bar{b}c - c\bar{b}a\right)\bar{a}\right\rangle_S \\ &= \frac{a\bar{a}}{2}\left\langle \bar{b}c - c\bar{b}\right\rangle_S = 0.\end{aligned} \tag{3.26}$$

Similarly, $\left\langle \left\langle a\bar{b}c\right\rangle_{\Im} \bar{b}\right\rangle_S = 0 = \left\langle \left\langle a\bar{b}c\right\rangle_{\Im} \bar{c}\right\rangle_S$. By the linearity of $\langle Tx\rangle_S$ in x, it follows that $\left\langle a\bar{b}c\right\rangle_{\Im}$ is orthogonal to any linear combination of a, b, and c. ■

Corollary 3.6 *The triparavector* $\left\langle a\bar{b}c\right\rangle_{\Im}$ *vanishes if and only if the three paravectors* a, b, c *are coplanar in paravector space.*

Proof: Antisymmetry ensures that $\left\langle a\bar{b}c\right\rangle_{\Im}$ vanishes if any of the paravectors is a linear combination of the other two. Let $c_\perp$ be that part of c with no components on a or b. Then $\left\langle a\bar{b}c\right\rangle_{\Im} = \left\langle a\bar{b}c_\perp\right\rangle_{\Im} = \left\langle \left\langle a\bar{b}\right\rangle_V c_\perp\right\rangle_{\Im}$. By Theorem 3.1, the real part of $\left\langle a\bar{b}\right\rangle_V c_\perp$ vanishes. Consequently, $\left\langle a\bar{b}c\right\rangle_{\Im} = 0$ implies $\left\langle a\bar{b}\right\rangle_V c_\perp = c_\perp \left\langle \bar{a}b\right\rangle_V = 0$. If $\left\langle a\bar{b}\right\rangle_V$ is invertible, $c_\perp$ must also vanish. Should $\left\langle a\bar{b}\right\rangle_V$ not be invertible, then the product $\left\langle a\bar{b}\right\rangle_V c_\perp$ could also vanish if a or b and $c_\perp$ were proportional to the same projector element of the form $\frac{1}{2}\left(1+\hat{\mathbf{e}}\right)$. However, then $c_\perp$ would lie in the plane of a and b, and this case was excluded by the definition of $c_\perp$. ■

The next theorem follows directly:

Theorem 3.7 *Let* B *be any simple biparavector. Then the planes represented by* B *and its dual* $^*\mathrm{B}$ *are orthogonal, that is every paravector in* B *is orthogonal to every paravector in* $^*\mathrm{B}$ *and* vice versa.

Proof: Let u be any paravector. Because a biparavector $\mathbf{B}$ and its dual $^*\mathbf{B}$ are related by $^*\mathrm{B} = -i\mathrm{B}$, the triparavector $\langle \mathbf{B}u \rangle_{\Im}$ vanishes if and only if $\langle {}^*\mathbf{B}u \rangle_{\Re}$ does. According to the last two theorems and their corollaries, this means that a paravector lies in the plane of B if and only if it is orthogonal to every paravector in $^*\mathbf{B}$. The argument can be repeated with B replaced by its dual $^*\mathrm{B}$. ■

When a corollary of Theorem (3.1) is substituted into the paravector part (3.21) of Maxwell's equation, we obtain the common component form of the inhomogeneous equations:

$$\begin{aligned}
\left\langle \bar{\partial}\mathbf{F} \right\rangle_{\Re} &= \frac{1}{2}\partial^{\lambda}\mathbf{F}^{\mu\nu}\left\langle \bar{\mathbf{e}}_{\lambda}\left\langle \mathbf{e}_{\mu}\bar{\mathbf{e}}_{\nu}\right\rangle_{V}\right\rangle_{\Re} \\
&= \frac{1}{2}\partial^{\lambda}\mathbf{F}^{\mu\nu}\left(\left\langle \bar{\mathbf{e}}_{\lambda}\mathbf{e}_{\mu}\right\rangle_{S}\bar{\mathbf{e}}_{\nu} - \left\langle \bar{\mathbf{e}}_{\lambda}\mathbf{e}_{\nu}\right\rangle_{S}\bar{\mathbf{e}}_{\mu}\right) \\
&= \partial^{\lambda}\mathbf{F}^{\mu\nu}\left\langle \bar{\mathbf{e}}_{\lambda}\mathbf{e}_{\mu}\right\rangle_{S}\bar{\mathbf{e}}_{\nu} = \left(c\varepsilon_{0}\right)^{-1} j^{\nu}\bar{\mathbf{e}}_{\nu}\,,
\end{aligned} \tag{3.27}$$

where we noted that $\mathbf{F}$ is antisymmetric in its indices. Since $\left\langle \mathbf{e}_{\lambda}\bar{\mathbf{e}}_{\mu}\right\rangle_{S} = \eta_{\lambda\mu}$, the inhomogeneous equations in component form become

$$\partial_{\mu}\mathbf{F}^{\mu\nu} = \left(c\varepsilon_{0}\right)^{-1} j^{\nu}. \tag{3.28}$$

The bar-conjugate of Theorem (3.4) together with the homogeneous part (3.22) of Maxwell's equation gives the component form

$$\partial^{\lambda}\mathbf{F}^{\mu\nu} + \partial^{\mu}\mathbf{F}^{\nu\lambda} + \partial^{\nu}\mathbf{F}^{\lambda\mu} = 0\,. \tag{3.29}$$

An alternative form of the homogeneous equation (3.22) can be written in terms of the dual field $^*\mathbf{F} = -i\mathbf{F}$:

$$\left\langle \bar{\partial}\mathbf{F} \right\rangle_{\Im} = \left\langle i\bar{\partial}\,{}^*\mathbf{F} \right\rangle_{\Im} = \left\langle \bar{\partial}\,{}^*\mathbf{F} \right\rangle_{\Re}$$

which, as above for the inhomogeneous equation, gives

$$\partial_{\mu}\,{}^*\mathbf{F}^{\mu\nu} = 0\,. \tag{3.30}$$

3.1.4 Interpretations of the Fields

In the previous chapter, we discussed the simple covariant electromagnetic field $\mathbf{F}$ and its split into real and imaginary parts $\mathbf{E}$ and $ic\mathbf{B}$, which can be viewed as projections of the field biparavector $\mathbf{F}$ onto a "vertical" hyperbolic (timelike) plane and a "horizontal" elliptic (spacelike) plane. The theorems of the last section allow an alternative view. We can write

$$\mathbf{E} = \langle \mathbf{F}\mathbf{e}_0 \rangle_{\Re} \tag{3.31}$$

$$c\mathbf{B} = {}^{*}\langle \mathbf{F}\mathbf{e}_0 \rangle_{\Im} \,. \tag{3.32}$$

These suggest that $\mathbf{E}$ is a *paravector* lying in the (biparavector) plane(s) of $\mathbf{F}$ and perpendicular to the (paravector) time axis $\mathbf{e}_0$. Since the volume orthogonal to the time axis is physical space at an instant of time, the electric field $\mathbf{E}$ is the spatial slice of the biparavector $\mathbf{F}$ at the given instant of its measurement. Since $\langle \mathbf{F}\mathbf{e}_0 \rangle_{\Re}$ is orthogonal to $\mathbf{e}_0$, the scalar part of the "paravector" $\mathbf{E}$ vanishes.

The magnetic field $c\mathbf{B}$, on the other hand, can be viewed as the paravector dual in spacetime to the triparavector $\langle \mathbf{F}\mathbf{e}_0 \rangle_{\Im}$. Since the triparavector contains the time axis $\mathbf{e}_0$, its dual must be orthogonal to $\mathbf{e}_0$. Thus, the scalar part of the "paravector" $c\mathbf{B}$, like that of $\mathbf{E}$, also vanishes. We will use this alternative view of $\mathbf{E}$ and $c\mathbf{B}$ to gain insight into the Lorentz-force equation in the following chapter.

3.2 Conservation of Charge

The continuity equation, which ensures the conservation of charge, follows easily from (3.8) by a single differentiation

$$\partial\bar{\partial}\mathbf{F} = (\varepsilon_0 c)^{-1}\,\partial\bar{j}\,. \tag{3.33}$$

The differential operator $\partial\bar{\partial} = c^{-2}\partial_t^2 - \nabla^2 \equiv \Box$ is a scalar operator called the *D'Alembertian.*[4] Since it is a scalar, whereas $\mathbf{F}$ is a vector, the left-hand side of (3.33) is a pure vector, and its scalar part vanishes, giving for the right-hand side

$$\langle \partial\bar{j} \rangle_S = 0 \tag{3.34}$$

[4] Caution is needed when referring to different texts: the D'Alembertian is frequently defined with an overall minus sign, and sometimes the symbol $\Box^2$ is used. We follow the usage in Jackson's text (J. D. Jackson, *Classical Electrodynamics*, 2nd edn., Wiley, New York, 1975), p.536.

which is the Lorentz-invariant continuity equation. It is often given in expanded form as

$$\left\langle \left(c^{-1}\partial_t - \nabla\right)(\rho c - \mathbf{j})\right\rangle_S = \partial_t \rho + \nabla \cdot \mathbf{j} = 0\,. \tag{3.35}$$

Its physical meaning in spacetime is analogous to that for the divergence-free condition of the magnetic field, in three-dimensional space. Recall that the Maxwell equation $\nabla \cdot \mathbf{B} = 0$ means that the field lines of $\mathbf{B}$ are continuous; they do not start or stop in any volume. With an extension of the familiar divergence theorem to spacetime, (3.34) may be integrated to give a vanishing net charge flowing through the boundary of any (four-dimensional) spacetime volume (see Fig. 3.2). In other words, the flow lines of j, which can be associated with world lines of typical discrete charges that constitute j, are continuous: they do not start or stop in any spacetime volume. The current is conserved, and net charge is neither created nor destroyed.

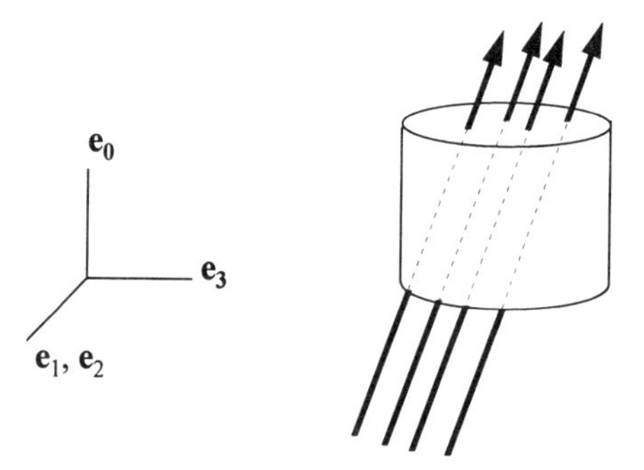

Fig. 3.2. The heavy arrows may be thought of as world lines of discrete charges or as flow lines of a current density j. The current is conserved if j satisfies the continuity equation $\langle \partial \bar{j} \rangle_S = 0$. The lines are then unbroken, and as much charge flows into any spacetime volume as flows out.

The conservation law can also be expressed on any spacelike slice through spacetime. For a given inertial observer, the noncovariant three-dimensional form (3.35) is integrated over any fixed volume V to obtain, with the help of the divergence theorem, the result

$$\frac{d}{dt}\int_V \rho\, d^3x = -\oint_{\partial V} \mathbf{j} \cdot \mathbf{n}\, d^2x\,. \tag{3.36}$$

This is the integral form of the common conservation statement. It equates the rate of increase of charge in V to the current flowing inward through the bounding surface ∂V.

3.3 Wave Equation

The vector part of (3.33) gives the *relativistic wave equation* for the electromagnetic field

$$\Box\mathbf{F} = (\varepsilon_0 c)^{-1} \langle \partial \bar{j} \rangle_V \,. \tag{3.37}$$

In regions of spacetime that are source-free ($j = 0$), the wave equation for the fields is simply

$$\Box\mathbf{F} = 0 \,, \tag{3.38}$$

which holds independently for each of the six component planes in spacetime. As we will see below, the same relation holds for the paravector potential A in source-free regions.

Many solutions to the wave equation exist. For example, any twice differentiable field $\mathbf{F}(s)$ is a solution to (3.37) if it depends on the spacetime position x only through the scalar $s = \langle k\bar{x} \rangle_S = \omega t - \mathbf{k}\cdot\mathbf{x}$, where $k = \omega/c + \mathbf{k}$, with $k\bar{k} = 0$, is any constant null paravector, and any linear superposition of such solutions with different k is also a solution. The investigation of these solutions occupies a major part of the following chapters.

3.4 Magnetic Monopoles

The four Maxwell equations given at the beginning of this chapter appear strangely asymmetric: they seem to beg for a "charge" density and current for magnetic fields in order to make them more symmetric. The inhomogeneous and homogenous Maxwell equations (3.21) and (3.22) similarly beg for a source term for the dual field. Indeed, the existence of a *magnetic monopole charge* has often been proposed as a way of "completing" Maxwell's equations. Dirac showed that if there were a fixed elementary magnetic charge corresponding to the electric charge on an electron, the product of the fundamental electric (e) and magnetic

(g) charges would be determined by the fundamental quantum of angular momentum, $\hbar$, to be

$$\frac{\mu_0}{4\pi} eg = \frac{1}{2} n\hbar \tag{3.39}$$

where n is an integer. Despite many searches for magnetic monopoles, no sighting has ever been confirmed, and from the existence of small magnetic fields in interstellar space, very low limits can be placed on the possible monopole flux ($< 10^{-11}\text{m}^{-2}\text{s}^{-1}$) in our galaxy.[5] (See example 3.7 below associated with an $e\,g$ pair.)

The Clifford-algebra expression (3.8) of Maxwell's equation has more geometrical content than the usual form and makes the experimental absence of magnetic monopoles more natural. The magnetic field has a distinct character from the electric field. Whereas the electric field is the intersection of a timelike plane in spacetime with a spatial surface, the magnetic field is the vector dual to a spatial plane. To be sure, magnetic monopoles could be introduced into the theory by adding an imaginary part to the current density j, changing it from a paravector to a triparavector and thereby making the magnetic-monopole current density a spacetime pseudovector rather than a vector. In particular, the magnetic-monopole charge density would have to change sign under spatial inversion or time reversal, and its nature would therefore be fundamentally different from that of electric charge density.

The basic idea of a magnetic monopole of strength g is that in its rest frame it would generate a static radial magnetic field

$$\mathbf{B} = \frac{\mu_0}{4\pi} \frac{g}{r^2} \hat{\mathbf{r}} \,.$$

The total magnetic flux passing through any closed surface surrounding the magnetic charge is therefore

$$\mu_0 g \,.$$

Such a magnetic field is not derivable as $\nabla \times \mathbf{A}$ from a continuous vector potential $\mathbf{A}$. This is verified by application of Stokes' theorem to a surface S with only one small opening ∂S:

$$\int_S d^2x\, \mathbf{n} \cdot \mathbf{B} = \int_S d^2x\, \mathbf{n} \cdot (\nabla \times \mathbf{A}) = \oint_{\partial S} d\mathbf{s} \cdot \mathbf{A} \,,$$

[5] M. S. Turner, E. N. Parker, and T. J. Bogdan, *Phys. Rev.* D**26**, 1296 (1982).

where $\mathbf{n}$ is the unit vector normal to the surface element and $d\mathbf{s}$ is a line element around the opening. As the surface S becomes closed, the line integral vanishes unless $\mathbf{A}$ is singular. Since the magnetic flux through any surface cannot vanish, the only apparent option is to have $\mathbf{A}$ singular on every surface surrounding the magnetic monopole.

Dirac's model of a magnetic monopole is a vanishingly thin solenoid called a *string*, one end of which is the monopole source. The other end of the solenoid is an infinite distance away. This monopole with a string does seem to satisfy the required symmetry conditions.

3.5 Electromagnetic Potential

It is useful to find the covariant extension of the familiar electrostatic relation

$$\mathbf{E} = -\nabla\phi, \text{ (static case)} \tag{3.40}$$

where ϕ is a scalar potential. Since the electromagnetic field $\mathbf{F}$ is a spacetime bivector, we may expect to express it as the vector part of the product of one spacetime vector with the spatial reversal of another. The desired generalization is

$$\mathbf{F} = c\left\langle \partial\bar{A}\right\rangle_V \tag{3.41}$$

where $A = \phi/c + \mathbf{A}$ is a paravector potential. The real vector part of (3.41) is

$$\mathbf{E} = -\partial_t\mathbf{A} - \nabla\phi \tag{3.42}$$

which reduces to (3.40) for static potentials, whereas the imaginary vector part gives

$$\mathbf{B} = \nabla\times\mathbf{A} \tag{3.43}$$

Example 3.3. The vector potential $A = -3x + 5y\left(7\mathbf{e}_2 + \mathbf{e}_3\right)$ leads to the compound biparavector field (3.41) $\mathbf{F} = c\left\langle \partial\bar{A}\right\rangle_V = 3c\mathbf{e}_1 + 5c\mathbf{e}_2\mathbf{e}_3 = c\mathbf{e}_1\left(3 + 5i\right)$ with parallel electric and magnetic fields: $\mathbf{E} = 3c\mathbf{e}_1$ and $\mathbf{B} = 5\mathbf{e}_1$.

Exercise 3.4. Show that the vector potential $A' = A + 1 + 2t + 3x\mathbf{e}_1 + 4y\mathbf{e}_2 + 5z\mathbf{e}_3$ gives the same field $\mathbf{F}$ as A.

As a paravector, the potential A transforms as a spacetime vector:

$$A \to A' = LAL^{\dagger}. \tag{3.44}$$

However, (3.44) may not tell the whole story of the transformation. A is generally a *field* that depends on the spacetime position $x = x^{\mu}\mathbf{e}_{\mu} : A = A(x)$ and x is affected by the same transformation. If we assume that the origins of the two reference frames of the transformation coincide, then for constant L,we can also write

$$x \to x' = LxL^{\dagger}. \tag{3.45}$$

We want to find $A'(x')$, but the transformation (3.44) does not change the functional dependence of A. We need to express x in terms of x' by applying the inverse transformation:

$$A'(x') = LA(x)\,L^{\dagger} = LA(\bar{L}x'\bar{L}^{\dagger})\,L^{\dagger}. \tag{3.46}$$

Example 3.5. Consider the paravector potential

$$A(x) = a\left[\left(x^1\right)^2\mathbf{e}_2 - \left(x^2\right)^2\mathbf{e}_1\right]. \tag{3.47}$$

The corresponding electromagnetic field is

$$\mathbf{F}(x) = c\left\langle\partial\bar{A}\right\rangle_V = 2ac\left(x^1 + x^2\right)\mathbf{e}_1\mathbf{e}_2\,. \tag{3.48}$$

By boosting the system from rest to proper velocity $u = \gamma + |\mathbf{u}|\,\mathbf{e}_1$ relative to the observer, we obtain

$$A' = LAL^{\dagger} = a\left[\left(x^1\right)^2\mathbf{e}_2 - \left(x^2\right)^2 u\mathbf{e}_1\right], \tag{3.49}$$

where

$$\begin{aligned} x &= \bar{L}x'\bar{L}^{\dagger} = \bar{u}\left(x'^0 + x'^1\mathbf{e}_1\right) + x'^2\mathbf{e}_2 + x'^3\mathbf{e}_3 \\ &= \left(\gamma x'^0 - |\mathbf{u}|\,x'^1\right) + \left(\gamma x'^1 - |\mathbf{u}|\,x'^0\right)\mathbf{e}_1 + x'^2\mathbf{e}_2 + x'^3\mathbf{e}_3\,. \end{aligned} \tag{3.50}$$

Therefore,

$$A'(x') = a\left[\left(\gamma x'^1 - |\mathbf{u}|\,x'^0\right)^2\mathbf{e}_2 - \left(x'^2\right)^2 u\mathbf{e}_1\right]. \tag{3.51}$$

The gradient of A' with respect to x' gives the transformed electromagnetic field $\mathbf{F}'$:

$$\begin{aligned}\mathbf{F}' &= c\left\langle \partial' \bar{A}'(x')\right\rangle_V \\ &= 2ac\left\langle \left(\gamma x'^1 - |\mathbf{u}|\, x'^0\right)\left(\gamma \mathbf{e}_1\mathbf{e}_2 + |\mathbf{u}|\, \mathbf{e}_2\right) - x'^2 \mathbf{e}_2\mathbf{e}_1\bar{u}\right\rangle_V \\ &= 2ac\left[\gamma x'^1 - |\mathbf{u}|\, x'^0 + x'^2\right] u\mathbf{e}_1\mathbf{e}_2\,.\end{aligned} \tag{3.52}$$

The result is identical with the transformation of (3.48)

$$\begin{aligned}\mathbf{F}' &= L\mathbf{F}\bar{L} = u\mathbf{F} \\ &= 2ac\left(x^1 + x^2\right) u\mathbf{e}_1\mathbf{e}_2\end{aligned} \tag{3.53}$$

after the components of x are written in terms of those of x'.

3.6 Gauge Invariance

Because the differential operator $\Box = \partial\bar{\partial} = \bar{\partial}\partial$ is a scalar, its vector part $\langle\Box\rangle_V = 0$. More concretely, if $\chi(x)$ is any scalar function, $\langle\Box\chi\rangle_V = 0$. Consequently, the addition of a term $\partial\chi$ to the paravector potential A, where $\chi(x)$ is a scalar function, leaves the field $\mathbf{F}$ (3.41) unaltered. Thus, $\mathbf{F}$ is invariant under the "gauge" transformation

$$A \to A + \partial\chi \tag{3.54}$$

In classical physics, the electromagnetic field $\mathbf{F}$ is usually taken to be the fundamental quantity, and the potential A is often viewed as being a handy mathematical artifact. However, the Aharonov-Bohm effect[6] in quantum mechanics has shown that the potential A can influence results even when $\mathbf{F} = 0$. A change in A generally induces a change in the phase of a quantum wave function, and this can be detected in some interference experiments. However, gauge transformations have no effect because in quantum theory they are always applied in conjunction with compensating phase shifts in the wave functions.

One constraint that is often convenient to place on the "gauge freedom" of the electromagnetic field is the Lorentz-invariant condition called the *Lorenz condition*[7]:

$$\left\langle \bar{\partial} A\right\rangle_S = 0 \tag{3.55}$$

[6] Y. Aharonov and D.Bohm, *Phys. Rev.* **115**, 485-491 (1959).

[7] Often misspelled as the "Lorentz condition," it is really a constraint suggested by the Danish physicist Ludwig V. Lorenz in 1867, when the Dutch physicist Hendrik Antoon Lorentz, famed for his transformations and for fundamental work on the electron, was only 14 years old. See E. T. Whittaker (1910) as well as notes in R. Penrose and W. Rindler (1984) and in P. Lounesto (1997).

Even with this condition, gauge transformations (3.54) can be made as long as $\Box\chi = 0$. When (3.55) is valid, Maxwell's equation can be written in the simple paravector form

$$\Box A = \mu_0 j \,. \tag{3.56}$$

In source-free space, equation (3.56) reduces to the vacuum wave equation:

$$\Box A = 0. \tag{3.57}$$

3.7 Energy and Momentum Density

The product $\mathbf{F}\mathbf{F}^{\dagger} = (\mathbf{E}^2 + c^2\mathbf{B}^2) + 2c\mathbf{E}\times\mathbf{B}$ gives both the energy density $\mathcal{E}$ and the energy-current density $\mathbf{S}$, known as the *Poynting vector*:

$$\frac{1}{2}\varepsilon_0 c\mathbf{F}\mathbf{F}^{\dagger} = \mathcal{E}c + \mathbf{S} \tag{3.58}$$

with

$$\mathcal{E} = \frac{1}{2}\varepsilon_0\left(\mathbf{E}^2 + c^2\mathbf{B}^2\right) \tag{3.59}$$

$$\mathbf{S} = \mathbf{E}\times\mathbf{B}/\mu_0 \,. \tag{3.60}$$

The quantity $\mathbf{S}$ is the *energy flow* and $\mathbf{S}/c^2 = \varepsilon_0\mathbf{E}\times\mathbf{B}$ is the momentum density. The identification of the electromagnetic expressions for $\mathcal{E}$ and $\mathbf{S}$ in terms of mechanical energy and energy flow is based largely on the following argument.

The scalar part of the derivative of (3.58) is

$$\frac{1}{2}\varepsilon_0 c\left\langle\bar{\partial}\left(\mathbf{F}\mathbf{F}^{\dagger}\right)\right\rangle_S = \varepsilon_0 c\left\langle\left(\bar{\partial}\mathbf{F}\right)\mathbf{F}^{\dagger}\right\rangle_{\Re S} = \partial_t\mathcal{E} + \nabla\cdot\mathbf{S} \tag{3.61}$$

$$= \left\langle\bar{j}\mathbf{F}^{\dagger}\right\rangle_{\Re S} = -\mathbf{j}\cdot\mathbf{E}\,, \tag{3.62}$$

where in the second line, Maxwell's equation (3.8) was applied. The resulting relation

$$\partial_t\mathcal{E} + \nabla\cdot\mathbf{S} + \mathbf{j}\cdot\mathbf{E} = 0\,, \tag{3.63}$$

called *Poynting's theorem*, ensures the conservation of energy: after integration over a small volume, the first term is the rate of increase in field energy, the second term gives the energy flow through the surrounding surface, and the third term, the integral of $\mathbf{j}\cdot\mathbf{E}$, gives the rate of change of mechanical energy in the volume. The first two terms are analogous to the continuity equation (3.34) for charge conservation; the third term accounts for an energy change of the current.

The energy density $\frac{1}{2}\varepsilon_0\mathbf{E}^2$ may be viewed as the interaction of the vacuum electric polarization $-\varepsilon_0\mathbf{E}$ with the electric field $\mathbf{E}$ that it generates. Similarly, $-\mathbf{B}/\mu_0$ is the vacuum magnetic polarization that interacts with the magnetic field $\mathbf{B}$ that it generates to give an energy density $\frac{1}{2}\mathbf{B}^2/\mu_0$. The factor of $\frac{1}{2}$ arises from the quadratic dependence on the field: a small change $d\mathbf{P} = -\varepsilon_0 d\mathbf{E}$ in the polarization changes the energy density by $-\mathbf{E}\cdot d\mathbf{P}$; integration over the field gives the total energy density and introduces the factor of $\frac{1}{2}$.

The statement (3.63) of energy conservation is not covariant, but is just the zero-component of a paravector conservation law that can be expressed covariantly. Even the expression $\mathbf{F}\mathbf{F}^\dagger$ does not form a covariant quantity. We return to this problem following the discussion in the next chapter of the Lorentz-force equation.

It is important to understand that the field energy is not to be added to the direct interaction of charges with each other or to the interaction of charges with fields; rather, it provides another way of computing the *same* interaction.

Example 3.6. Consider the interaction of two point charges, a charge q_1 at the origin and a charge q_2 at $\mathbf{R}$. Coulomb's law gives a direct interaction

$$W = \frac{q_1 q_2}{4\pi\varepsilon_0}\frac{1}{R}. \tag{3.64}$$

The energy W can also be interpreted as the interaction of charge q_2 in the electrostatic potential $q_1\,(4\pi\varepsilon_0 R)^{-1}$ of charge q_1. Of course it can also be viewed as the interaction of charge q_1 in the field of q_2 .It is *not* equal to the *sum* of the interactions of each charge in the field of the others; such a sum would double count the interaction. Now we find the interaction part of the field energy:

$$\begin{aligned} W' &= \frac{1}{2}\varepsilon_0\int d^3r\left[(\mathbf{E}_1{+}\mathbf{E}_2)^2 - \mathbf{E}_1^2 - \mathbf{E}_2^2\right] = \varepsilon_0\int d^3r\,\mathbf{E}_1\cdot\mathbf{E}_2 \\ &= \varepsilon_0\frac{q_1q_2}{(4\pi\varepsilon_0)^2}\int d^3r\frac{(\mathbf{r}-\mathbf{R})\cdot\hat{\mathbf{r}}}{|\mathbf{r}-\mathbf{R}|^3 r^2} \\ &= 2\pi\varepsilon_0\frac{q_1q_2}{(4\pi\varepsilon_0)^2}\int_{-1}^{1} d\cos\theta\int_0^\infty dr\,\frac{r - R\cos\theta}{(r^2+R^2-2rR\cos\theta)^{3/2}} \end{aligned} \tag{3.65}$$

Now change the last dummy integration variable from r to $s = r^2 - 2rR\cos\theta$. Since

$ds = 2\,(r - R\cos\theta)\,dr$, then

$$\begin{aligned} W' &= \frac{q_1 q_2}{(16\pi\varepsilon_0)} \int_{-1}^{1} d\cos\theta \int_0^{\infty} ds\,\left(R^2 + s\right)^{-3/2} \\ &= \frac{q_1 q_2}{4\pi\varepsilon_0 R} = W\,. \end{aligned} \qquad (3.66)$$

The interaction energy is the same. There are several ways to calculate the interaction. Which is the *right* way? They are evidently all valid. Then where does the interaction energy reside? There is no clear answer to this question. The different calculations of the interaction energy suggest different locations. Apparently, only a change in the total interaction energy is physically significant, not the apparent location of the energy.

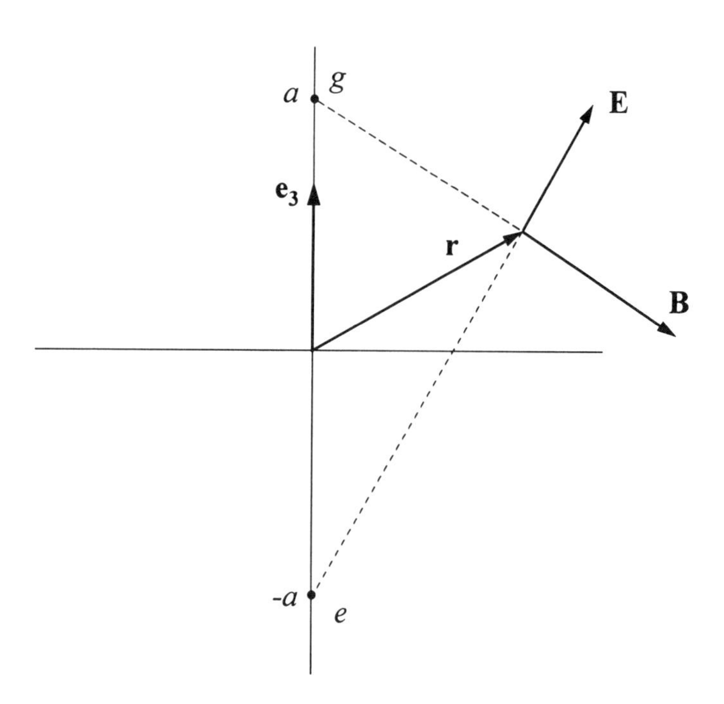

Fig. 3.3. The combination of an electric charge and a magnetic monopole gives an electromagnetic field with circular energy flow and angular momentum.

Example 3.7. Calculate the energy flow associated with an electric charge e at $\mathbf{r} = -a\mathbf{e}_3$

and a magnetic monopole g at $\mathbf{r} = a\mathbf{e}_3$ (see Fig. 3.3). The electric and magnetic fields are

$$\begin{aligned}\mathbf{E} &= \frac{e(\mathbf{r}+a\mathbf{e}_3)}{4\pi\varepsilon_0 |\mathbf{r}+a\mathbf{e}_3|^3} \\ \mathbf{B} &= \frac{\mu_0 g(\mathbf{r}-a\mathbf{e}_3)}{4\pi |\mathbf{r}-a\mathbf{e}_3|^3}.\end{aligned}$$

The energy flow is

$$\begin{aligned}\mathbf{S} &= \mathbf{E}\times\mathbf{B}/\mu_0 \\ &= \frac{2aeg\mathbf{e}_3\times\mathbf{r}}{(4\pi)^2\varepsilon_0 |\mathbf{r}-a\mathbf{e}_3|^3 |\mathbf{r}+a\mathbf{e}_3|^3}.\end{aligned}$$

It is natural to use cylindrical coordinates: $\mathbf{r} = z\mathbf{e}_3 + \rho\hat{\boldsymbol{\rho}}$ where $\hat{\boldsymbol{\rho}} = \mathbf{e}_1\cos\phi + \mathbf{e}_2\sin\phi$. Then

$$\begin{aligned}\mathbf{S} &= \frac{2aeg\rho\hat{\boldsymbol{\phi}}}{(4\pi)^2\varepsilon_0\left[(z-a)^2+\rho^2\right]^{3/2}\left[(z+a)^2+\rho^2\right]^{3/2}} \\ &= \frac{aeg\rho\hat{\boldsymbol{\phi}}}{8\pi^2\varepsilon_0\left[(z^2-a^2)^2+2(z^2+a^2)\rho^2+\rho^4\right]^{3/2}}. \qquad (3.67)\end{aligned}$$

There is thus an energy flow in the direction $\hat{\boldsymbol{\phi}} = \mathbf{e}_2\cos\phi - \mathbf{e}_1\sin\phi$ circulating around the line joining e and g.

Example 3.8. Find the total angular momentum $\mathbf{J}$ associated with the circulating flow in the previous example.
Solution: We want

$$\begin{aligned}\mathbf{J} &= \int d^3x\,\mathbf{r}\times\mathbf{S}/c^2 \\ &= \frac{\mu_0 aeg}{8\pi^2}\int_{-\infty}^{\infty}dz\int_0^{\infty}\rho d\rho\int_0^{2\pi}d\phi\frac{\rho^2\mathbf{e}_3 - z\rho\hat{\boldsymbol{\rho}}}{\left[(z^2-a^2)^2+2(z^2+a^2)\rho^2+\rho^4\right]^{3/2}} \\ &= \frac{\mu_0 aeg\mathbf{e}_3}{4\pi}\int_{-\infty}^{\infty}dz\int_0^{\infty}\rho d\rho\frac{\rho^2}{\left[(z^2-a^2)^2+2(z^2+a^2)\rho^2+\rho^4\right]^{3/2}} \\ &= \frac{\mu_0 aeg\mathbf{e}_3}{4\pi}\int_0^{\infty}dz\,\left(z^2+a^2+\left|z^2-a^2\right|\right)^{-1} \qquad (3.68) \\ &= \frac{\mu_0}{4\pi}eg\mathbf{e}_3. \qquad (3.69)\end{aligned}$$

Note that the result is independent of the separation $2a$ of the electric and magnetic charges! Dirac's quantization condition (3.39) is equivalent to setting $J = n\hbar/2$.

3.8 Problems

1. **Coulomb gauge.**

   (a) Show that the common Coulomb gauge, defined by

   $$\nabla \cdot \mathbf{A} = 0$$

   satisfies the Lorenz condition (3.55) if and only if ϕ is constant.

   (b) Show that more generally, the Coulomb gauge together with the scalar part of Maxwell's equation implies Poisson's equation

   $$\nabla^2 \phi = -\rho/\varepsilon_0$$

2. Find the electromagnetic field associated with the paravector potential

   $$A = ax^3 + b\left(x^1\mathbf{e}_2 - x^2\mathbf{e}_1\right) + d$$

   where a, b, d are constants. Show that the field is unchanged if A is replaced by $A + \partial\left[\left(x^1\right)^2 + \left(x^2\right)^2\right]$.

3. Consider a time-independent electromagnetic spacetime potential given in the lab frame by

   $$A(r) = a\left(z\mathbf{e}_0 + xy\mathbf{e}_1\right)$$

   where $r = ct\mathbf{e}_0 + x\mathbf{e}_1 + y\mathbf{e}_2 + z\mathbf{e}_3$ is the spacetime position and a is a constant scalar.

   (a) What is the electric field $\mathbf{E}$? The magnetic field $\mathbf{B}$?

   (b) Show that the fields are the same under the gauge transformation that replaces $A(r)$ by $A(r) + bx\mathbf{e}_1$, where b is a constant scalar.

   (c) Show that A is transformed by any boost in the $\mathbf{e}_3$ direction, but that the field $\mathbf{F}$ is unchanged under such a boost. Verify that $\left\langle c\partial'\bar{A}'(r')\right\rangle_V = \mathbf{F}'(r')$, where the primes indicate transformed quantities.

   (d) Use Maxwell's equation to find the current density paravector j in the lab frame.

4. **Null spacetime planes.**

   (a) Show that the paravector $P = \frac{1}{2}\left(1 + \mathbf{e}_3\right)$ is null and orthogonal to itself.

(b) Demonstrate that the spacetime plane represented by the biparavector $\mathbf{F} = P\bar{\mathbf{e}}_1$ is null.

(c) Prove that P lies in both $\mathbf{F}$ and its dual $^*\mathbf{F}$ and is orthogonal to both.

5. **Reflections in spacetime planes.** As pointed out in Chapter 2, the reflection of a four-dimensional spacetime vector in a spatial mirror corresponds to a reflection in a three-dimensional subspace of spacetime, since it results in a change of sign of only the one component that is orthogonal to the two spatial dimensions in the mirror and to the time axis. However, we can also consider reflections in spacetime planes, represented by simple biparavectors, by which we mean the transformation that changes the signs of paravector components orthogonal to vectors in the plane. Consider the simple biparavector $\langle p\bar{q}\rangle_V$ that represents the plane spanned by the real paravectors p and q. Let x be another real paravector.

   (a) Use the theorems of this chapter to prove that if x is orthogonal to p and q that

$$\langle p\bar{q}\rangle_V x = -x \langle p\bar{q}\rangle_V^\dagger . \tag{3.70}$$

   (b) Similarly show that if x lies in the $\langle p\bar{q}\rangle_V$ plane, then

$$\langle p\bar{q}\rangle_V x = x \langle p\bar{q}\rangle_V^\dagger . \tag{3.71}$$

   (c) Combine these results to prove that the reflection of an arbitrary real paravector in $\langle p\bar{q}\rangle_V$ is expressed by the transformation

$$x \longrightarrow \frac{\langle p\bar{q}\rangle_V x \langle p\bar{q}\rangle_V^\dagger}{\langle p\bar{q}\rangle_V^2} . \tag{3.72}$$

   (d) Verify for the case $\langle x\rangle_S = 0$ and $p\bar{q} = \mathbf{e}_1\bar{\mathbf{e}}_2$ that the result of part (c) correctly gives the reflection of the spatial vector $x = \mathbf{x}$ in the $\mathbf{e}_1\bar{\mathbf{e}}_2$ plane.

6. **Electron self-energy.** Find the total electromagnetic-field energy of a stationary sphere of radius r_e with the charge e uniformly distributed on its surface. Show that this energy is the same as the mechanical energy required to assemble the charge on the sphere from infinity. At what value of r_e is this energy equal to half the rest energy, $mc^2 = 0.511\,\text{MeV}$, of the electron? (The radius r_e is called the *classical electron radius.*) What can you say about the location of the energy of the electron?

7. **Maxwell stress tensor.** Let $p = p^0 + \mathbf{p}$ be a real spacetime vector. Show that $\mathbf{F}p\mathbf{F}^\dagger$ also transforms as a spacetime vector and can be written

$$\frac{1}{2}\varepsilon_0 c\mathbf{F}p\mathbf{F}^\dagger = p^0\left(\mathcal{E}c + \mathbf{S}\right) + \mathbf{p}\cdot\overleftrightarrow{\mathbf{T}}c - \mathbf{p}\cdot\mathbf{S} \tag{3.73}$$

where $\overleftrightarrow{\mathbf{T}}$ is the Maxwell stress tensor and $\mathbf{p}\cdot\overleftrightarrow{\mathbf{T}}$ is defined to be a spatial vector. Find a dyadic expression for $\overleftrightarrow{\mathbf{T}}$ in terms of the electric and magnetic fields. [Hint: a *dyadic expression* such as $\overleftrightarrow{\mathbf{AB}}$ is defined so that the inner product of any vector $\mathbf{a}$ with either side gives a vector, namely

$$\mathbf{a}\cdot\overleftrightarrow{\mathbf{AB}} = (\mathbf{a}\cdot\mathbf{A})\,\mathbf{B}$$
$$\overleftrightarrow{\mathbf{AB}}\cdot\mathbf{a} = \mathbf{A}\,(\mathbf{B}\cdot\mathbf{a}) .$$

The unit dyad $\overleftrightarrow{\mathbf{1}}$ is defined so that for any vector $\mathbf{v}$, $\mathbf{v}\cdot\overleftrightarrow{\mathbf{1}} = \overleftrightarrow{\mathbf{1}}\cdot\mathbf{v}$. It can be written

$$\overleftrightarrow{\mathbf{1}} = \overleftrightarrow{\mathbf{e}_1\mathbf{e}_1} + \overleftrightarrow{\mathbf{e}_2\mathbf{e}_2} + \overleftrightarrow{\mathbf{e}_3\mathbf{e}_3} .$$

The stress tensor $\overleftrightarrow{\mathbf{T}}$ is a linear combination of such terms and the dot product $\mathbf{p}\cdot\overleftrightarrow{\mathbf{T}}$ is therefore a vector. You can assume that $\mathbf{S}$ and $\overleftrightarrow{\mathbf{T}}$ are both independent of p. The expression can be simplified by noting that $\mathbf{F}p\mathbf{F}^{\dagger}$ is real.]

8. Consider a long coaxial conductor with positive charge evenly distributed on the outer surface of the inner cylinder at radius b and an equal but opposite charge on the inner surface of the outer cylinder at radius a. Let the voltage between the cylinders be V. Find λ, the charge per unit length on each conductor. Show that the total field energy per unit length for an observer for whom the cable is at rest is $\frac{1}{2}\lambda V$, and find the factor by which this result is changed for another observer for whom the cable is moving with velocity $\mathbf{v}$ along its axis.

9. Derive the **Biot-Savart law** (3.2) by superimposing the field due to a moving positive charge dq on one due to a stationary negative charge of the same magnitude, and then taking the nonrelativistic limit.

10. Use the Biot-Savart law (3.2) to derive the magnetic field arising from a current I through a long straight wire [see equation (3.1)].

11. Consider the **circular wire loop** of radius a centered at the origin and carrying a current I shown in Fig. 3.4.

    (a) Show that the magnetic field at $z\mathbf{e}_3$ on the axis of the loop is

    $$\mathbf{B} = \frac{\mu_0 I}{2}\mathbf{e}_3\frac{a^2}{(z^2+a^2)^{3/2}} .$$

    (b) Consider a long **solenoid** of radius a with n loops per unit length carrying current I. Integrate the result of part (a) to demonstrate that the magnetic field inside the solenoid is $\mu_0 In$.

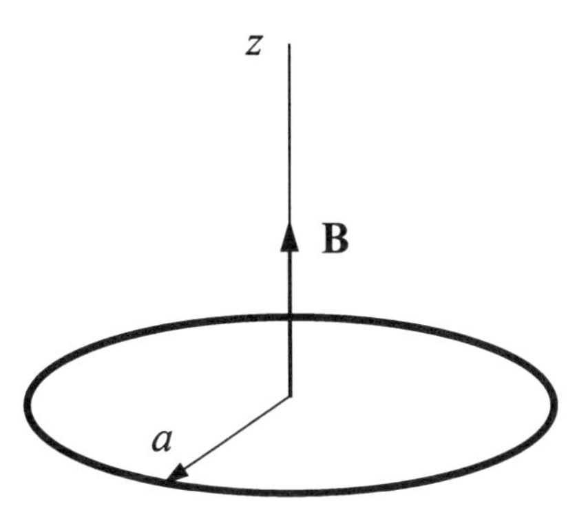

Fig. 3.4. A wire loop of radius a centered at the origin carries a current I and generates a magnetic field $\mathbf{B}$ on the $\mathbf{e}_3$ axis.

(c) Show that the field inside a long solenoid can also be found from **Ampere's law** and cylindrical symmetry if is assumed that the field outside the solenoid is negligible compared to that inside.

12. **Electromagnetic field tensor.** Define the tensor elements

$$F^{\mu\nu} = c\left(\partial^\mu A^\nu - \partial^\nu A^\mu\right) \tag{3.74}$$

and show that they are related to the electromagnetic field $\mathbf{F} = c\left\langle \partial \bar{A} \right\rangle_V$ by

$$\mathbf{F} = \frac{1}{2} F^{\mu\nu} \left\langle \mathbf{e}_\mu \bar{\mathbf{e}}_\nu \right\rangle_V . \tag{3.75}$$

Use the above result together with the reality of $F^{\mu\nu}$ and the linear independence of the three spacetime bivector elements $\mathbf{e}_0\bar{\mathbf{e}}_1 = -i\mathbf{e}_2\bar{\mathbf{e}}_3$, $\mathbf{e}_0\bar{\mathbf{e}}_2 = -i\mathbf{e}_3\bar{\mathbf{e}}_1$, and $\mathbf{e}_0\bar{\mathbf{e}}_3 = -i\mathbf{e}_1\bar{\mathbf{e}}_2$ to verify that the tensor elements $F^{\mu\nu}$ are given in terms of the components of $\mathbf{E}$ and $\mathbf{B}$ by

$$\left(F^{\mu\nu}\right) = \begin{pmatrix} 0 & -E_x & -E_y & -E_z \\ E_x & 0 & -cB_z & cB_y \\ E_y & cB_z & 0 & -cB_x \\ E_z & -cB_y & cB_x & 0 \end{pmatrix} . \tag{3.76}$$

13. **Dual field tensor.** Define the dual tensor elements by

$$^*F_{\alpha\beta} = \frac{1}{2} F^{\mu\nu} \epsilon_{\mu\nu\alpha\beta} , \tag{3.77}$$

where $\epsilon_{\mu\nu\alpha\beta}$ is the fully antisymmetric tensor of rank 4 with $\epsilon_{1230} = 1$. Thus, for example,

$$\begin{aligned}\epsilon_{1230} &= \epsilon_{2103} = \epsilon_{2310} = \epsilon_{3120} = 1 \\ \epsilon_{0123} &= \epsilon_{0231} = \epsilon_{1320} = \epsilon_{1203} = -1\,.\end{aligned}$$

Show

$$^*\mathbf{F} := -i\mathbf{F} = \frac{1}{2}{}^*F_{\alpha\beta}\left\langle \mathrm{e}^{\alpha}\bar{\mathrm{e}}^{\beta}\right\rangle_V, \tag{3.78}$$

and verify that the elements are given by

$$({}^*F_{\alpha\beta}) = \begin{pmatrix} 0 & cB_x & cB_y & cB_z \\ -cB_x & 0 & E_z & -E_y \\ -cB_y & -E_z & 0 & E_x \\ -cB_z & E_y & -E_x & 0 \end{pmatrix}, \tag{3.79}$$

and that using the metric tensor $\eta^{\alpha\rho}$ (see subsections 1.4.9 and 2.1.1) to obtain ${}^*F^{\rho\sigma} = \eta^{\alpha\rho}\eta^{\beta\sigma}\,{}^*F_{\alpha\beta}$, one can also write

$$({}^*F^{\rho\sigma}) = \begin{pmatrix} 0 & -cB_x & -cB_y & -cB_z \\ cB_x & 0 & E_z & -E_y \\ cB_y & -E_z & 0 & E_x \\ cB_z & E_y & -E_x & 0 \end{pmatrix}. \tag{3.80}$$

The following problems provide additional practice with Lorentz transformations.

14. Consider a charged medium with a static charge density of $\rho_0 = 3 \times 10^{-8}\,\mathrm{Cm}^{-3}$. Find the current density j in units of A/m^2 in the lab frame where the medium is moving with velocity $\mathbf{v} = 0.6\,c\mathbf{e}_1$.

15. **Joule heat.** Model a resistor of R ohms resistance by a uniform cylinder of radius a and length l. Calculate the Poynting vector at the surface of the resistor when a current I is flowing through. Show that the total field energy entering the resistor per second is just I^2R.

16. Consider a current $\mathbf{j}$ that in the lab frame consists of equal and opposite flows at constant velocity $\pm v$ of positive and negative charge densities, $\pm\rho/2$. Find the force acting on an isolated charge e moving with the same velocity $+v$ as the positively charged current, by calculating the force in the rest frame of e. [Hint: note that because of Lorentz contraction, the electric field is nonvanishing in this frame.] Compare to the Lorentz force of the moving charge in the lab-frame magnetic field, as determined by Ampere's law.

17. An electron moves through a pure electric field $\mathbf{E} = E_0\mathbf{e}_2$ with velocity $\mathbf{v} = 0.8\,c\mathbf{e}_1$, where E_0 is 180 kV/m. Find the electromagnetic field in the frame of the electron. Verify explicitly that $E^2 - c^2B^2$ and $\mathbf{E}\cdot\mathbf{B}$ are invariant under the boost. Sketch the spacetime diagram showing the electromagnetic field as spacetime planes in the two reference frames.

18. A cylindrical capacitor comprises two long concentric tubes with opposite charges, $\pm q$, which generate a radial electric field (see Fig. 3.5). The two cylinders, which are free to rotate independently about their common axis, are placed in a uniform magnetic field. When the magnetic field is turned off, the cylinders begin to rotate.

    (a) Use Faraday's law of induction to calculate the force on the charges $\pm q$ resulting from a time-rate of change $\dot{B}$ of the magnetic field strength, and find the torque on the cylinders. Integrate the result over time to obtain the total angular momentum J transferred to the cylinders.[Ans. $\mathbf{J} = -\frac{1}{2}q\mathbf{B}\left(b^2 - a^2\right)$.]

    (b) Calculate the electromagnetic energy flow inside the capacitor when the magnetic field $\mathbf{B}$ is present. Use this to determine the angular momentum of the fields and show that this equals the angular momentum picked up by the cylinders.

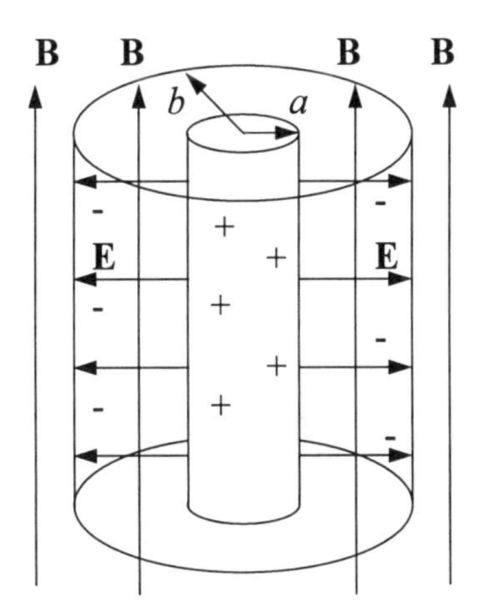

Fig. 3.5. When the magnetic field is turned off, the cylinders begin to rotate in opposite directions. Although the effect can be explained by Faraday induction, the conservation of angular momentum implies the existence of momentum in the static fields.

Chapter 4

Lorentz-Force Equation

The foundations of classical electrodynamics are summarized in just two equations: Maxwell's equation and the Lorentz-force equation. Maxwell's equation, as discussed in the previous chapter, relates the electromagnetic field to the sources. The Lorentz-force equation describes the motion of charged particles in external electromagnetic fields. It is probably most familiar in its vector form

$$d\mathbf{p}/dt = e\mathbf{E} + e\mathbf{v} \times \mathbf{B} \tag{4.1}$$

where e is the charge, $\mathbf{v} = \mathbf{dr}/dt$ the velocity, $\mathbf{p} = mc\mathbf{u} = \gamma m\mathbf{v}$ the momentum, t the coordinate time, and $\mathbf{E}$ and $\mathbf{B}$ the electric and magnetic fields. Equation (4.1) is relativistically correct but not covariant: the scalar time t, the velocity $\mathbf{v}$, and the split of the field into electric and magnetic parts are all observer-dependent.

4.1 Covariant Form

A common covariant form which also includes a component representing the change in energy is

$$\dot{p}^{\mu} = eF^{\mu\nu}u_{\nu}\,, \tag{4.2}$$

where $F^{\mu\nu}$ is the antisymmetric, second-rank electromagnetic field tensor and the dot indicates differentiation with respect to the proper time τ of the charge: $cd\tau = \langle \bar{u}dx \rangle_S$. In most cases, the physical and geometrical interpretation of the tensor-component form (4.2) is less obvious than that of the vector form (4.1).

In the Pauli algebra $C\ell_3$, we are able to combine the covariance of the tensor form with the component-free simplicity of the vector form

$$\dot{p} = \langle e\mathbf{F}u \rangle_{\Re} , \tag{4.3}$$

where $\mathbf{F} = \mathbf{E} + ic\mathbf{B}$ is the Faraday, that is, the electromagnetic field. The vector part of (4.3) is just γ times the vector equation (4.1), whereas the scalar part gives the proper rate at which the field performs work on the charge:

$$\dot{p}^0 = e\mathbf{E} \cdot \mathbf{u} , \tag{4.4}$$

where $p^0 c$ is the energy of the charge. Furthermore, by expanding (4.3) in basis paravectors and their products (see theorem 3.1),

$$\begin{aligned} \dot{p}^\lambda \mathbf{e}_\lambda &= \frac{1}{2} e F^{\mu\nu} u^\rho \left\langle \langle \mathbf{e}_\mu \bar{\mathbf{e}}_\nu \rangle_V \mathbf{e}_\rho \right\rangle_{\Re} \\ &= \frac{1}{2} e F^{\mu\nu} u^\rho \left(\mathbf{e}_\mu \langle \bar{\mathbf{e}}_\nu \mathbf{e}_\rho \rangle_S - \mathbf{e}_\nu \langle \bar{\mathbf{e}}_\mu \mathbf{e}_\rho \rangle_S \right) \\ &= \frac{1}{2} e F^{\mu\nu} u^\rho \left(\mathbf{e}_\mu \eta_{\nu\rho} - \mathbf{e}_\nu \eta_{\mu\rho} \right) \\ &= e F^{\mu\nu} u_\nu \mathbf{e}_\mu , \end{aligned} \tag{4.5}$$

the tensor component (4.2) form is readily obtained.

The covariant form (4.3) itself has a straightforward geometric interpretation. When $\mathbf{F}$ is a simple biparavector, the electromagnetic field is a single spacetime plane and $\langle \mathbf{F}u \rangle_{\Re}$ is a paravector whose direction in spacetime lies in the spacetime plane of $\mathbf{F}$ and is orthogonal to u (see the previous chapter). Since u is the time axis of the frame commoving with the charge at proper velocity u, $\langle \mathbf{F}u \rangle_{\Re}$ is the spatial slice of $\mathbf{F}$ at a given instant in the moving frame. The force (more precisely: the proper time rate of change of the momentum) in the instantaneous rest frame of the charge is the paravector $\langle e\mathbf{F}_{rest}\mathbf{e}_0 \rangle_{\Re}$, and the lab frame value $\langle e\mathbf{F}u \rangle_{\Re}$ is just that value transformed to the lab.

If $\mathbf{F}$ is compound, it is the sum of two mutually orthogonal planes and $\langle \mathbf{F}u \rangle_{\Re}$ is the sum of two paravectors both of which are orthogonal to u. This geometric picture is an extension to spacetime of the familiar result that if $\mathbf{F}$ is purely magnetic (for a given observer), then the acceleration is perpendicular to the velocity v and, in agreement with equation (4.4), there is no change in the particle energy. The more general orthogonality of $\dot{p} = mc\dot{u}$ with u ensures that the square lengths $u\bar{u}$ and $p\bar{p}$ are invariant under the action of $\mathbf{F}$.

Of course, $u\bar{u} = 1$ and $p\bar{p} = mc^2$ are Lorentz scalars, invariant under Lorentz transformations. It is clear that they should also be invariant under the Lorentz force. The relationship of the Lorentz-force law to Lorentz transformations is formalized in the *spinorial form* of the Lorentz-force equation, which is introduced in this chapter. It is often easier to interpret and solve than equation (4.3) for $\dot{p}$, and its spinor solution is closely related to a quantum wave function.

4.2 Eigenspinors

The motion and orientation of a particle is determined by its *eigenspinor* Λ, which is just the special Lorentz transformation L that relates the rest-frame of the charge to the lab frame: properties such as a spacetime-vector q_r known in the rest frame of the particle are transformed by Λ to the lab frame: $q = \Lambda q_r \Lambda^\dagger$. In particular, the time axis $\mathbf{e}_0 = 1$ in the rest frame is transformed to the proper velocity (in units of c) in the lab frame

$$u = \Lambda\Lambda^\dagger. \tag{4.6}$$

More generally, elements of the rest-frame tetrad $\{\mathbf{e}_\mu\}$ of the particle become

$$\mathbf{u}_\mu = \Lambda \mathbf{e}_\mu \Lambda^\dagger \tag{4.7}$$

in the lab frame, with $\mathbf{u}_0 \equiv u$.

The eigenspinor Λ is often referred to as a SL(2,C) ("spinorial") form[1] of the Lorentz transformation between the frames. As with the algebraic form L of Lorentz transformations discussed in chapter 2, it can be written as the exponential of a complex vector:

$$\Lambda = \exp\left[\frac{1}{2}\mathbf{W}\right], \quad \mathbf{W} = \mathbf{w} - i\boldsymbol{\theta}. \tag{4.8}$$

It induces a spatial rotation by angle θ in the plane $i\widehat{\boldsymbol{\theta}}$ (about the direction $\widehat{\boldsymbol{\theta}}$) if $\mathbf{w} = 0$, and describes a pure boost (velocity transformation) of rapidity $\mathbf{w}$ if $\boldsymbol{\theta} = 0$. Together, the six real parameters $w^j, \theta^k \in \mathbb{R}$ of the complex vector $\mathbf{w} - i\boldsymbol{\theta}$ determine the *spacetime rotation* of the charge as observed in the lab. As mentioned previously, $\mathbf{W}$ is a biparavector with the covariant expansion

$$\mathbf{W} = \frac{1}{2} W^{\mu\nu} \langle \mathbf{e}_\mu \bar{\mathbf{e}}_\nu \rangle_V . \tag{4.9}$$

[1] The designation SL(2,C) comes from the smallest faithful representation, which is formed from elements in a linear space of *special* (that is, unimodular) 2×2 complex matrices.

Here we will show below that $\mathbf{W}$ should indeed transform as a spacetime bivector, and we will reconcile the use of the noncovariant form $\mathbf{W} = \mathbf{w} - i\boldsymbol{\theta}$ when transforming between inertial reference frames.

The eigenspinor of a particle is different for different observers. Suppose Λ_{AP} is the eigenspinor of the charged particle P as seen by observer A. Observer A thus sees the charge moving with proper velocity

$$u_{AP} = \Lambda_{AP}\,\Lambda_{AP}^{\dagger}\,. \tag{4.10}$$

Let L_{BA} transform properties from the rest frame of the observer (either from frame A as seen by observer A or from frame B as seen by observer B) to those of frame A as viewed by observer B. Then observer B will see frame A moving with spacetime velocity

$$u_{BA} = L_{BA}\,L_{BA}^{\dagger}\,. \tag{4.11}$$

The proper velocity of P as seen by observer B is found by transforming u_{AP} :

$$u_{BP} = L_{BA}u_{AP}L_{BA}^{\dagger} = L_{BA}\Lambda_{AP}\Lambda_{AP}^{\dagger}L_{BA}^{\dagger}\,. \tag{4.12}$$

The eigenspinor of the particle for observer B can thus be taken to be

$$\Lambda_{BP} = L_{BA}\Lambda_{AP}. \tag{4.13}$$

The transformation of the eigenspinor thus takes the form

$$\Lambda \to L\Lambda\,, \tag{4.14}$$

which is a *spinor*-type transformation. The single mathematical form is used for several physical transformations: it expresses an active transformation ($B \to A$) of the *system* as viewed by a given observer, but it also formally describes a "backwards" transformation ($A \to B$) of the *observer* for a given observed system, as well as many possible combinations in which the frames of both the observer and the observed system are changed. The mathematics depends only on the relative motion of the two frames, that is, on the motion of one frame with respect to (*i.e.*, as viewed from) the other. While spinors have been defined in different ways, for us the defining property is its transformation behavior (4.14). The 'eigen" part of the term "eigenspinor" means "self", that is the particle's "own" spinor.

The spinor-type transformation (4.14) is simpler than either the paravector ($p \to LpL^{\dagger}$) or biparavector ($\mathbf{F} \to L\mathbf{F}\bar{L}$) type transformations, and in fact, the paravector and biparavector transformations can be derived from it. For example, let q_r be a paravector in the rest frame of the particle. The corresponding paravector $q = \Lambda q_r \Lambda^\dagger$ in the lab frame transforms as

$$q = \Lambda q_r \Lambda^\dagger \to L\Lambda q_r \Lambda^\dagger L^\dagger = LqL^\dagger. \tag{4.15}$$

Spinors like Λ, are said to be carriers of a representation of the group SL(2,C), the universal covering group of restricted Lorentz transformations.[2]

Although the eigenspinors Λ are Lorentz transformations, they warrant the special designation of eigenspinor by virtue of three special properties:

1. They relate inequivalent frames: an observed-object frame and an observer frame. Further Lorentz transformations are applied to the two frames differently. In particular, active transformations act only on the frame of the observed object whereas passive ones act only on the observer frame.[3]

2. The spinor-type transformation (4.14) is distinct. A Lorentz transformation L_{BA} that relates equivalent frames transforms instead by the biparavector type of similarity transformation: $L_{BA} \to LL_{BA}\bar{L}$. This can be derived by inverting relation (4.13) to give the Lorentz transformation L_{BA} relating A to B in terms of the eigenspinors of P:

   $$L_{BA} = \Lambda_{BP}\bar{\Lambda}_{AP}. \tag{4.16}$$

   Application of the spinor transformation (4.14) then gives a similarity transformation of L_{BA} [see also the section below on SL(2,C) Diagrams].

3. Whereas Lorentz transformations L_{BA} relate inertial frames and are generally time independent, the eigenspinor describes the orientation and velocity of the observed object as it changes in time. It is therefore time dependent. We will usually express it as an explicit function $\Lambda(\tau)$ of the proper time τ of the observed object. At each τ it is the Lorentz transformation relating the inertial frame instantaneously commoving with the particle at τ to the inertial frame of the observer.

[2]Formally, the spinor transformations (4.14), in which L acts from the left, make the eigenspinors elements of the *left regular representation* of the geometric algebra restricted to the group SL(2,C).

[3]In both cases, L is the transformation as viewed by the initial observer. An important invariance of the form of L is discussed in more detail later in this chapter.

The spinor transformation (4.14) can also be expressed in terms of the transformation L_r of the charge as seen from the frame P, the former *rest* frame of the charge. Thus, $L_r L_r^\dagger$ is the proper velocity of the transformed charge as viewed from frame P, and L_r is its eigenspinor as seen in P. To express properties of the transformed charge (as seen from the lab) in terms of the eigenspinor Λ of its former rest frame, we must apply L_r to relate the new and former rest frames of the charge. The effect on the eigenspinor of the transformation is[4]

$$\Lambda \to \Lambda L_r \,. \tag{4.17}$$

4.3 Invariant Planes and Directions

Any rotation in the $\mathbf{e}_1\bar{\mathbf{e}}_2$ plane changes vectors in the plane but leaves the plane itself invariant:

$$\mathbf{e}_1\bar{\mathbf{e}}_2 \to R\mathbf{e}_1 R^\dagger \bar{R}^\dagger \bar{\mathbf{e}}_2 \bar{R} = R\mathbf{e}_1\bar{\mathbf{e}}_2\bar{R} = \mathbf{e}_1\bar{\mathbf{e}}_2 \tag{4.18}$$

where $R = \exp(\mathbf{e}_1\bar{\mathbf{e}}_2\theta/2)$. The time axis $\mathbf{e}_0$ and the rotation axis $\mathbf{e}_3$ are also invariant, and it follows that the spacetime plane $\mathbf{e}_3\bar{\mathbf{e}}_0$ as well as any paravector in it is also invariant. A more general rotation $R = \exp(-i\boldsymbol{\theta}/2)$ leaves the rotation plane $i\hat{\boldsymbol{\theta}}$ and the spacetime directions $\mathbf{e}_0$ and $\hat{\boldsymbol{\theta}}$ (and any linear combination of them) invariant. A boost $B = \exp(\mathbf{w}/2)$ leaves the timelike spacetime plane $\hat{\mathbf{w}}\bar{\mathbf{e}}_0$ and paravectors in spatial directions perpendicular to $\hat{\mathbf{w}}$ invariant. It follows that the spatial plane $i\hat{\mathbf{w}}$ is also invariant.

More generally, any simple Lorentz transformation

$$L = \exp(\mathbf{W}/2) \tag{4.19}$$

produces a rotation in the spacetime plane $\mathbf{W}$ and leaves the dual biparavectors $\mathbf{W}$ and $-i\mathbf{W}$ invariant. Even compound transformations have invariant dual planes in spacetime. To find them for an arbitrary $\mathbf{W} = \mathbf{w} - i\boldsymbol{\theta}$, we define

$$\begin{aligned} \tan 2\alpha &= \frac{-2\mathbf{w}\cdot\boldsymbol{\theta}}{\mathbf{w}^2 - \boldsymbol{\theta}^2} \\ \mathbf{a} &= \mathbf{w}\cos\alpha - \boldsymbol{\theta}\sin\alpha \\ \mathbf{b} &= \boldsymbol{\theta}\cos\alpha + \mathbf{w}\sin\alpha \end{aligned} \tag{4.20}$$

[4]This relation makes Λ an element of the *right regular representation* of the geometric algebra restricted to SL(2,C).

and then find that we can expand $\mathbf{W}$ in the form

$$\mathbf{W} = \mathbf{w} - i\boldsymbol{\theta} = e^{i\alpha}(\mathbf{a} - i\mathbf{b}) , \qquad (4.21)$$

where $\mathbf{a}\cdot\mathbf{b} = 0$ and α is a real phase angle. Thus, $\mathbf{a} - i\mathbf{b}$ and its dual $-i(\mathbf{a}-i\mathbf{b})$ are simple biparavectors representing spacetime planes that are invariant under $L = \exp(\mathbf{W}/2)$:

$$\mathbf{a} - i\mathbf{b} \to L(\mathbf{a} - i\mathbf{b})\bar{L} = e^{-i\alpha}L\mathbf{W}\bar{L} = e^{-i\alpha}\mathbf{W} = \mathbf{a} - i\mathbf{b}\,.$$

4.3.1 Transformations of Transformations

As seen in the previous section, Lorentz transformations transform in the same way as biparavectors. However, they themselves are *not* biparavectors since they generally have scalar as well as complex vector parts. Nevertheless, they may always be written as the exponential of a biparavector:

$$L_{BA} = \exp(\mathbf{W}_{BA}/2) , \qquad (4.22)$$

with the transformation

$$L_{BA} \to LL_{BA}\bar{L} = \exp\left(\frac{1}{2}L\mathbf{W}_{BA}\bar{L}\right). \qquad (4.23)$$

The last equality is one that will prove very useful. It is easily verified by expanding the exponential and observing that intermediate factors of $\bar{L}_{BA}L_{BA}$ cancel.

Example 4.1. The product $B(\mathbf{w})\,R$ of a boost $B(\mathbf{w}) = \exp(\mathbf{w}/2)$ and a rotation R can also be written in the alternative forms

$$\begin{aligned} B(\mathbf{w})\,R &= R\bar{R}B(\mathbf{w})\,R = RB(\mathbf{w}') \\ &= R^{1/2}R(\mathbf{w}'')\,R^{1/2}, \end{aligned} \qquad (4.24)$$

where $\mathbf{w}' = \bar{R}\mathbf{w}R$ and $\mathbf{w}'' = \bar{R}^{1/2}\mathbf{w}R^{1/2}$.

Since biparavectors generally have different coefficients (and split into different real and imaginary parts) when expressed in different frames, when we transform *between* frames, how do we know which frame to use in the expression of L? Consider $L_{BA} = \exp\left[\frac{1}{4}W_{BA}^{\mu\nu}\left\langle \mathbf{e}_\mu\bar{\mathbf{e}}_\nu\right\rangle_V\right]$, which transforms from A to B in the sense that given properties with respect to A are transformed by

L_{BA} into ones with respect to B. The orthonormal biparavectors $\langle \mathbf{e}_\mu \bar{\mathbf{e}}_\nu \rangle_V$ are those at rest relative to the observer whereas those at rest in A as seen by B are $L_{BA} \langle \mathbf{e}_\mu \bar{\mathbf{e}}_\nu \rangle_V \bar{L}_{BA} =: \langle \mathbf{u}_\mu \bar{\mathbf{u}}_\nu \rangle_V$. The transformation

$$\begin{aligned} L_{BA} &= \exp\left[\frac{1}{4} W_{BA}^{\mu\nu} \langle \mathbf{e}_\mu \bar{\mathbf{e}}_\nu \rangle_V\right] \\ &= L_{BA} L_{BA} \bar{L}_{BA} = \exp\left[\frac{1}{4} W_{BA}^{\mu\nu} L_{BA} \langle \mathbf{e}_\mu \bar{\mathbf{e}}_\nu \rangle_V \bar{L}_{BA}\right] \\ &= \exp\left[\frac{1}{4} W_{BA}^{\mu\nu} \langle \mathbf{u}_\mu \bar{\mathbf{u}}_\nu \rangle_V\right] \end{aligned} \tag{4.25}$$

has the same coefficients $\frac{1}{4} W_{BA}^{\mu\nu}$ with respect to either biparavector basis, and hence with respect to observers in either frame. This important result states that it does not matter which of the two frames we use to express the transformation between the frames, the scalar coefficients are the same. It is a consequence of the invariance of the biparavector $\mathbf{W}_{BA}$ under the Lorentz transformation $L_{BA} = \exp(\mathbf{W}_{BA}/2)$.

Exercise 4.2. As an example of the important relation (4.25), consider the rotation R composed of a 90° rotation about $\mathbf{e}_3$ (in the $\mathbf{e}_1\mathbf{e}_2 = i\mathbf{e}_3$ plane) followed by a 90° rotation about $\mathbf{e}_2$ (see Fig. 4.1). Perform this rotation on a book in the $\mathbf{e}_1\mathbf{e}_3$ plane and sketch the result. Repeat for the same two 90° rotations acting in the opposite order, and show thereby that rotations about different axes generally do not commute. Next consider how the basis triad $\{\mathbf{e}_1, \mathbf{e}_2, \mathbf{e}_3\}$ is related to the transformed triad $\{\mathbf{u}_1, \mathbf{u}_2, \mathbf{u}_3\}$ that results from the rotation R described above applied to $\{\mathbf{e}_1, \mathbf{e}_2, \mathbf{e}_3\}$. Show that the rotation composed of a 90° rotation about $\mathbf{u}_3$ followed by a 90° rotation about $\mathbf{u}_2$ has the same effect on the book as R. (Try some other rotations to convince yourself that this result does not depend on the particular angles chosen.) Finally, note that if the *hybrid* ("body-fixed") rotation axes $\{\mathbf{e}_2, \mathbf{u}_3\}$ are used, the order of rotations must be reversed to produce the same result: first rotate about $\mathbf{e}_2$ and then about $\mathbf{u}_3$, each by 90°. See Fig. 4.1.

The relation (4.25) can be understood for rotations by recognizing that any product of rotations can be combined into a single rotation R in a plane $-i\hat{\boldsymbol{\theta}}$ by an angle θ,and that the plane itself is invariant under the rotation: the axis of rotation $\hat{\boldsymbol{\theta}}$ normal to the plane has the same components in the initial triad $\{\mathbf{e}_1, \mathbf{e}_2, \mathbf{e}_3\}$ as in the rotated basis $\{\mathbf{u}_1, \mathbf{u}_2, \mathbf{u}_3\}$,and the bivector $-i\boldsymbol{\theta}$ has the same components in the bivector basis $\{\langle \mathbf{e}_j \bar{\mathbf{e}}_k \rangle_V\}$ as in the rotated basis $\{\langle \mathbf{u}_j \bar{\mathbf{u}}_k \rangle_V\}$. Both the initial triad $\{\mathbf{e}_1, \mathbf{e}_2, \mathbf{e}_3\}$ and the final rotated triad $\{\mathbf{u}_1, \mathbf{u}_2, \mathbf{u}_3\}$ are space-fixed bases, with elements related by the rotation: $\mathbf{u}_k = R\mathbf{e}_k R^\dagger$ (see Fig. 4.1). It is also possible to

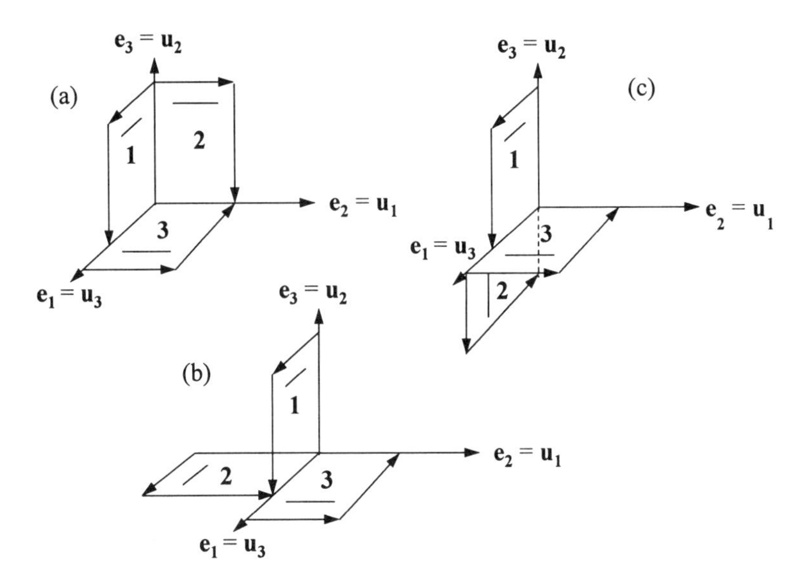

Fig. 4.1. The rotation R is a rotation of 90° about e_3 followed by a rotation of 90° about e_2, as shown in (a). The planes labeled 1,2,3, show successive orientations of a book as it is rotated by R. The rotation R takes the basis $\{e_1, e_2, e_3\}$ into $\{u_1, u_2, u_3\}$,and the same rotation with respect to the transformed basis, namely a rotation of 90° about u_3 followed by one of 90° about u_2, gives the same result, as seen in (b). Hybrid axes, consisting of the axes as they are rotated (the "body-fixed axes") can also be used, but then the order of the rotations must be reversed, as shown in (c): a rotation of 90° about e_2 followed by one of 90° about u_3 is equivalent to R.

choose a hybrid, "body-fixed" basis in which the order of rotations in the product must be reversed (see problem 1).

One can generalize the invariance of the rotation plane to an arbitrary physical Lorentz transformation L: as we will see in more detail below, for any such L, one can identify two invariant paravector planes that are dual to each other ($i\hat{\boldsymbol{\theta}}$ and $\hat{\boldsymbol{\theta}}\bar{e}_0$ for the rotation; more generally the plane of the transformation and its dual plane for a simple transformation). These invariant planes have the same biparavector components in the initial as in the rotated bases.

4.4 The Group SL(2,C)

The restricted Lorentz transformations in spinorial form, L, constitute a *group*. The group operation is multiplication, which as we have seen, is associative but generally noncommutative (*nonabelian*). The required identity element is the unit 1, obtained from (4.19) when $\mathbf{w} = \boldsymbol{\theta} = 0$. The inverse of any element L is $\bar{L}$, obtained from (4.19) by changing the sign of the exponent. The group property of *closure* ensures that if L_{AB} and L_{BC} are restricted Lorentz transformations, so is their product $L_{AC} = L_{AB}L_{BC}$. This is the appropriate generalization to non-commuting transformations of the velocity composition rule $u_{AC} = u_{AB}u_{BC}$ for aligned boosts (see chapter 2, problem 8). The group is unimodular (*special*) and, as noted above, is called SL(2,C). Since both L and $-L$ induce the same Lorentz transformation of a spacetime vector, SL(2,C) is said to be the universal two-fold covering group of $SO_+(1,3)$, the group of restricted Lorentz transformations of spacetime vectors. Elements L of SL(2,C) that are also members of the even subalgebra of $\mathcal{C}\ell_3$ are unitary $\left(L^{-1} = L^\dagger\right)$ and form the three-parameter subgroup SU(2), which is the two-fold covering group of rotations SO(3) in 3-dimensional Euclidean space. The three parameters that specify a rotation may be expressed as the three Euler angles α, β, γ, when the elements of SO(3) are written as products of rotations first around $\mathbf{e}_3$ by α, then about $\mathbf{e}_2$ by β, and finally about $\mathbf{e}_3$ again, this time by γ.

The *real* elements of SL(2,C) represent boosts, but they do *not* form a group since the product of real elements is not generally real. In other words, the product of boosts is not generally a pure boost but may contain a rotational component. Various abelian (commutative) subgroups of SL(2,C) can be identified, such as the group of rotations in any single plane, the boosts in any given direction, or the *rifling subspace* formed by rotations in a spatial plane combined with boosts normal to the plane.

Example 4.3. The product of boosts $B_1 = (2 + \mathbf{e}_1)/\sqrt{3}$ and $B_2 = (2 + \mathbf{e}_2)/\sqrt{3}$ in orthogonal directions is

$$L = B_1 B_2 = (4 + 2\mathbf{e}_1 + 2\mathbf{e}_2 + \mathbf{e}_1\mathbf{e}_2)/3\,. \tag{4.26}$$

Since L is not real, it is not a boost. But it can be expressed as the product $L = BR$ of a boost

$$B = \left(LL^\dagger\right)^{1/2} = \frac{17 + 10\mathbf{e}_1 + 6\mathbf{e}_2}{3\sqrt{17}} \tag{4.27}$$

and a rotation by the angle $2\arctan(1/4) \approx 28°$ in the $\mathbf{e}_1\mathbf{e}_2$ plane:

$$R = \frac{\langle L\rangle_+}{\langle B\rangle_S} = \frac{4+\mathbf{e}_1\mathbf{e}_2}{\sqrt{17}}. \tag{4.28}$$

Exercise 4.4. Let $B_1 = \exp(\mathbf{w}_1/2)$ and $B_2 = \exp(\mathbf{w}_2/2)$ be two arbitrary boosts. Show that when their product B_1B_2 is expressed as a product BR, the rotation $R = \exp(-i\boldsymbol{\theta}/2)$ is in the plane of $\mathbf{w}_1$ and $\mathbf{w}_2$ and indeed

$$-i\hat{\boldsymbol{\theta}}\tan\frac{\theta}{2} = \frac{\langle\hat{\mathbf{w}}_1\hat{\mathbf{w}}_2\rangle_V \tanh\frac{w_1}{2}\tanh\frac{w_2}{2}}{1+\langle\hat{\mathbf{w}}_1\hat{\mathbf{w}}_2\rangle_S \tanh\frac{w_1}{2}\tanh\frac{w_2}{2}}. \tag{4.29}$$

4.4.1 SL(2,C) Diagrams

Every physically accessible inertial frame is uniquely determined relative to an arbitrary origin by a restricted Lorentz transformation, and thus by the six real parameters[5] $\left(w^k,\theta^k\right)$, $k = 1,2,3$, in the domains $w^k \in \mathbb{R}$ and $0 \le \theta^k < 2\pi$. Every point in *frame space* (the 6-dimensional parameter space restricted to these domains) thus represents an inertial frame. Every ordered pair of points corresponds, within a sign (a factor of ±1), to a member $L \in$ SL(2,C) which relates the two inertial frames. In the abstract depiction of the parameter space in figure 4.2, points A and B represent the inertial frames of observers A and B, and the line connecting them represents (to within a sign) both the transformation L_{AB}, which gives the motion of B as seen by A, and L_{BA}, which gives the motion of A relative to B. The two transformations, which are inverses of one another, are distinguished by placing an arrow on the line: an arrow from A to B denotes L_{AB}, whereas one from B to A represents L_{BA}. The direction of the arrow makes sense from the standpoint of an active transformation on the frame tetrad: L_{AB} applied to a tetrad $\{\mathbf{e}_\mu\}$ initially at rest in the observer's frame A transforms it to one $\{\mathbf{u}_\mu\}$ commoving with B but observed by A. In this sense, one may describe L_{AB} as being the transformation from A to B. Unfortunately, in a passive sense, L_{AB} describes the transformation (of an observer) from B to A. These contrary ways of describing the same transformation can easily lead to confusion. It is usually helpful to specify the meaning of the proper velocity $u_{AB} = L_{AB}L_{AB}^\dagger$, which gives the velocity of frame B as seen from frame A.

[5]Or equivalently by the antisymmetric second-rank tensor $W^{\mu\nu} = -W^{\nu\mu}$ described in section 2.4. The fact that there are six parameters reflects the six coefficients needed to specify a point in the six-dimensional linear space of biparavectors (= spacetime bivectors).

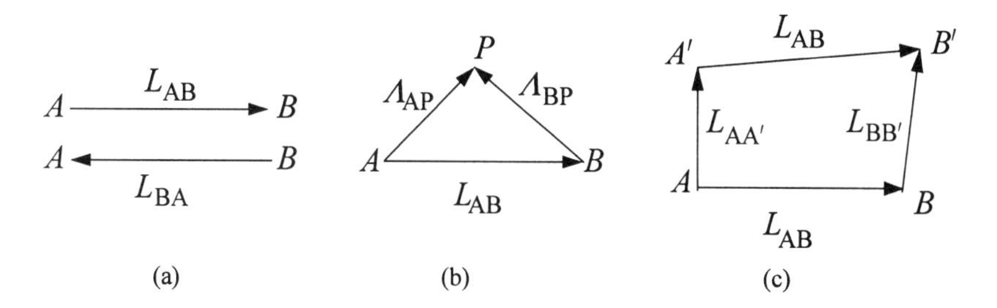

Fig. 4.2. Lorentz transformations in frame space. The transformation L_{AB} specifies the motion and orientation of frame B as viewed from frame A. The product of transformations on a closed loop is ± 1.

Moving from point to point in frame space corresponds to applying successive Lorentz transformations, that is, to multiplying the spinorial elements L together. The multiplication order is from left to right as one proceeds in the direction of the arrows.[6] The product of successive transformations corresponding to a closed loop in frame space must equal ± 1. In the simplest case, we consider only two frames A and B, and $L_{AB}L_{BA} = \pm 1$ [see Fig. 4.2(a)]. Thus, within a sign, $L_{BA} = \bar{L}_{AB}$. If we consider three inertial frames, say two observers A and B and a particle P [see Fig. 4.2(b)], then $L_{BA}\Lambda_{AP}\bar{\Lambda}_{BP} = \pm 1$, which implies (4.13) to within a sign. With four inertial frames, A, A', B, B' one finds $L_{AA'}L_{A'B'} = L_{AB}L_{BB'}$, and if B' is defined to be related to A' by the same transformation that relates B to A, that is, if $L_{AB} =: L_{A'B'}$, then $L_{BB'} = L_{BA}L_{AA'}\bar{L}_{BA}$ (see Fig. 4.2(c)), which gives the result of the transformation at A of $L_{AA'}$ as seen by observer B. It has just the form of the similarity transformation used for biparavectors.

The fact that generally $L_{BB'} \neq L_{AA'}$ even when $L_{A'B'}$ is defined equal to L_{AB}, is a manifestation of the nonlinearity of frame space. A consequence of this nonlinearity is that vectors, representing Lorentz transformations, cannot be freely translated from one position to another in the space. The lack of translation invariance is indicated in Fig. 4.2(c) by the two vectors labeled L_{AB} : they have different effects (lengths and directions) when applied at the two different frames A and A'. (Needless to say, our 2-dimensional depiction of 6-dimensional frame space is schematic, at best. The rotational subspace spanned by the three imaginary (spatial bivector) directions is the 3-dimensional surface of a 4-dimensional

[6] The actual order of *application* is reversed, that is in the right-to-left order. This is necessary since each observer will apply each transformation from his/her frame.

sphere. The rifling subspace formed by one rotation axis that coincides with the boost direction has the geometry of a cylinder.)

4.5 Time Evolution of Eigenspinor

For any given observer, the eigenspinor depends on the proper time τ of the charge, and the motion of the charge is determined by the dependence $\Lambda(\tau)$. The "time-evolution" operator relating eigenspinors at different proper times is itself a Lorentz transformation:

$$\Lambda(\tau_2) = L(\tau_2, \tau_1)\Lambda(\tau_1) \tag{4.30}$$

with $L(\tau_2, \tau_1) = \Lambda(\tau_2)\,\bar{\Lambda}(\tau_1)$.

The proper time-rate of change of the eigenspinor can always be written

$$\dot{\Lambda} = \frac{1}{2}\Omega\Lambda \tag{4.31}$$

where $\Omega := 2\dot{\Lambda}\bar{\Lambda}$. The spatial reversal of (4.31) is

$$\bar{\dot{\Lambda}} = \frac{d}{d\tau}\bar{\Lambda} = \frac{1}{2}\bar{\Lambda}\bar{\Omega}. \tag{4.32}$$

Note that since the proper time is a scalar, the proper-time derivative of a spatially reversed element is the same as the spatial reversal of the proper-time derivative. Since $\Lambda\bar{\Lambda} = 1$,

$$\frac{d}{d\tau}\Lambda\bar{\Lambda} = \dot{\Lambda}\bar{\Lambda} + \Lambda\bar{\dot{\Lambda}} = \frac{1}{2}\left(\Omega + \bar{\Omega}\right) = 0 \tag{4.33}$$

and thus the scalar part of Ω vanishes. It is therefore a pure complex vector and represents a spacetime bivector. Since the proper time τ is a Lorentz scalar, Ω also transforms as a covariant spacetime bivector:

$$\begin{aligned} \Omega &= 2\dot{\Lambda}\bar{\Lambda} \\ &\to L\Omega\bar{L}\,. \end{aligned} \tag{4.34}$$

Physically, Ω may be interpreted as the spacetime rotation rate of the particle frame. This interpretation is justified by expanding the eigenspinor $\Lambda(\tau + d\tau)$ about τ:

$$\begin{aligned} \Lambda(\tau + d\tau) &= \Lambda(\tau) + \dot{\Lambda}(\tau)\,d\tau = \left(1 + \tfrac{1}{2}\Omega d\tau\right)\Lambda(\tau) \\ &= \exp\left(\tfrac{1}{2}\Omega d\tau\right)\Lambda(\tau) \end{aligned} \tag{4.35}$$

to lowest order in $d\tau$. Thus, from τ to $\tau + d\tau$, the eigenspinor picks up an additional spacetime rotation of $\Omega d\tau$.

If (4.31) can be solved, then the time evolution of any spacetime vector or bivector that is fixed in the particle frame is readily found. For example, the momentum p has the fixed value m in the rest frame of the charge. Its proper-time derivative is

$$\begin{aligned}
\dot{p} &= \dot{\Lambda} mc \Lambda^{\dagger} + \Lambda mc \dot{\Lambda}^{\dagger} \\
&= \frac{1}{2}\Omega \Lambda mc \Lambda^{\dagger} + \Lambda mc \Lambda^{\dagger} \Omega^{\dagger} \\
&= \frac{1}{2}\left(\Omega p + p \Omega^{\dagger}\right) \\
&= \langle \Omega p \rangle_{\Re}
\end{aligned} \tag{4.36}$$

Exercise 4.5. Show that in the special case that $\Omega = -i\omega$ is purely imaginary, (4.36) describes a pure rotation at the angular rate ω and reduces to the familiar form

$$\dot{\mathbf{p}} = \boldsymbol{\omega} \times \mathbf{p}\, , \dot{E} = 0\, . \tag{4.37}$$

Exercise 4.6. Let the biparavector $\Omega = \left\langle a\bar{b}\right\rangle_V$, where a and b are real paravectors. Recall that

$$\langle \Omega p \rangle_{\Re} = a \left\langle \bar{b} p \right\rangle_S - b \left\langle \bar{a} p \right\rangle_S \, , \tag{4.38}$$

and show that this vanishes if p is orthogonal to the paravector plane spanned by a and b. Show that as a result, such a paravector p is invariant under the time evolution (4.36).

4.6 Thomas Precession

As an application of the evolution equation (4.31), we derive the precession of an accelerated frame. It is possible to describe the motion of a point by an eigenspinor that is a pure boost, that is, with $\Lambda = B$. Since

$$B = u^{1/2} = \frac{u + 1}{\sqrt{2(\gamma + 1)}}\, , \tag{4.39}$$

the spacetime rotation rate (4.34) is

$$\begin{aligned}
\Omega &= 2\left\langle \dot{B}\bar{B}\right\rangle_V = \frac{\langle \dot{u}\left(\bar{u} + 1\right)\rangle_V}{\gamma + 1} \\
&= \dot{\mathbf{u}} - \frac{\left(\dot{\mathbf{u}} \cdot \mathbf{u}\right)\mathbf{u}}{\gamma\left(\gamma + 1\right)} - i\frac{\dot{\mathbf{u}} \times \mathbf{u}}{\gamma + 1}\, .
\end{aligned} \tag{4.40}$$

The negative imaginary part gives the spatial rotation rate of the accelerated frame; it is known as the (proper) Thomas precession rate:

$$\omega_{Th} = i \left\langle \Omega \right\rangle_{\Im} = \frac{\dot{\mathbf{u}} \times \mathbf{u}}{\gamma + 1} . \tag{4.41}$$

Thomas precession was important in the history of the concept of electron spin, since it explained in the context of the nonrelativistic theory why an electron g factor of $g = 2$, as required by observed Zeeman splittings, did not require twice the observed fine structure intervals in hydrogen. Note that the Thomas precession is nonzero only if the acceleration is not collinear with the velocity. In such cases, successive boosts of the eigenspinor are in different directions and do not commute. The physical origin of the precession can thus be traced to the fact that nonaligned boosts do not commute and that their composition leads to a boost times a rotation in the plane of the boosts. Note also that the Thomas precession is not a covariant quantity; it is, instead, the space-like part of the covariant rotation rate biparavector Ω.

4.7 Spinorial Form of Lorentz-Force Equation

A comparison of (4.36) with the Lorentz-force equation (4.3) shows that the spacetime rotation rate Ω of a charged particle may be identified with the electromagnetic field $\mathbf{F}$ at the position of the charge

$$\Omega = \frac{e}{mc} \mathbf{F} , \tag{4.42}$$

and that the Lorentz-force equation itself can be expressed in the form

$$\dot{\Lambda} = \frac{e}{2mc} \mathbf{F} \Lambda . \tag{4.43}$$

These relations specify explicitly the nature of the physical action of a simple electromagnetic-field biparavector $\mathbf{F}$: it induces a rotation at the rate Ω in the spacetime plane of $\mathbf{F}$.

Since the field $\mathbf{F}$ is related to the field $\mathbf{F}_r$ in the inertial frame commoving with the charge by

$$\mathbf{F} = \Lambda \mathbf{F}_r \bar{\Lambda} \tag{4.44}$$

the spinorial form of the Lorentz-force equation can also be written

$$\dot{\Lambda} = \frac{e}{2mc}\Lambda \mathbf{F}_r \,. \tag{4.45}$$

Note that the spinorial form (4.43) is invariant under a fixed Lorentz transformation to the particle frame: $\Lambda \to \Lambda \bar{L}$, where $\dot{L} = 0$. Consequently, the charge does not really need to be at rest in its reference frame; it is sufficient if it is moving there at constant velocity. On the other hand, a change in the observer, $\Lambda \to \Lambda' = L\Lambda$ where L is constant, will transform the spinorial force equation (4.43) to

$$\dot{\Lambda}' = \frac{e}{2mc}\mathbf{F}'\Lambda', \quad \mathbf{F}' = L\mathbf{F}\bar{L}\,,$$

which has the same form as (4.43) but uses the field $\mathbf{F}'$ seen by the new observer. Of course the field in the particle reference frame is unchanged by changes in the observer frame, and indeed the Lorentz-force equation in form (4.45) is invariant under $\Lambda \to L\Lambda$, with L constant.

4.8 Motion in Constant Fields

According to the form (4.34) of the Lorentz transformation for spacetime bivectors, Ω^2 and hence $\mathbf{F}^2$ are Lorentz invariants. Now

$$\mathbf{F}^2 = \left(\mathbf{E}^2 - c^2\mathbf{B}^2\right) + 2ic\mathbf{E}\cdot\mathbf{B}\,, \tag{4.46}$$

so that the invariance of $\mathbf{F}^2$ implies the invariance independently of both $\mathbf{E}^2 - c^2\mathbf{B}^2$ and $\mathbf{E}\cdot\mathbf{B}$. If $\langle\Omega^2\rangle_{\Im} = 0$, then Ω is simple and $\mathbf{F}$ generates a rotation in a single spacetime plane. If furthermore $\langle\Omega^2\rangle_{\Re} > 0$, the rotation occurs in a timelike plane and is predominantly a boost, whereas if $\langle\Omega^2\rangle_{\Re} < 0$, the spacetime rotation is about a spacelike plane and is predominantly a spatial rotation.

In any case, if the electromagnetic field $\mathbf{F}$ at the position of the charge is constant, the spinorial Lorentz-force equation (4.43) is easily integrated to give

$$\Lambda\left(\tau\right) = \exp\left(\frac{e\mathbf{F}}{2mc}\tau\right)\Lambda(0)\ . \tag{4.47}$$

Of course our choice of the initial proper time is arbitrary and we can equally well write

$$\Lambda(\tau) = \exp\left(\frac{e\mathbf{F}}{2mc}\left(\tau - \tau_0\right)\right)\Lambda(\tau_0)\ . \tag{4.48}$$

For every initial position $r(0)$ and every initial proper velocity $u(0) = \Lambda(0)\,\Lambda^\dagger(0)$, the solution (4.47) generates a new trajectory:

$$r(\tau) = r(0) + \int_0^\tau d\tau' u(\tau') \ , \tag{4.49}$$

where

$$\begin{aligned} u(\tau) &= \Lambda(\tau')\,\Lambda^\dagger(\tau') \\ &= \exp\left(\frac{e\mathbf{F}}{2mc}\tau\right) u(0) \exp\left(\frac{e\mathbf{F}^\dagger}{2mc}\tau\right) . \end{aligned} \tag{4.50}$$

The solution (4.47) therefore represents a family of trajectories, each one of which satisfies the Lorentz-force equation.

Exercise 4.7. Prove that for any constant, uniform field $\mathbf{F}$, the motion of a charge in that field always leaves invariant the field $\mathbf{F}_r$ in the instantaneous rest frame of charge.

Let's consider solutions for several special field configurations.

4.8.1 Pure Electric Field

In a pure constant electric field $\mathbf{F} = \mathbf{E}$, a charge released from rest $[\Lambda\,(0) = 1]$ accelerates and experiences, during its acceleration, a constant field in its rest frame given by

$$\mathbf{F}_r = \bar{\Lambda}\mathbf{F}\Lambda = \mathbf{F} = \mathbf{E}\,, \tag{4.51}$$

since $\Lambda = \exp[e\mathbf{F}\tau/\,(2mc)]$ commutes with $\mathbf{F}$. Its rest-frame acceleration is therefore the constant $\mathbf{a} := (e/m)\,\mathbf{F} = (e/m)\,\mathbf{E}$. The proper velocity in the lab frame is

$$u = \Lambda\Lambda^\dagger = \exp(\mathbf{a}\tau/c) \ , \tag{4.52}$$

and its integral gives the spacetime position

$$r(\tau) = c\int u\,d\tau = r(0) + c^2\mathbf{a}^{-1}\left[\exp(\mathbf{a}\tau/c) - 1\right] . \tag{4.53}$$

It is convenient to choose an origin of coordinates so that $r(0) = c^2\mathbf{a}^{-1}$. Then (4.53) reduces to

$$r(\tau) = c^2\mathbf{a}^{-1}\exp(\mathbf{a}\tau/c) \ , \tag{4.54}$$

with a spatial-vector part

$$\mathbf{r}(\tau) = c^2\mathbf{a}^{-1}\cosh(a\tau/c)\ , \tag{4.55}$$

and a scalar part

$$t(\tau) = ca^{-1}\sinh(a\tau/c) \tag{4.56}$$

where a is the length of $\mathbf{a}$. Since $-r\bar{r} = \mathbf{r}^2 - c^2t^2 = c^4a^{-2}$, a spacetime plot of $\mathbf{r}(t)$ gives a hyperbola. For example, in units of the length c^2/a, with $\mathbf{a} = c^2\mathbf{e}_1$, $[x, ct] = [\cosh c\tau, \sinh c\tau]$ gives the plot of Fig. (4.3). The particle is seen to approach the origin from positive x, slowing down as it comes. It comes instantaneously to rest where the world line crosses the x axis and then accelerates away. The motion is called *hyperbolic*.

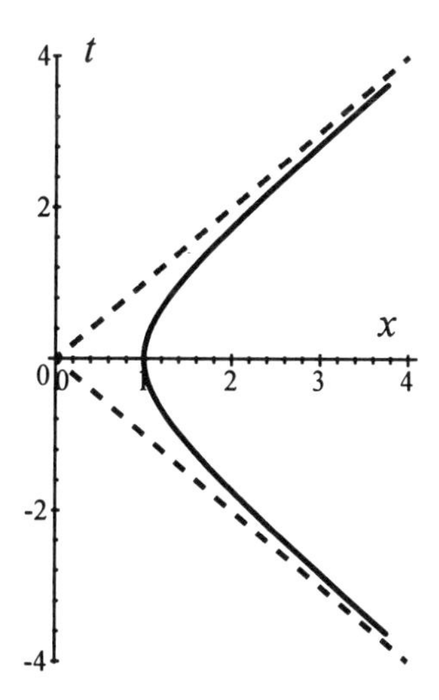

Fig. 4.3. A spacetime diagram of hyperbolic motion, showing a particle with constant rest-frame acceleration in the x direction.

4.8.2 Pure Magnetic Field

On the other hand, in a constant magnetic field $\mathbf{F} = ic\mathbf{B}$, the Lorentz transformation Λ is a pure spatial rotation, and the velocity of the charge rotates at proper angular velocity $\boldsymbol{\omega}_c = -e\mathbf{B}/m$, which is also known as the proper cyclotron frequency:

$$u(\tau) = e^{-i\omega_c\tau/2}u(0)\, e^{i\omega_c\tau/2}. \tag{4.57}$$

Because the length of a vector is unchanged by a spatial rotation, the speed and hence the energy of the charge is constant in the magnetic field. The particle follows a helix in what is called cyclotron motion, with the component $\mathbf{v}_\parallel$ of the coordinate velocity $\mathbf{v}$ parallel to $\mathbf{B}$ remaining fixed and the perpendicular component $\mathbf{v}_\perp$ rotating in a circle:

$$\begin{aligned} \mathbf{v}(t) &= e^{-i\omega_c t/2\gamma}\mathbf{v}(0)\, e^{i\omega_c t/2\gamma} = \mathbf{v}_\parallel(0) + e^{-i\omega_c t/\gamma}\mathbf{v}_\perp(0) \\ \gamma &= \sqrt{1+\mathbf{u}^2} = 1/\sqrt{1-\mathbf{v}^2/c^2}\,. \end{aligned} \tag{4.58}$$

The particle is locked to the magnetic field; its path twists around the field lines. It cannot move far in the plane perpendicular to $\mathbf{B}$ unless other forces (or collisions) are present. Note that the proper angular velocity $\boldsymbol{\omega}_c$ (and therefore the period $2\pi/\omega_c$) is independent of the speed, and consequently higher energy particles follow larger orbits. The path is found by integrating (4.58):

$$\begin{aligned} \mathbf{r}(t) &= \mathbf{r}_0 + \mathbf{v}_\parallel(0)\, t + e^{-i\omega_c t/\gamma}\mathbf{r}_c \\ \mathbf{r}_c &\equiv \gamma\mathbf{v}(0) \times \boldsymbol{\omega}_c^{-1}, \end{aligned} \tag{4.59}$$

where $\mathbf{r}_0 + \mathbf{v}_\parallel(0)\, t$ is the center of the orbit and $\mathbf{r}_c$ is the initial radius vector of the helix.

Exercise 4.8. Show that the magnitude of the helical cyclotron radius is

$$r_c = \frac{|\mathbf{p}_\perp|}{eB}, \tag{4.60}$$

where $\mathbf{p}_\perp$ is the component of the momentum perpendicular to the magnetic field $\mathbf{B}$.

4.8.3 Electric and Magnetic Fields

If both electric and magnetic fields are nonzero constants, the solution (4.47) can still be used. It describes a mixture of accelerating and rotating motion that is generally difficult to picture. However, if $\mathbf{F}^2 \neq 0$, a "drift frame" can be found in which the electric and magnetic parts of $\mathbf{F}_{drift}$ are parallel (or in which one of them vanishes if $\mathbf{E} \cdot \mathbf{B} = 0$) and the motion (4.47) is relatively simple. In the drift frame the motion is rifle-like: there is a rotation about the direction of the boost and the rotation and boost parts commute (see problem 11). In the lab frame, the rifle-like motion is compounded with the drift velocity and "drifts" through the frame.

4.8.4 Simple Fields

If $\mathbf{F}$ is simple, $\mathbf{E} \cdot \mathbf{B} = 0$ and one of $\mathbf{E}_{drift}$ or $\mathbf{B}_{drift}$ vanishes. We can align the axes so that $\mathbf{E} = E\mathbf{e}_1$ and $\mathbf{B} = B\bar{\mathbf{e}}_3$, and then

$$\mathbf{F} = E\mathbf{e}_1\bar{\mathbf{e}}_0 + cB\mathbf{e}_1\bar{\mathbf{e}}_2 \,. \tag{4.61}$$

If $\mathbf{E}^2 > c^2\mathbf{B}^2$, the field $\mathbf{F}$ can be factored

$$\begin{aligned} \mathbf{F} &= E\mathbf{e}_1\left(\bar{\mathbf{e}}_0 + \frac{cB}{E}\bar{\mathbf{e}}_2\right) \\ &= \mathbf{F}_{drift}\bar{u}_{drift} = u_{drift}\mathbf{F}_{drift} \,, \end{aligned} \tag{4.62}$$

into the boost to proper velocity $u_{drift} = \gamma_{drift}\left(\mathbf{e}_0 + cB\mathbf{e}_2/E\right)$ of a pure electric field $\mathbf{F}_{drift} = \gamma_{drift}^{-1}E\mathbf{e}_1$, where $\gamma_{drift} = E/\sqrt{E^2 - c^2B^2}$. In this case, the proper velocity u_{drift} of the drift frame lies in the plane of the field $\mathbf{F}$ (see Fig. 4.4 for an example.)

In the drift frame, $\mathbf{F}$ in this case is rotated in spacetime to the "vertical" plane $\mathbf{F}_{drift}$ with no spatial (magnetic) component. A boost $\bar{u}_{drift}^{1/2}$ can be found for any timelike (predominantly electric) simple field $\mathbf{F}$ that rotates it into a purely electric field. However, if $\mathbf{E}^2 < c^2\mathbf{B}^2$, the field is spacelike (predominantly magnetic) and the magnetic field cannot be eliminated by any boost. In this case, a drift frame can be found in which the electric component vanishes.

Exercise 4.9. Factor the field $\mathbf{F} = \mathbf{e}_1\bar{\mathbf{e}}_0 + 2\mathbf{e}_1\bar{\mathbf{e}}_2$ (in units of kV/m) into $\mathbf{F}_{drift}\bar{u}_{drift}$ where the field $\mathbf{F}_{drift}$ in the drift frame is purely magnetic. Find $\mathbf{F}_{drift}$ and u_{drift}. Sketch the fields in the lab frame and indicate the transformed unit paravectors of the drift frame.

4.8.5 Compound Fields

Compound fields are not single spacetime planes. Since for such fields the Lorentz invariant $\mathbf{E} \cdot \mathbf{B} \neq 0$, there is no frame in which either the electric or magnetic

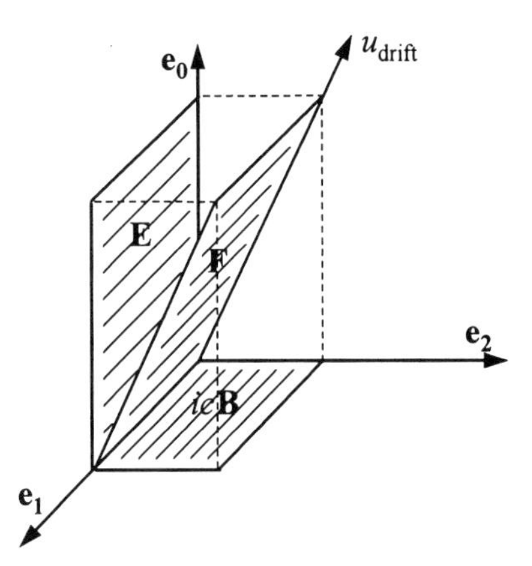

Fig. 4.4. The simple electromagnetic field $\mathbf{F} = 2\mathbf{e}_1\bar{\mathbf{e}}_0 + \mathbf{e}_1\bar{\mathbf{e}}_2$ (in units of kV/m) has electric and magnetic components in the lab frame but is purely electric in the drift frame, which moves with proper velocity $u_{drift} = \gamma_{drift}\left(\mathbf{e}_0 + \mathbf{e}_2/2\right)$. Note that the coordinates of this diagram are distinct from position coordinates. In general, there is a field diagram such as this at every point (event) in spacetime.

component of the field vanishes. However, there is a drift frame in which $\mathbf{E}_{drift}$ and $\mathbf{B}_{drift}$ are parallel and hence on orthogonal (dual) spacetime planes.

Since any boost along the aligned fields $\mathbf{E}_{drift}$ and $\mathbf{B}_{drift}$ gives another drift frame, there are an infinite number of such frames. One can always choose a drift frame whose proper velocity u_{drift} has a vector part perpendicular to $\mathbf{F}$, that is to both $\mathbf{E}$ and $\mathbf{B}$ as seen in the lab frame. The field in this drift frame is

$$\mathbf{F}_{drift} = \bar{u}_{drift}^{1/2}\mathbf{F}u_{drift}^{1/2} = \bar{u}_{drift}\mathbf{F}\,, \tag{4.63}$$

which gives a (real) proper drift velocity

$$u_{drift} = \mathbf{F}\mathbf{F}_{drift}^{-1} \tag{4.64}$$

whose square is

$$u_{drift}^2 = \frac{\mathbf{F}\mathbf{F}^\dagger}{\mathbf{F}_{drift}\mathbf{F}_{drift}^\dagger} = \frac{\mathbf{F}\mathbf{F}^\dagger}{\left|\mathbf{F}^2\right|} = \frac{\mathcal{E} + \mathbf{S}/c}{\frac{1}{2}\varepsilon_0\left|\mathbf{F}^2\right|}, \tag{4.65}$$

where $\mathcal{E} = \frac{1}{2}\varepsilon_0 (\mathbf{E}^2 + c^2\mathbf{B}^2)$ is the energy density of the field, $\mathbf{S} = \mathbf{E} \times \mathbf{B}/\mu_0$ is the Poynting vector (see chapter 3), and we noted that since $\mathbf{F}_{drift}$ and $\mathbf{F}^\dagger_{drift}$ are parallel, their product is the real invariant scalar $|\mathbf{F}^2_{drift}|$. (On the other hand, generally $\mathbf{F}$ and $\mathbf{F}^\dagger$ are *not* parallel and $\mathbf{F}\mathbf{F}^\dagger$is not a scalar!) Most texts give the coordinate velocity $\mathbf{v}$ of the drift frame in cases where $\mathbf{E} \cdot \mathbf{B} = 0$ and hence $\mathbf{F}^2$ is real. However, we can easily find a more general expression. Since $u\bar{u} = 1$,

$$u^2 + 1 = u(u + \bar{u}) = 2\gamma u = 2\gamma^2 (1 + \mathbf{v}/c) \,. \tag{4.66}$$

Thus $\mathbf{v}/c$ is the vector part of $u^2 + 1$ divided by its scalar part. In the case of the drift velocity,

$$\mathbf{v}_{drift} = \frac{c \left\langle u^2_{drift} \right\rangle_V}{\left\langle u^2_{drift} \right\rangle_S + 1} = \frac{\mathbf{S}}{\mathcal{E} + \frac{1}{2}\varepsilon_0 |\mathbf{F}^2|} \,. \tag{4.67}$$

(For an alternative solution, see problem 14). When $\mathbf{E} \cdot \mathbf{B} = 0$, $\mathbf{v}_{drift}/c$ reduces to $\mathbf{E} \times c\mathbf{B}$ divided by the larger of $\mathbf{E}^2$ or $c^2\mathbf{B}^2$ (see problem 9) and the motion in the drift frame is either uniformly accelerated (if $\mathbf{E}^2 > c^2\mathbf{B}^2$) or purely rotational (if $c^2\mathbf{B}^2 > \mathbf{E}^2$).

In the case of a constant simple null field $\mathbf{F}^2 = 0$, the drift frame moves at the speed of light and cannot be reached by massive charges or observers. However, the solution (4.47) has a simple expansion in this case:

$$\Lambda(\tau) = \left[1 + \frac{e}{2mc}\mathbf{F}\tau\right] \Lambda(0) \,, \qquad \mathbf{F}^2 = 0, \tag{4.68}$$

and

$$\begin{aligned} u(\tau) &= \Lambda(\tau)\,\Lambda^\dagger(\tau) \\ &= u(0) + \frac{e\tau}{mc}\left\langle \mathbf{F}u(0) \right\rangle_{\Re} + \left(\frac{e\tau}{2mc}\right)^2 \mathbf{F}u(0)\,\mathbf{F}^\dagger. \end{aligned} \tag{4.69}$$

Exercise 4.10. Find the spacetime momentum $p = mcu$ of a charge e released from rest at $\tau = 0$ in a constant simple null field. Express the part quadratic in τ in terms of the energy and momentum density of the field.

4.9 Motion in Varying Fields

If the electromagnetic field $\mathbf{F}$ is not constant in spacetime, the Lorentz-force equation generally becomes nonlinear and difficult to solve. The problem is that the field $\mathbf{F}(x)$ depends on the position x of the charge, but to know $x(\tau)$ we must have already solved the equation! However, there are still a number of important cases that we can solve.

If the field at the position of the charge is not a constant, but its "direction" is

$$\mathbf{F}(\tau) = \mathbf{F}_0 f(\tau), \tag{4.70}$$

where $f(\tau)$ is a complex scalar function, then the equation of motion (4.43) can still be integrated to give

$$\Lambda(\tau) = \exp(\mathbf{F}_0 s/2)\,\Lambda(0)\ , \tag{4.71}$$

where $s = e/(mc)\int_0^\tau d\tau'\, f(\tau')$. Since f is complex, the field $\mathbf{F}$ can still undergo duality rotations that rotate $\mathbf{E}$ into $c\mathbf{B}$ and $c\mathbf{B}$ into $-\mathbf{E}$. Such solutions prove useful in chapter 6 where we consider the motion of charges in an electromagnetic plane wave. (See also problem 16.)

4.9.1 Coulomb Field

Another important case is the field of a fixed point charge. Consider the relativistic motion of a charge e at $x = ct + r\hat{\mathbf{r}}$ in the Coulomb field

$$\mathbf{F} = -\frac{Ze}{4\pi\varepsilon_0}\frac{\hat{\mathbf{r}}}{r^2}\,. \tag{4.72}$$

The spinorial Lorentz-force equation (4.43) is then

$$\dot{\Lambda} = -\frac{\alpha\hat{\mathbf{r}}}{2mr^2}\Lambda\,, \tag{4.73}$$

where $\alpha = Ze^2/(4\pi\varepsilon_0 c)$. Energy conservation is guaranteed by relation (4.4). In addition. because the force is radial, the orbital angular momentum

$$\mathbf{l} = -i\langle p\bar{x}\rangle_{\Im} = \hat{\mathbf{l}}mr^2\dot{\phi} \tag{4.74}$$

is conserved: since $p\bar{u} = mc$ is real,

$$\begin{aligned}
\frac{d}{d\tau}\mathbf{l} &= -i\langle\dot{p}\bar{x}\rangle_{\Im} = -i\langle e\langle\mathbf{F}u\rangle_{\Re}\bar{x}\rangle_{\Im} \\
&= i\frac{\alpha c}{r^2}\langle\langle\hat{\mathbf{r}}u\rangle_{\Re}\bar{x}\rangle_{\Im} = i\frac{\alpha c}{r^2}\langle(\hat{\mathbf{r}}\gamma + \hat{\mathbf{r}}\cdot\mathbf{u})\bar{x}\rangle_{\Im} \\
&= 0\,.
\end{aligned} \tag{4.75}$$

Expressing the eigenspinor Λ as a function of the orbital angle ϕ, we can reformulate the Lorentz-force equation (4.73) to read

$$\frac{d}{d\phi}\Lambda = \dot{\phi}^{-1}\dot{\Lambda} = -\frac{\alpha\hat{\mathbf{r}}}{2l}\Lambda\,. \tag{4.76}$$

Of course $\hat{\mathbf{r}}$ depends on ϕ :

$$\hat{\mathbf{r}}\,(\phi) = \exp\left(-i\phi\hat{\mathbf{l}}/2\right)\,\hat{\mathbf{r}}_0 \exp\left(i\phi\hat{\mathbf{l}}/2\right)\,. \tag{4.77}$$

Let Λ_{rot} be the eigenspinor with respect to the frame rotating with $\mathbf{r}$:

$$\Lambda = e^{-i\hat{\mathbf{l}}\phi/2}\Lambda_{rot}\,. \tag{4.78}$$

It obeys

$$\frac{d}{d\phi}\Lambda_{rot} = \frac{i}{2}\mathbf{h}_0\Lambda_{rot}\,, \tag{4.79}$$

where $\left\{\hat{\mathbf{r}},\hat{\boldsymbol{\phi}},\hat{\mathbf{l}}\right\}$ form a right-handed orthonormal basis, the constant biparavector $\mathbf{h}_0$ is

$$\mathbf{h}_0 = \hat{\mathbf{l}} + i\frac{\alpha}{l}\hat{\mathbf{r}}_0 = \left(1+\frac{\alpha}{l}\hat{\boldsymbol{\phi}}_0\right)\hat{\mathbf{l}}\,, \tag{4.80}$$

and the zero subscripts refer to the $\phi = 0$ directions. Integration then gives

$$\begin{aligned}\Lambda(\phi) &= e^{-i\hat{\mathbf{l}}\phi/2}e^{i\mathbf{h}_0\phi/2}\Lambda(0)\\ &= e^{i\mathbf{h}\phi/2}e^{-i\hat{\mathbf{l}}\phi/2}\Lambda(0)\end{aligned} \tag{4.81}$$

where

$$\mathbf{h} = e^{-i\hat{\mathbf{l}}\phi/2}\mathbf{h}_0 e^{i\hat{\mathbf{l}}\phi/2} = \left(1+\frac{\alpha}{l}\hat{\boldsymbol{\phi}}\right)\hat{\mathbf{l}}\,. \tag{4.82}$$

The proper velocity is

$$\begin{aligned}u(\phi) &= \Lambda(\phi)\,\Lambda^{\dagger}(\phi)\\ &= e^{i\mathbf{h}\phi/2}e^{-i\hat{\mathbf{l}}\phi/2}u(0)\,e^{i\hat{\mathbf{l}}\phi/2}e^{-i\mathbf{h}^{\dagger}\phi/2}\,.\end{aligned} \tag{4.83}$$

We write the initial proper velocity in terms of components:

$$u(0) = \gamma(0) + u_r(0)\,\hat{\mathbf{r}}_0 + u_\phi(0)\,\hat{\boldsymbol{\phi}}_0 \tag{4.84}$$

and then rotate the initial unit vectors:

$$e^{-i\hat{\mathbf{l}}\phi/2}u(0)\,e^{i\hat{\mathbf{l}}\phi/2}=\gamma(0)+u_r(0)\,\hat{\mathbf{r}}+u_\phi(0)\,\hat{\boldsymbol{\phi}}\,. \tag{4.85}$$

The solution is thus

$$u\det(\phi)=e^{i\mathbf{h}\phi/2}\left[\gamma(0)+u_r(0)\,\hat{\mathbf{r}}+u_\phi(0)\,\hat{\boldsymbol{\phi}}\right]e^{-i\mathbf{h}^\dagger\phi/2}. \tag{4.86}$$

By expanding $\exp(i\mathbf{h}\phi/2)=\cos\phi h/2+i\mathbf{h}\left(\sin\phi h/2\right)/h$, where $h^2=\mathbf{h}^2=\mathbf{h}^{\dagger 2}=1-\alpha^2/l^2$, we find that for an arbitrary paravector p,

$$e^{i\mathbf{h}\phi/2}pe^{-i\mathbf{h}^\dagger\phi/2}=p+\frac{1}{2}\left(\mathbf{h}p\mathbf{h}^\dagger-ph^2\right)\frac{(1-\cos\phi h)}{h^2}+i\left\langle\mathbf{h}p\right\rangle_{\Im}\frac{\sin\phi h}{h}\,. \tag{4.87}$$

If the initial condition $u_r(0)=0$ is imposed, the solution thus reduces to

$$\begin{aligned}\gamma(\phi) &= \gamma(0)+C\frac{\alpha}{lh^2}\left(1-\cos\phi h\right) &(4.88)\\ u_r(\phi) &= -C\frac{\sin\phi h}{h} &(4.89)\\ u_\phi(\phi) &= u_\phi(0)+C\left(1-\cos\phi h\right)/h^2, &(4.90)\end{aligned}$$

where $C=\alpha\gamma(0)/l-u_\phi(0)$. We can apply the conservation of angular momentum to obtain the orbital path:

$$r=\frac{r_0u_\phi(0)}{u_\phi(\phi)}=r_0\left[1+\frac{C}{h^2u_\phi(0)}\left(1-\cos\phi h\right)\right]^{-1}. \tag{4.91}$$

The shape of the orbit is a precessing ellipse

$$r=r_0\frac{1-\varepsilon}{1-\varepsilon\cos\phi h} \tag{4.92}$$

of eccentricity

$$\varepsilon=\frac{C}{h^2u_\phi(0)+C}\,. \tag{4.93}$$

Exercise 4.11. Show that the total energy E is given by

$$\begin{aligned}E &= \gamma(0)\,mc^2-\alpha c/r_0\\ &= mc^2\left[\gamma(0)-u_\phi(0)\,\alpha/l\right]. &(4.94)\end{aligned}$$

Use this result together with $u_\phi(0) = \sqrt{\gamma^2(0) - 1}$ to express the eccentricity in the form

$$\varepsilon = \frac{l}{\alpha}\sqrt{1 - (hmc^2/E)^2}. \tag{4.95}$$

Show that in the nonrelativistic limit $c \to \infty$, this takes the form

$$\varepsilon \to \varepsilon_{nr} = \sqrt{1 + \frac{2l^2 E_{nr}}{\alpha^2 mc^2}}, \tag{4.96}$$

where $E_{nr} = E - mc^2$ and the product αc is finite.

The orbit $r(\phi)$ repeats itself when ϕ increases by $2\pi/h = 2\pi/\sqrt{1 - \alpha^2/l^2} > 2\pi$. The bound relativistic motion is therefore not a closed ellipse as in the nonrelativistic case, but a precessing ellipse. If $\alpha > l$, the factor $h \equiv i\kappa$ is imaginary and the orbit becomes a spiral that wraps around the origin an infinite number of times in a finite time interval (see problem 17).[7]

4.10 Field Momenta

In chapter 3, we derived Poynting's theorem for the conservation of energy. This theorem gave us expressions for the energy density and energy flow of electromagnetic fields. We now extend these ideas to momentum conservation.

The change in the momentum of a charge e interacting with an external field $\mathbf{F}$ between times $t(1)$ and $t(2)$ is found by integrating the Lorentz-force equation (4.3):

$$p(2) - p(1) = \int_1^2 d\tau \, \langle e\mathbf{F}u\rangle_\Re, \tag{4.97}$$

where $uc = dr/d\tau$ is the tangent vector to the path $r(\tau)$. When there are many charges, each with its own path $r_n(\tau)$, the total change in particle momentum is given by the sum of such expressions, where one notes that the proper time is generally different for each particle:

$$P(2) - P(1) = \sum_n e_n \int_1^2 d\tau_n \, \langle \mathbf{F}\left[r_n(\tau_n)\right] u_n(\tau_n)\rangle_\Re \tag{4.98}$$

[7] Although this surprising solution is not well known, it certainly is known. See for example L. D. Landau and E. M. Lifshitz (1975), p. 93.

It is convenient to define a charge current (density)

$$j(x) = \sum_n c^2 e_n \int_{-\infty}^{\infty} d\tau\, u_n(\tau)\, \delta^{(4)}\left[x - r_n(\tau)\right] \tag{4.99}$$

by which we avoid problems associated with distinct proper times.

Example 4.12. Suppose all of the charges have the same constant proper velocity u. The charge current is then

$$j(x) = \rho(x)\, cu/\gamma\,, \tag{4.100}$$

where $\rho(x) = \sum_n e_n \delta\left(\mathbf{x} - \mathbf{r}(t/\gamma)\right)$ is the charge density. If the charge density is uniform, there will be a rest frame in which

$$j_{rest} = \rho_{rest} c\,. \tag{4.101}$$

Since a boost of j_{rest} to the frame moving with proper velocity u gives

$$j = j_{rest} u = \rho_{rest} cu\,, \tag{4.102}$$

one finds that the moving density is larger than when at rest:

$$\rho = \gamma \rho_{rest}\,. \tag{4.103}$$

Of course, this is expected because of the Lorentz contraction of spatial volumes.

The change in particle momentum can thus be written

$$P(2) - P(1) = c^{-2} \int d^4x\, \langle \mathbf{F}(x)\, j(x) \rangle_{\Re}\,, \tag{4.104}$$

where the four-dimensional volume of integration is the physical volume observed bounded by times $t(1)$ and $t(2)$. Evidently, $\langle \mathbf{F}(x)\, j(x) \rangle_{\Re}$ is the density of the time-rate-of change of the particle momentum. Applying the bar-dagger, $\bar{\partial}\mathbf{F}^{\dagger} = -\left(\varepsilon_0 c\right)^{-1} j$, of Maxwell's equation, we obtain the covariant relation

$$\langle \mathbf{F}(x)\ j(x) \rangle_{\Re} = -\varepsilon_0 c \left\langle \mathbf{F}(x)\ \bar{\partial}\mathbf{F}^{\dagger}(x) \right\rangle_{\Re}\,, \tag{4.105}$$

which can be expanded in a given inertial frame to give

$$\mathbf{j} \cdot \mathbf{E} + \rho \mathbf{E} c + \mathbf{j} \times \mathbf{B} c + \partial_t \left(\mathcal{E} + \mathbf{S}/c\right) - \nabla \cdot \overleftrightarrow{\mathbf{T}} c + \nabla \cdot \mathbf{S} = 0\,, \tag{4.106}$$

where

$$\overleftrightarrow{\mathbf{T}} = \varepsilon_0 \left(\overleftrightarrow{\mathbf{EE}} + c^2 \overleftrightarrow{\mathbf{BB}} \right) - \mathcal{E} \overleftrightarrow{\mathbf{1}} \tag{4.107}$$

is the Maxwell stress dyad (see Chapt. 3, problem 7). The scalar part is Poynting's theorem derived in the last chapter; it specifies energy conservation. The vector part is the analogous statement for momentum conservation.

For a covariant component expression of the conservation law, note

$$\begin{aligned} \langle \mathbf{F}(x)\, j(x) \rangle_{\Re} &= \frac{1}{2} F^{\mu\nu} j^{\lambda} \left\langle \langle \mathbf{e}_\mu \bar{\mathbf{e}}_\nu \rangle_V \mathbf{e}_\lambda \right\rangle_{\Re} \\ &= F^{\mu\nu} j_\nu \mathbf{e}_\mu \end{aligned} \tag{4.108}$$

and

$$\begin{aligned} \varepsilon_0 \left\langle \mathbf{F} \partial \mathbf{F}^\dagger \right\rangle_{\Re} &= \frac{\varepsilon_0}{2} \left[\mathbf{F} \mathbf{e}^\lambda \left(\partial_\lambda \mathbf{F}^\dagger \right) + \left(\partial_\lambda \mathbf{F} \right) \mathbf{e}^\lambda \mathbf{F}^\dagger \right] \\ &= \frac{\varepsilon_0}{2} \partial_\lambda \left(\mathbf{F} \mathbf{e}^\lambda \mathbf{F}^\dagger \right) \\ &= \partial_\lambda T^{\mu\lambda} \mathbf{e}_\mu \end{aligned} \tag{4.109}$$

with the symmetric *electromagnetic stress energy-momentum tensor* $T^{\mu\lambda}$ given by

$$T^{\mu\lambda} \mathbf{e}_\mu = \frac{\varepsilon_0}{2} \mathbf{F} \mathbf{e}^\lambda \mathbf{F}^\dagger . \tag{4.110}$$

The conservation theorem can thus be written

$$F^{\mu\nu} j_\nu + c \partial_\lambda T^{\mu\lambda} = 0. \tag{4.111}$$

We can apply theorems of chapter 3 to derive other expressions for $T^{\mu\lambda}$. First, we note that the whole expression is real, and then, we expand the factor $\mathbf{F}\mathbf{e}^\lambda$ in real and imaginary parts:

$$\begin{aligned} \mathbf{F} \mathbf{e}^\lambda \mathbf{F}^\dagger &= \left\langle \mathbf{F} \mathbf{e}^\lambda \mathbf{F}^\dagger \right\rangle_{\Re} \\ &= \left\langle \left\langle \mathbf{F} \mathbf{e}^\lambda \right\rangle_{\Re} \mathbf{F}^\dagger \right\rangle_{\Re} + \left\langle \left\langle \mathbf{F} \mathbf{e}^\lambda \right\rangle_{\Im} \mathbf{F}^\dagger \right\rangle_{\Re} . \end{aligned} \tag{4.112}$$

Now

$$\begin{aligned} \left\langle \mathbf{F} \mathbf{e}^\lambda \right\rangle_{\Re} &= \frac{1}{2} F^{\mu\nu} \left\langle \langle \mathbf{e}_\mu \bar{\mathbf{e}}_\nu \rangle_V \mathbf{e}^\lambda \right\rangle_{\Re} \\ &= \frac{1}{2} F^{\mu\nu} \left(\mathbf{e}_\mu \delta_\nu^\lambda - \mathbf{e}_\nu \delta_\mu^\lambda \right) \\ &= F^{\mu\lambda} \mathbf{e}_\mu . \end{aligned} \tag{4.113}$$

Therefore the first term of equation (4.112) is

$$\begin{aligned}\left\langle\left\langle \mathbf{F}\mathbf{e}^{\lambda}\right\rangle_{\Re}\mathbf{F}^{\dagger}\right\rangle_{\Re} &= F^{\mu\lambda}\left\langle \mathbf{e}_{\mu}\mathbf{F}^{\dagger}\right\rangle_{\Re} \\ &= F^{\mu\lambda}\left\langle \mathbf{F}\mathbf{e}_{\mu}\right\rangle_{\Re} \\ &= F_{\mu}^{\nu}F^{\mu\lambda}\mathbf{e}_{\nu}\,. \end{aligned} \tag{4.114}$$

In the second term of equation (4.112), we replace $\mathbf{F}$ by $i\,{}^{*}\mathbf{F}$ to get

$$\begin{aligned}\left\langle\left\langle \mathbf{F}\mathbf{e}^{\lambda}\right\rangle_{\Im}\mathbf{F}^{\dagger}\right\rangle_{\Re} &= \left\langle\left\langle i\,{}^{*}\mathbf{F}\mathbf{e}^{\lambda}\right\rangle_{\Im}\left(-i\,{}^{*}\mathbf{F}^{\dagger}\right)\right\rangle_{\Re} \\ &= \left\langle\left\langle {}^{*}\mathbf{F}\mathbf{e}^{\lambda}\right\rangle_{\Re}{}^{*}\mathbf{F}^{\dagger}\right\rangle_{\Re}\,, \end{aligned} \tag{4.115}$$

which is just the first term with $\mathbf{F}$ replaced by its dual. Consequently,

$$\mathbf{F}\mathbf{e}^{\lambda}\mathbf{F}^{\dagger} = \left(F_{\mu}^{\nu}F^{\mu\lambda} + {}^{*}F_{\mu}^{\nu}\,{}^{*}F^{\mu\lambda}\right)\mathbf{e}_{\nu}\,. \tag{4.116}$$

Direct calculation verifies

$$\begin{aligned}{}^{*}F_{\mu}^{\nu}\,{}^{*}F^{\mu\lambda} &= F_{\mu}^{\nu}F^{\mu\lambda} - \eta^{\lambda\nu}\left(\mathbf{E}^{2} - c^{2}\mathbf{B}^{2}\right) \\ &= F_{\mu}^{\nu}F^{\mu\lambda} + \frac{1}{2}\eta^{\lambda\nu}F^{\alpha\beta}F_{\alpha\beta}\,. \end{aligned} \tag{4.117}$$

Putting this together we finally obtain

$$\begin{aligned}T^{\nu\lambda}\mathbf{e}_{\nu} &= \frac{\varepsilon_{0}}{2}\mathbf{F}\mathbf{e}^{\lambda}\mathbf{F}^{\dagger} \\ &= \varepsilon_{0}\left(F_{\mu}^{\nu}F^{\mu\lambda} + \frac{1}{4}\eta^{\lambda\nu}F^{\alpha\beta}F_{\alpha\beta}\right)\mathbf{e}_{\nu}\,. \end{aligned} \tag{4.118}$$

Exercise 4.13. Prove that $T_{\lambda}^{\ \lambda} \equiv T^{\nu\lambda}\left\langle \mathbf{e}_{\nu}\bar{\mathbf{e}}_{\lambda}\right\rangle_{S} = 0$. (See also problem 18.)

Returning to the total change in particle momentum, we have

$$\begin{aligned}P(2) - P(1) &= -\frac{\varepsilon_{0}}{c}\int d^{4}x\,\left\langle \mathbf{F}(x)\,\partial\mathbf{F}^{\dagger}(x)\right\rangle_{\Re} \\ &= -\frac{\varepsilon_{0}}{2c}\int d^{4}x\partial_{\lambda}\left(\mathbf{F}\mathbf{e}^{\lambda}\mathbf{F}^{\dagger}\right)\,. \end{aligned} \tag{4.119}$$

For a fixed spatial volume $\mathcal{V}$, we can integrate by parts to get the paravector

$$P(2) - P(1) = -c^{-1}\int_{\mathcal{V}} d^{3}x\ \left(\mathcal{E} + \mathbf{S}/c\right)\Big|_{t_1}^{t_2} + \int_{t_1}^{t_2} dt \oint_{\partial\mathcal{V}} d\boldsymbol{\sigma}\cdot\left(\overleftrightarrow{\mathbf{T}} - \mathbf{S}/c\right) \tag{4.120}$$

where $d\boldsymbol{\sigma}$ is an outward-pointing surface element. The last term represents the flow of spacetime momentum inward through the surface $\partial\mathcal{V}$ bounding the volume $\mathcal{V}$.

4.11 Problems

1. Consider the product $R_{12} = R_1 R_2$ of two **noncommuting rotations** $R_1 = R\left(\theta_1^j \mathbf{e}_j\right)$ and $R_2 = R\left(\theta_2^k \mathbf{e}_k\right)$, and let $\mathbf{u}_j = R_{12}\mathbf{e}_j R_{12}^\dagger$ be the basis vectors in the rotated frame. Show that one can also express R_{12} as $R_{12} = R\left(\theta_2^k \mathbf{u}_k\right) R\left(\theta_1^j \mathbf{e}_j\right)$. [Hint: recall equation (4.23) and insert factors of $L\bar{L}$ where needed.]

2. Show that when the **product of boosts** $B_j = a + b\mathbf{e}_j$ is re-expressed in the form $B_2 B_1 = BR$, the rotation represented by R is by an angle $\phi = 2\arctan\left(b^2/a^2\right)$ about $\mathbf{e}_3$.

3. **Spinning particles.** If a particle is spinning rapidly, it may be convenient to describe the motion of the particle with the eigenspinor of a non-spinning reference frame. Thus, we factor the eigenspinor Λ
$$\Lambda = \Lambda_s \exp(-i\omega\tau/2)$$
into the spin motion $\exp(-i\omega\tau/2)$ and the eigenspinor Λ_s for the non-spinning particle frame, where ω is the angular velocity of the spin. Compare the equations of motion (4.45) for the eigenspinors Λ and Λ_s ,expressed in terms of the electromagnetic fields $\mathbf{F}_r$ and $\mathbf{F}_s$ in the particle reference frames. Show that the field $\mathbf{F}_s$ in the non-spinning particle frame (the *frame* is not spinning, but the particle is spinning with respect to this frame) is related to the field $\mathbf{F}_r$ in the particle rest frame by a rotation of the field plus the addition of an effective magnetic field proportional to the angular velocity ω of the spin. Find this field.

4. **Lagrangian and conjugate momentum.** The motion of a charge e in a paravector potential A can be determined from the covariant Lagrangian
$$L = \frac{1}{2} mcu\bar{u} + e\left\langle \bar{A}u\right\rangle_S , \tag{4.121}$$
where the tangent vector $u = dx/ds$ and spacetime position x are taken as independent variables. The *conjugate momentum* is defined to have components
$$\pi^\mu = \frac{\partial L}{\partial u_\mu} = eA^\mu + mcu^\mu . \tag{4.122}$$
In paravector form, $\pi = p + eA$, where $p = mcu$. Here, s is a scalar path parameter that is identified by $ds = cd\tau$ with the proper time τ only for the path that maximizes the action
$$\int_1^2 L\, ds \tag{4.123}$$

between the fixed end points labeled 1 and 2. This path is given by Euler-Lagrange equation of motion:

$$\frac{d\pi}{d\tau} = ec\partial \left\langle \bar{A}u \right\rangle_S . \tag{4.124}$$

Show that this gives the Lorentz-force equation. [Hint: note the *convective derivative* $dA/d\tau = c \left\langle u\bar{\partial} \right\rangle_S A(x)$ and treat the proper velocity u and the spacetime position x as independent variables. Note also that $dp/d\tau$ is real.]

5. **Hamiltonian.** In the Hamiltonian formulation, the conjugate momentum π and the spatial position $\mathbf{x}$ are taken to be independent variables. Show that the Lorentz-force equation follows from the equation of motion

$$\frac{d\pi}{d\tau} = \gamma \partial H , \tag{4.125}$$

where the Hamiltonian is

$$H = e\phi + \sqrt{(\boldsymbol{\pi} - e\mathbf{A})^2 c^2 + m^2c^4} . \tag{4.126}$$

[Hint: recall the convective derivative of the previous problem.]

6. **Hyperbolic motion.** Let a charge e start from rest at time $t = 0$ in a constant electric field $\mathbf{E}$. Find the proper velocity as a function of the proper time τ. Show that the proper time can be expressed

$$\tau = \alpha^{-1} \ln\left(\alpha t + \sqrt{\alpha^2 t^2 + 1}\right) \tag{4.127}$$

and determine the constant α.

7. **Wien velocity filter.** Consider the motion of an electric charge e in a constant electromagnetic field

$$\mathbf{F} = E\left(\mathbf{e}_1 + 2i\mathbf{e}_2\right) \tag{4.128}$$

What condition on the velocity of the charge must be fulfilled in order that the force vanishes. Check your result by determining the electromagnetic field in the rest frame of the charge.

8. Consider the motion of a charge e in the lab frame with an electromagnetic field $\mathbf{F} = E\left(3\mathbf{e}_1 + 5i\mathbf{e}_2\right)$, where $E = \dot{3}00$ kV/m (we use the short-hand: $\dot{3} \equiv 2.99792458$).

   (a) Factor $\mathbf{F}$ into a proper velocity and a pure magnetic field.

(b) Identify the velocity $\mathbf{v}$ of the drift frame ($\mathbf{v}$ is assumed perpendicular to $\mathbf{E}$ and $\mathbf{B}$) and show that the charge experiences no force if it is traveling at the drift velocity.

(c) Find the electromagnetic field in the drift frame and verify that the electric field vanishes there. Use the appropriate SI units for the magnetic field.

(d) Use the field invariants to provide an independent calculation of the magnitude of the magnetic field in the frame in which the electric field vanishes.

(e) Show that

$$\Lambda = \frac{1}{2\sqrt{2}}\left(3 + \mathbf{e}_3\right) \tag{4.129}$$

is the eigenspinor of a particle moving at the drift velocity.

(f) Find the proper time rate of change of Λ for a charge moving in $\mathbf{F}$ and combine it with $\bar{\Lambda}$ to obtain the electromagnetic field in both the lab and drift frames.

9. Verify that when the constant fields $\mathbf{E}$ and $\mathbf{B}$ are perpendicular, the biparavector $\mathbf{F}$ is a single spacetime plane. Show that the drift velocity $\mathbf{v}_{drift}$ (4.67) reduces to $c^2\mathbf{E}\times\mathbf{B}$ divided by the larger of $\mathbf{E}^2$ or $c^2\mathbf{B}^2$, and that the transformation to the drift frame rotates $\mathbf{F}$ so that it becomes purely electric or purely magnetic.

10. **Magnetic brake.** A conducting plate moves through a pure magnetic field $\mathbf{F} = ic\mathbf{B}$ with a velocity $\mathbf{v}$ perpendicular to $\mathbf{B}$. Find the field $\mathbf{F}'$ in the frame of the plate. Show that the drift velocity in the frame of the plate is $-\mathbf{v}$. Interpret.

11. Consider an electromagnetic field with parallel and constant $\mathbf{E}$ and $\mathbf{B}$.

(a) Show explicitly in this case that the energy density $\mathcal{E}$ is related to the Lorentz invariant $\mathbf{F}^2$ by $\mathcal{E} = \frac{1}{2}\varepsilon_0\left|\mathbf{F}^2\right|$.

(b) Show that the eigenspinor can be expressed in the form

$$\Lambda(\tau) = BR\Lambda(0) \tag{4.130}$$

where B is a pure boost, R is a pure rotation, and both depend on the proper time.

(c) Demonstrate further that the proper velocity u is

$$u(\tau) = B^2\left[\gamma(0) + \mathbf{u}_{\parallel}(0)\right] + R^2\mathbf{u}_{\perp}(0)\ , \tag{4.131}$$

where $\mathbf{u}_{\parallel}$ and $\mathbf{u}_{\perp}$ are the vector parts of u parallel and perpendicular to the field direction, respectively.

(d) Show that the velocity of a particle starting from rest in such a field is independent of the magnitude of the magnetic field $\mathbf{B}$.

12. **Limiting boost.** Consider the boost $B = \exp(\mathbf{w}/2)$ in the limit of large $w = |\mathbf{w}|$. Show that it can be expressed by

$$B = \exp(w/2)\, P_{\hat{\mathbf{w}}} + \exp(-w/2)\, \bar{P}_{\hat{\mathbf{w}}}\,, \tag{4.132}$$

which in the limit of large w approaches

$$B \to \exp(w/2)\, P_{\hat{\mathbf{w}}}\,, \tag{4.133}$$

where $P_{\hat{\mathbf{w}}} = \frac{1}{2}(1 + \hat{\mathbf{w}})$. Show that the paramomentum of a particle of small mass m, when boosted from its rest frame by B, is in this limit

$$p = mc \exp(w)\, P_{\hat{\mathbf{w}}}\,. \tag{4.134}$$

Compare this form to the paramomentum of a photon, $p = \hbar\omega\left(1 + \hat{\mathbf{k}}\right)$.

13. **Constant compound field.** Consider the compound field $\mathbf{F} = 3\mathbf{e}_1 + \mathbf{e}_2 + i\mathbf{e}_2 - 3i\mathbf{e}_1$ (in units of kV/m).

(a) Show that $\mathbf{F}$ can be written in the form $(1 - i)\,\mathbf{F}_s$ where $\mathbf{F}_s$ is a simple field, given by a single spacetime plane. Find $\mathbf{F}_s$.

(b) Factor $\mathbf{F}_s$ in the form $u_{drift}\mathbf{E}_{drift}$ and interpret the factors.

(c) Find $\mathbf{F}^2$ and verify that it is the same in the lab as in the drift frame.

(d) Show that the coordinate velocity $\mathbf{v}_{drift}$ determined from part (b) is the same as that given by equation (4.67).

14. Express the constant electromagnetic field $\mathbf{F} = \mathbf{E} + ic\mathbf{B}$ as

$$\mathbf{F} = e^{i\alpha}(\mathbf{a} + i\mathbf{b})\,, \tag{4.135}$$

where

$$\begin{aligned} \tan 2\alpha &= \frac{2c\mathbf{E}\cdot\mathbf{B}}{\mathbf{E}^2 - c^2\mathbf{B}^2} \\ \mathbf{a} &= \mathbf{E}\cos\alpha + c\mathbf{B}\sin\alpha \\ \mathbf{b} &= c\mathbf{B}\cos\alpha - \mathbf{E}\sin\alpha\,. \end{aligned} \tag{4.136}$$

(a) Show that the biparavector $\mathbf{a} + i\mathbf{b}$ is simple and represents a single spacetime plane.

(b) Prove that the spacetime plane $\mathbf{a}+i\mathbf{b}$ and its dual are invariant under the spacetime rotations induced by $\mathbf{F}$.

(c) Demonstrate that one can always choose the quadrant of α so that $\mathbf{a}^2 \leq \mathbf{b}^2$.

(d) Show that if $\mathbf{a}^2 < \mathbf{b}^2$, an inertial *drift frame* exists, moving with proper velocity $u_{drift} = \gamma_{drift}\left(1+\mathbf{v}_{drift}/c\right)$, in which the transformed plane

$$\bar{u}_{drift}^{1/2}\left(\mathbf{a}+i\mathbf{b}\right)u_{drift}^{1/2} = \left(\mathbf{a}+i\mathbf{b}\right)u_{drift} \tag{4.137}$$

is imaginary and hence a spatial bivector:

$$\left(\mathbf{a}+i\mathbf{b}\right)\gamma_{drift}\left(1+\mathbf{v}_{drift}/c\right). \tag{4.138}$$

Find $\mathbf{v}_{drift}$ in terms of $\mathbf{a}$ and $\mathbf{b}$.

(e) Find the motion of a charge e in the drift frame.

15. Integrate (4.69) to find the path $x(\tau)$ of a charge e released from the origin at $\tau = 0$ in a constant simple null field.

16. Find the proper velocity of a charge e, initially at rest, after the **field pulse**

$$\mathbf{F}(\tau) = \mathbf{F}_0 f(\tau) \tag{4.139}$$

passes, where $\tau f(\tau) \to 0$ as $\tau \to \pm\infty$ and

$$\int_{-\infty}^{\infty} d\tau\, f(\tau) = \tau_1 . \tag{4.140}$$

(Note that this solution is of rather limited use since if $\mathbf{F}$ is not constant, it must satisfy Maxwell's equation, and it is unlikely that it will be known as a function of the proper time τ of the charge. The method is important for solving the relativistic motion of charges in pulsed plane waves, however, as we see in chapter 6.)

17. **Capture solutions.** Consider the relativistic motion of a charge e in the **attractive Coulomb field** of a charge $-Ze$ when $l < \alpha = Ze^2/\left(4\pi\varepsilon_0 c\right)$.

(a) Show that with the initial condition $u_r(0) = 0$, the motion is given by

$$\gamma(\phi) = \gamma(0) + C\frac{\alpha}{l\kappa^2}\left(\cosh\phi\kappa - 1\right) \tag{4.141}$$

$$u_r(\phi) = -C\frac{\sinh\phi\kappa}{\kappa} \tag{4.142}$$

$$u_\phi(\phi) = u_\phi(0) + C\left(\cosh\phi\kappa - 1\right)/\kappa^2, \tag{4.143}$$

where $C = \alpha\gamma(0)/l - u_\phi(0)$ and $\kappa = \sqrt{\alpha^2/l^2 - 1}$.

(b) Use these results to calculate the orbital shape $r(\phi)$. For simplicity, let $u_\phi(0) = \gamma(0)\, l/\alpha$, and show that in this case $u_\phi(0)\,\kappa = 1$ and

$$\begin{aligned}\gamma(\phi) &= \gamma(0)\cosh\phi\kappa && (4.144)\\ u_\phi(\phi) &= u_\phi(0)\cosh\phi\kappa && (4.145)\\ u_r(\phi) &= -\sinh\phi\kappa && (4.146)\\ r(\phi) &= r_0/\cosh\phi\kappa\,. && (4.147)\end{aligned}$$

(c) Sketch the orbit.

(d) Calculate the proper time τ and the coordinate time t for the charge to spiral into the origin, using

$$\begin{aligned}\tau &= \int_0^\infty \frac{d\phi}{\dot\phi} = \frac{m}{l}\int_0^\infty d\phi\, r^2 && (4.148)\\ t &= \frac{m}{l}\int_0^\infty d\phi\, r^2\gamma\,. && (4.149)\end{aligned}$$

18. **Explicit matrices.** Starting with the matrix $(F^{\mu\nu})$ (see Chapt. 3, problem 12), determine $\left(F^{\mu}_{\ \lambda}\right) = (\eta_{\lambda\nu}F^{\mu\nu})$ and the product matrix $\left(F^{\mu}_{\ \lambda}F^{\lambda\nu}\right)$. Write down the dual equivalent of the product matrix by making the substitutions $E \to cB$ and $cB \to -E$ and find the difference $\left(F^{\mu}_{\ \lambda}F^{\lambda\nu}\right) - \left({}^*F^{\mu}_{\ \lambda}\,{}^*F^{\lambda\nu}\right)$. Use this to show that the stress-energy-momentum tensor is

$$(T^{\mu\nu}) = \begin{pmatrix} \mathcal{E} & S_x/c & S_y/c & S_z/c \\ S_x/c & & & \\ S_y/c & & \left(\overleftrightarrow{\mathbf{T}}\right) & \\ S_z/c & & & \end{pmatrix} \qquad (4.150)$$

where $\overleftrightarrow{\mathbf{T}}$ is the Maxwell stress tensor and the elements of the 3×3 matrix $\left(\overleftrightarrow{\mathbf{T}}\right)$ are

$$\left(\overleftrightarrow{\mathbf{T}}\right)^{jk} = \mathbf{e}^j\cdot\overleftrightarrow{\mathbf{T}}\cdot\bar{\mathbf{e}}^k = \delta^{jk}\mathcal{E} - \varepsilon_0\left(E^jE^k + c^2B^jB^k\right). \qquad (4.151)$$

Chapter 5

Electromagnetic Waves in 1-D

Before we turn to wave solutions of Maxwell's equation in three-dimensional physical space, it is useful to review the properties of waves in one dimension. In this chapter, we look at transverse waves on a flexible string and electromagnetic waves in a transmission line. We derive the wave equation for these cases and investigate solutions to it. We can emphasize the significance of linear equations of motion by studying linear superpositions, including standing waves and Fourier decomposition.

Since geometry plays a minor role when there is only one spatial dimension, there is relatively little use of Clifford algebra in this chapter. However, in the following chapter, where we treat waves in three dimensions, the geometric-algebra approach will prove indispensable.

5.1 Wave Equation

The one-dimensional wave equation for the motion of a disturbance V in some property along the x axis is

$$c^2 \frac{\partial^2}{\partial x^2} V = \frac{\partial^2}{\partial t^2} V, \tag{5.1}$$

where x is the longitudinal position, t, the time, and c, the speed of the wave in the medium. Such equations describe the lossless, linear propagation of waves in one dimension in many physical systems: small-amplitude transverse waves in a string, pressure waves in the fluid medium inside a cylindrical pipe, or electrical waves (voltage or current) in a transmission line. In each case, $V(x, t)$ is the displacement of the measured property from its equilibrium value: the transverse position of the string, the pressure of the fluid, or the voltage in the line.

Two general solutions have the form

$$V(x,t) = f_1(ct - x)\,,\ V(x,t) = f_2(ct + x)\,, \tag{5.2}$$

where f_1 and f_2 are *any* twice-differentiable functions. Since $f_1\,(ct - x)$ depends only on the difference $ct - x$, its value is unchanged if both ct and x increase by the same amount. Evidently the function $f_1(ct - x)$ describes a solution that propagates in the $+x$ direction at speed c. Similarly, the solution $f_2(ct + x)$ propagates in the $-x$ direction at the same speed.

Exercise 5.1. Set $z = ct - x$ and use the chain rule

$$\frac{\partial f(z)}{\partial t} = \frac{\partial z}{\partial t}\frac{df(z)}{dz} =: \frac{\partial z}{\partial t} f'(z) \tag{5.3}$$

to prove that $f\,(z)$ really is a solution of the wave equation (5.1). Convince yourself that $z = ct + x$ works as well.

The wave equation (5.1) is *real* and *linear*. Its linearity means that if V_1 and V_2 are any two solutions, then any linear combination

$$V = a_1 V_1 + a_2 V_2\,, \tag{5.4}$$

where a_1 and a_2 are complex constants, is also a solution. Its reality means that if $V\,(x,t)$ is a solution, so is the complex conjugate $V^*(x,t)$. The combination of reality and linearity means that if $V\,(x,t)$ is a solution, then so is its real part[1]

$$\langle V(x,t)\rangle_{\Re} = \frac{1}{2}\left[V(x,t) + V^*(x,t)\right]\,. \tag{5.5}$$

In most physical situations, we are interested in real solutions to the wave equation (5.1). We will often find it convenient to write down complex solutions, but it is usually understood that the physical solution is the real part.

The result of the linearity of the wave equation is known as the *superposition principle:* any superposition (= linear combination) of solutions is also a solution.

[1] Its imaginary part is also a solution, but the imaginary part of V is the real part of $-iV$, so it's not necessary to specify separately.

In particular, waves can propagate in both forward and backward directions at once:

$$V(x,t) = f_1(ct - x) + f_2(ct + x) \ . \tag{5.6}$$

At each point (x, t) the values of the waves f_1 and f_2 simply add. The two waves simply pass through one another, without affecting the shape of the other.

5.2 Waves in a String

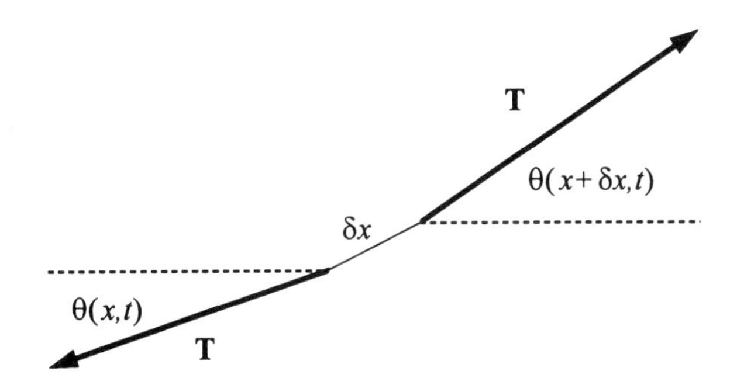

Fig. 5.1. The force on a segment δx of string arises from unbalanced tension at the two ends of the segment.

As an example of the wave equation, consider the equation of motion for small-amplitude vibrations of a flexible string. Let T be the tension of the string, ρ its mass per unit length, and $y(x, t)$ the transverse displacement of the string from equilibrium. The net transverse force δF_y on a segment δx of the string arises from the unbalanced y components of the tension pulling on the opposite ends of the segment. Let $\theta(x, t)$ be the angle between the string and its orientation at equilibrium. Then for small θ,

$$\begin{aligned} \delta F_y &= T\sin\theta(x+\delta x,t) - T\sin\theta(x,t) \\ &\approx T\tan\theta(x+\delta x,t) - T\tan\theta(x,t) \\ &= T\frac{\partial y(x+\delta x,t)}{\partial x} - T\frac{\partial y(x,t)}{\partial x} \\ &= T\frac{\partial^2 y(x,t)}{\partial x^2}\delta x \, . \end{aligned} \tag{5.7}$$

Newton's second law equates this force to the mass $\rho\delta x$ of the segment times its acceleration in the y direction:

$$T\frac{\partial^2 y(x,t)}{\partial x^2}\delta x = \rho\delta x\frac{\partial^2 y}{\partial t^2}, \tag{5.8}$$

and this is recognized as the wave equation with $\sqrt{T/\rho}$ identified as the wave speed c in the string. Derivations will be given below for electrical potential and current in a transmission line.

5.3 Separation of Variables

Particular solutions of the wave equation (5.1) can be obtained in the form of a product

$$V_p(x,t) = X(x)\,T(t) \tag{5.9}$$

of a function $X(x)$ of position only with a function $T(t)$ of time only. Substitution into the wave equation (5.1) and division by $V(x,t)$ separates the independent variables:

$$c^2\frac{X''(x)}{X(x)} = \frac{T''(t)}{T(t)}. \tag{5.10}$$

Since the left-hand side is independent of the variable t, its partial with respect to t must vanish:

$$\frac{\partial}{\partial t}\left[c^2\frac{X''(x)}{X(x)}\right] = \frac{\partial}{\partial t}\left[\frac{T''(t)}{T(t)}\right] = 0\,, \tag{5.11}$$

and consequently, the right-hand side, which depends only on t, must be constant. We put

$$\frac{T''(t)}{T(t)} = -\omega^2, \tag{5.12}$$

where ω is constant. It may in principle be complex, but we will mainly be concerned with persistent and bounded wave solutions, for which ω is real. The separation of variables has reduced the original partial-differential equation to a pair of ordinary differential equations, each of which is the equation of motion for a simple harmonic oscillator.

One solution to (5.12) is

$$T(t) = e^{-i\omega t}. \tag{5.13}$$

Another possible solution is the complex conjugate, but since we consider only the real part in physics applications, we do not need to treat the complex conjugate separately. Waves with such a simple time dependence are said to be *harmonic.* The corresponding solutions for $X(x)$ are

$$X(x) = e^{\pm ikx}, \tag{5.14}$$

where the *wave number* $k = 2\pi/\lambda = \omega/c$ for a lossless line, where λ is the wavelength. We must now retain both signs, since the relative sign of ωt and kx is important, as we saw above. Thus the total separation-of-variables solution can be written

$$V_p(x,t) = e^{\pm ikx}e^{-i\omega t}. \tag{5.15}$$

The complex solution (5.15), which gives real (and hence physical) solutions $\langle e^{i\phi}V_p(x,t)\rangle_{\Re} = \cos(\omega t \mp kx - \phi)$, where the phase angle ϕ is a real constant, represents a sinusoidal wave oscillating at angular frequency $\omega = 2\pi\nu$ and traveling with velocity $c = \pm\omega/k$.

Although most solutions of the wave equation (5.1) cannot be expressed as a simple product of the form (5.15), the separation-of-variable solutions form a *complete set*: any solution can be expressed as a linear combination of (5.15) with coefficients determined by initial conditions. This is formulated mathematically at the end of the next section.

5.4 Fourier Series

If boundary conditions are imposed, solutions are generally restricted to a discrete set of frequencies. For example, the boundary conditions

$$V(0,t) = V(L,t) = 0 \tag{5.16}$$

restrict the separated-variable solutions to standing waves

$$V_n(x,t) = e^{-i\omega_n t}\sin k_n x \tag{5.17}$$

at frequencies determined by

$$k_n = \omega_n / c = \frac{\pi n}{L} . \tag{5.18}$$

These solutions represent the *normal (or natural) modes* of the system. They are the only ways that the system can vibrate together sinusoidally and in phase. The boundary conditions have quantized the permissible frequencies.

To understand the quantization physically, consider the boundary condition at $x = 0$. The only way to satisfy the condition is for every wave $\exp[-i\,(kx + \omega t)]$ approaching the boundary from the right to be accompanied by one of the opposite amplitude from the left: $-\exp[i\,(kx - \omega t)]$, so that their sum vanishes at $x = 0$. The boundary has caused a *reflection* in the wave. The boundary at $x = L$ causes a similar reflection, and the doubly reflected wave will be superimposed on the original. If the doubly reflected wave is not in phase with the original wave, then its phase difference with the original will increase with successive round trips, and the net effect after enough cycles will be *destructive interference* with the original wave. The only waves that can survive many reflections are those for which the doubly reflected wave is in phase with the original wave. This is precisely the condition of the normal modes.

All general solutions that satisfy the boundary conditions can be expressed as linear superpositions of the normal modes:

$$V(x,t) = \sum_n a_n V_n(x,t) = \sum_n a_n e^{-i\omega_n t} \sin k_n x \tag{5.19}$$

with complex coefficients a_n determined by the initial conditions

$$V(x,t_0) = \sum_n a_n e^{-i\omega_n t_0} \sin k_n x \tag{5.20}$$

and by the orthogonality of the normal modes at $t = t_0$, namely

$$\int_0^L dx \, \sin k_n x \, \sin k_m x = \frac{L}{2} \delta_{nm} , \tag{5.21}$$

to be

$$a_n e^{-i\omega_n t_0} = \frac{2}{L} \int_0^L dx \, V(x,t_0) \, \sin k_n x . \tag{5.22}$$

The simple relation (5.22) is sufficient if the complex initial shape (5.20) is known. Usually, it is only the physical part $\langle V(x,t_0)\rangle_{\Re}$ that is known, and this is sufficient for finding the real parts $\langle a_n e^{-i\omega_n t_0}\rangle_{\Re}$. To find the imaginary parts, we can use the time-rates of change $\partial V(x,t)/\partial t$ at $t = t_0$:

$$\left.\frac{\partial V(x,t)}{\partial t}\right|_{t=t_0} = -i\sum_n \omega_n a_n e^{-i\omega_n t_0}\sin k_n x\,. \tag{5.23}$$

This gives the real part of $-i\omega_n a_n e^{-i\omega_n t_0}$ and thus the imaginary part of $a_n e^{-i\omega_n t_0}$. Combining the results, we obtain

$$a_n e^{-i\omega_n t_0} = \frac{2}{L}\int_0^L dx\left[\langle V(x,t_0)\rangle_{\Re} + \frac{i}{\omega_n}\left\langle \left.\frac{\partial V(x,t)}{\partial t}\right|_{t=t_0}\right\rangle_{\Re}\right]\sin k_n x\,. \tag{5.24}$$

The important point is that the complete motion of the system is given by the *Fourier series* (5.19) once the discrete set $\{a_n\}$ of complex coefficients, sometimes called the *vibration recipe*, is known. These coefficients are in turn determined by the initial conditions. Because the system is linear, each normal mode progresses independently of the others, and *any* complicated motion of the system is simply a linear superposition of such sinusoidally varying modes. The particular initial conditions (5.16) above gave a Fourier sine series. Other initial conditions can give a cosine series or a mixture of sine and cosine terms.

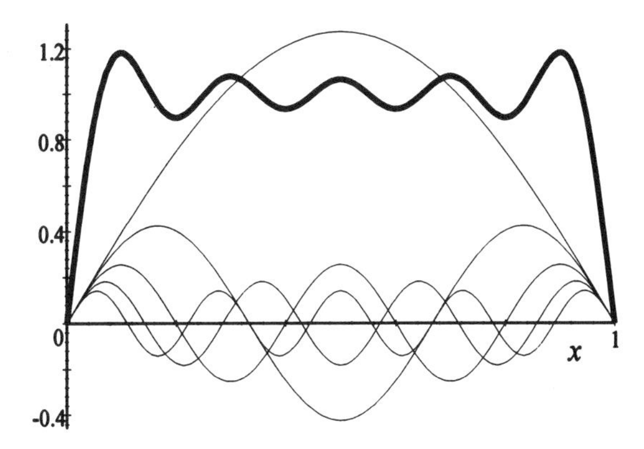

Fig. 5.2. The Fourier components of the square wave $y(x) = 1,\ 0 < x < 1$, with $n = 1,3,5,7,9$ and their sum. The components with even n vanish for the square wave.

Example 5.2. Find the Fourier series for a string with boundary conditions $y = 0$ at $x = 0$ and $x = L$, with an initial square-wave shape $y(x,0) = 1$, for $0 < x < L$ and $\partial y/\partial t = 0$ at $t = 0$. The amplitude a_n of the nth harmonic is given by (5.24) with $t_0 = 0$ and $k_n = n\pi/L$:

$$a_n = \frac{2}{L}\int_0^L dx \sin k_n x = \frac{2}{k_n L}(1 - \cos k_n L) \tag{5.25}$$

$$= \begin{cases} 0, & n \text{ even} \\ \frac{4}{n\pi}, & n \text{ odd} \end{cases} . \tag{5.26}$$

(See Figure 5.2.)

5.5 Fourier Integrals

If a pair of boundary conditions are not imposed, the wave equation (5.1) supports waves (5.15) at all frequencies. A linear superposition of such waves can form a wave *pulse* or *packet*

$$V(x,t) = \int_{-\infty}^{\infty} dk\, a(k)\, e^{ikx} e^{-i\omega t}, \quad \omega = |k|\, c \tag{5.27}$$

that is also a solution of the wave equation. All components travel at the same speed *c*, but those with $k > 0$ travel in the $+x$ direction whereas those with $k < 0$ travel in the $-x$ direction. The coefficients of the linear combination are given by the complex amplitude function $a(k)$, which is once again determined by the initial conditions:

$$a(k)\, e^{-i\omega t_0} = \frac{1}{2\pi}\int_{-\infty}^{\infty} dx\, V(x,t_0)\; e^{-ikx} . \tag{5.28}$$

In terms of real values, this is equivalent to

$$a(k)\, e^{-i\omega t_0} = \frac{1}{2\pi}\int_{-\infty}^{\infty} dx \left[\langle V(x,t_0)\rangle_{\Re} + \frac{i}{\omega}\left\langle \left.\frac{\partial V(x,t)}{\partial t}\right|_{t=t_0} \right\rangle_{\Re} \right] e^{-ikx} . \tag{5.29}$$

The integrals (5.27) and (5.28) are *Fourier integrals*. One says that $V(x,t)$ is the Fourier transform of $a(k)\, e^{-i\omega t}$, and $a(k)\, e^{-i\omega t_0}$ is the *inverse Fourier transform* of $V(x,t_0)$. To appreciate the relation between the Fourier transform and its inverse, we find it is useful to introduce Dirac's delta "function" $\delta(x)$.

5.5.1 Dirac Delta Function

The plane-wave solution $V(x,t) = \exp(ik_0 x - i\omega_0 t)$ is a special case of the wave packet (5.27) when only one frequency is present. Its amplitude function $a(k)$ must be extremely peaked; it must vanish for all k except $k = k_0$ but its integral must be unity. These are the basic properties of the *Dirac delta function* $\delta(k - k_0)$. Its properties are too singular for it to be accepted as a normal function, and many mathematicians invoke distribution theory in order to come to terms with it. For our purposes, the Dirac delta "function" $\delta(K)$ may be considered the limit of innumerable families of normalized, increasingly peaked functions that in the limit obey

$$\delta(K) = 0,\ K \neq 0 \tag{5.30}$$

and, for any bounded function $g(k)$,

$$\int dk'\, g(k')\, \delta(k - k') = g(k) \tag{5.31}$$

as long as the domain of integration includes a neighborhood of the point $k' = k$. For example, the rectangular steps

$$\Delta_w(K) = \begin{cases} 0, & |K| > w/2 \\ w^{-1}, & |K| \leq w/2 \end{cases} \tag{5.32}$$

defines a function of unit area that becomes increasingly peaked in a decreasing neighborhood of $K = 0$ as $w \to 0$ (see Figure 5.3). We can take

$$\delta(K) = \lim_{w \to 0} \Delta_w(K)\,. \tag{5.33}$$

Exercise 5.3. Write the Dirac delta function $\delta(K)$ as a limit of symmetric triangular functions peaked at $K = 0$.

The integral of a Dirac delta function is a step function. Using the form (5.32), we can write

$$\int_{-\infty}^{k} dK\, \delta(K) = \lim_{w \to 0} \int_{-\infty}^{k} dK\, \Delta_w(K) \tag{5.34}$$

$$= \lim_{w \to 0} \begin{cases} 0, & k < -w/2 \\ w^{-1}(k + w/2)\,, & -w/2 \leq k \leq w/2 \\ 1, & w/2 < k \end{cases} \tag{5.35}$$

$$= \theta(k)\,, \tag{5.36}$$

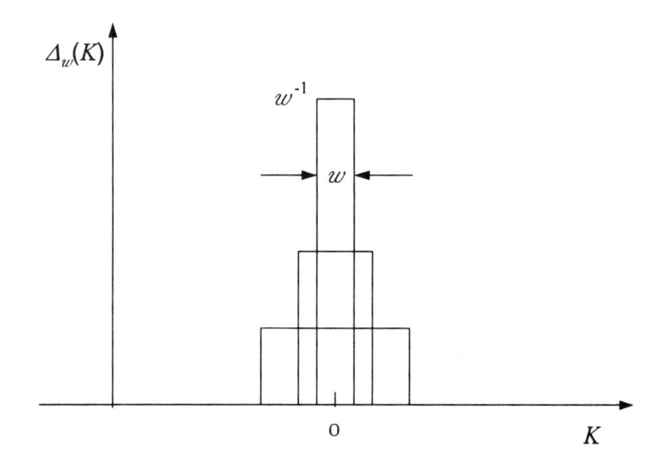

Fig. 5.3. The Dirac delta function can be viewed as the limit of a family of rectangular functions of unit area as they become more peaked.

which defines the *Heaviside step function* $\theta(k)$.

A common form of the Dirac delta function is

$$\begin{aligned}\delta(K) &= \lim_{X\to\infty}\Delta_X(K) \\ \Delta_X(K) &= \frac{1}{2\pi}\int_{-X}^{X} dx\,\exp(iKx) \\ &= \frac{1}{\pi}\frac{\sin(KX)}{K},\end{aligned} \tag{5.37}$$

which with $K = k - k'$ can be written simply as

$$\delta(k-k') = \frac{1}{2\pi}\int_{-\infty}^{\infty} dx\,\exp(ikx)\exp(-ik'x) \tag{5.38}$$

(see problem 10).

It is equation (5.38) that allows us to invert a Fourier transform. Any wave packet

$$V(x,0) = \int_{-\infty}^{\infty} dk\,a(k)\,e^{ikx}, \quad \omega = |k|\,c \tag{5.39}$$

can be multiplied by $\exp(-ik'x)$ and integrated over x to obtain

$$\int_{-\infty}^{\infty} dx\, V(x,0)\, e^{-ik'x} = \int_{-\infty}^{\infty} dk\, a(k) \int_{-\infty}^{\infty} dx\, e^{ikx} e^{-ik'x} \tag{5.40}$$

$$= 2\pi\, a(k')\ , \tag{5.41}$$

which gives the amplitude $a(k)$ (5.28) in terms of the packet shape at $t = 0$.

The relation (5.38) ensures that any bounded function of x can be expanded in the plane waves e^{ikx} and is therefore called the *completeness relation* for such waves.

Exercise 5.4. As a simple example of delta functions, use Maxwell's equation $\bar{j} = \varepsilon_0 c \bar{\partial} \mathbf{F}$ to show that the spacetime current j that gives the electromagnetic field

$$\mathbf{F}(x) = E_0 \mathbf{e}_1 \theta(x^1) \tag{5.42}$$

is the charge density (times c)

$$j(x) = \varepsilon_0 E_0 c \delta(x^1)\ . \tag{5.43}$$

Relate the result to an ideal parallel-plate capacitor. See also problem 11.

5.5.2 Group Velocity

Consider the wave packet (5.27) for an amplitude function that is concentrated in the $k > 0$ region. It can be approximated by

$$V(x,t) = \int_0^{\infty} dk\, a(k)\, e^{ikx - ikct} = V(x - ct, 0)\ . \tag{5.44}$$

The wave packet is seen to move forward without changing shape at the constant wave speed $c = \omega/k$. However, the result assumes a dispersion-free medium: one in which the wave speed ω/k is the same at all frequencies.

In the presence of *dispersion*, the relationship $\omega = \omega(k)$ is more complicated. Assuming the wave packet is sharply peaked around $k = k_0$, we approximate the dependence of $\omega(k)$ by expanding it around k_0: $\omega(k) \simeq \omega_0 + (k - k_0)\, d\omega/dk|_0$. The wave packet is then given by

$$\begin{aligned} V(x,t) &\simeq \int_0^{\infty} dk\, a(k)\, e^{ikx - i[\omega_0 + (k-k_0)d\omega/dk|_0]t} \\ &= e^{-i\omega_0 t + ik_0 d\omega/dk|_0\, t} \int_0^{\infty} dk\, a(k)\, e^{ik(x - d\omega/dk|_0 t)} \\ &= e^{-i\omega_0 t + ik_0 d\omega/dk|_0\, t} V(x - d\omega/dk|_0\, t, 0)\ . \end{aligned} \tag{5.45}$$

Apart from the complex phase factor, the wave shape propagates forward at about the *group velocity*

$$v_g = \frac{d\omega}{dk} \,. \tag{5.46}$$

(Note that v_g is usually the speed at which energy is transferred, provided that the expansion of $k(\omega)$ is sufficiently accurate to first order. However, $d\omega/dk$ can in some cases be greater than the speed of light. It can then not represent the speed of energy or signal propagation.)

If v_g is different at different frequencies, different frequency components of the packet will travel at different velocities and the packet will usually undergo *spreading* proportional to the second derivative $d^2\omega/dk^2$ and inversely proportional to the width of the packet (see problem 7).

5.5.3 Frequency Distributions

One of the most important uses of Fourier transforms is in associating an observed frequency spectrum with the time dependence of fluctuations in the source. Many phenomena in physics are characterized by a time-dependent amplitude, such as the potential V or current I in electronic circuits, the electromagnetic field $\mathbf{F}$ or parapotential A at a given field point in electromagnetic radiation, and the pressure or velocity in acoustics, and the power generated is proportional to the square of this amplitude.

Let $A(t)$ be a scalar amplitude, so normalized that the power is

$$P(t) = |A(t)|^2 \,. \tag{5.47}$$

(In most cases, a physical amplitude will be real and the absolute-magnitude sign is unnecessary, but its presence will allow us to generalize the result.) Define a Fourier transform of the amplitude

$$\mathcal{A}(\omega) = \frac{1}{\sqrt{2\pi}} \int_{-\infty}^{\infty} A(t)\, e^{-i\omega t} dt \,, \tag{5.48}$$

with the factor of 2π distributed to emphasize the symmetry with the inverse transform,

$$A(t) = \frac{1}{\sqrt{2\pi}} \int_{-\infty}^{\infty} \mathcal{A}(\omega)\, e^{i\omega t} d\omega \,. \tag{5.49}$$

Note, however, that if $A(t)$ is real, $\mathcal{A}(\omega)$ is generally complex with

$$\mathcal{A}(-\omega) = \mathcal{A}^*(\omega) \ . \tag{5.50}$$

We assume a finite process with

$$\lim_{t \to \pm\infty} tA(t) \to 0 \tag{5.51}$$

to ensure that the Fourier integral (5.48) is well defined.

The total energy from the process is

$$W = \int_{-\infty}^{\infty} P(t)\ dt \,. \tag{5.52}$$

Inserting the inverse transform (5.49) we find

$$W = \frac{1}{2\pi} \int_{-\infty}^{\infty} dt \int_{-\infty}^{\infty} d\omega \int_{-\infty}^{\infty} d\omega' \left[\mathcal{A}(\omega)\, \mathcal{A}^*(\omega')\right] e^{i(\omega-\omega')t}. \tag{5.53}$$

By changing the order of integration (usually safe for well-behaved, finite integrals) and noting the appearance of the Dirac delta function

$$\delta(\omega - \omega') = \frac{1}{2\pi} \int_{-\infty}^{\infty} e^{i(\omega-\omega')t} dt \,, \tag{5.54}$$

the total energy W is seen also to be

$$W = \int_{-\infty}^{\infty} |\mathcal{A}(\omega)|^2 \, d\omega \,. \tag{5.55}$$

The quantity

$$S(\omega) = |\mathcal{A}(\omega)|^2 \tag{5.56}$$

gives the frequency distribution of the energy and is known as the *spectral distribution* or *spectrum* of the process.

Exercise 5.5. Show that the spectrum, expressed in terms of the time-dependent amplitudes $A(t)$, is the Fourier transform

$$S(\omega) = \frac{1}{2\pi} \int_{-\infty}^{\infty} ds\, \Phi(s)\, e^{i\omega s} \tag{5.57}$$

of what is called the *autocorrelation function*

$$\Phi(s) = \int_{-\infty}^{\infty} dt\, A(t)\, A^*(t+s) \ . \tag{5.58}$$

5.6 Transmission Lines

A transmission line is a pair of conductors of constant cross section and with a length comparable or larger than the wavelength of the waves it transmits. Typical configurations have the parallel conductors either as side-by-side wires (a parallel-wire line) or as concentric cylinders (a coaxial cable). In both cases, we often refer to the two conductors as the two *sides* of the line. Since the wave speed in conductors is roughly the speed of light, the wave behavior of electrical signals in transmission lines of meter lengths is important for frequencies higher than roughly 100 MHz. In longer lines, the effects are significant at lower frequencies.

In elementary circuit theory, one ignores the delay of signals across conductors, and that is an excellent approximation as long as the time to traverse the conductor is small compared to the characteristic periods with which the circuit is driven. In this chapter, we consider cases in which this approximation breaks down.

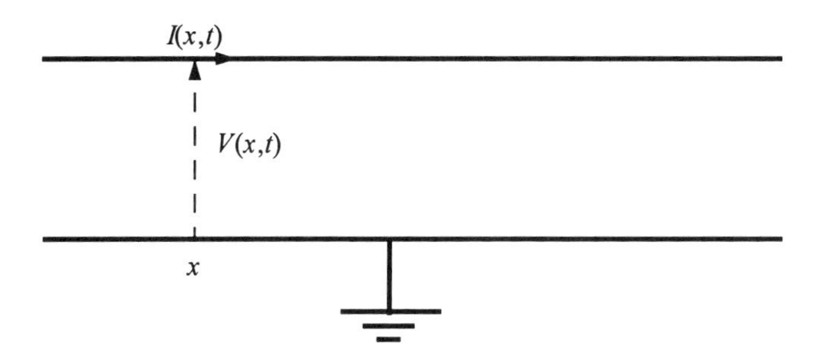

Fig. 5.4. The parameters of a transmission line.

Let's derive the wave equation for the voltage V and current I in a transmission line. If one of the conductors is charged up, it will tend to pull the opposite charge onto the other conductor. There is therefore a capacitance between the conductors which should be considered at high frequencies. Every transmission line is characterized by a characteristic capacitance per unit length, C. Similarly, the line, together with impedance loads at the two ends, forms a circuit that can develop a voltage (an emf) in response to changing magnetic flux linking the circuit. The resulting self-inductance is also proportional to the length of the two conductors, so that the transmission line is also characterized by an inductance L

per unit length.

Consider a short length δx of transmission line (see Figure 5.4). The charge δq held in δx by the voltage difference $V(x,t)$ between the two sides of the line is $\delta q = CV\delta x$. The time-rate of change of this charge gives the net current inflow into δx:

$$\frac{\partial(\delta q)}{\partial t} = C\frac{\partial V}{\partial t}\delta x = I(x,t) - I(x+\delta x, t) = -\frac{\partial I}{\partial x}\delta x\,.$$

Therefore,

$$C\frac{\partial V}{\partial t} = -\frac{\partial I}{\partial x}\,. \tag{5.59}$$

On the other hand, an increasing current induces a back emf (electromotive force) that resists the increase and thereby forces a compensatory drop in the line voltage[2] across δx of

$$\delta V = \left(\frac{\partial V}{\partial x}\right)\delta x = -L\delta x\frac{\partial I}{\partial t}\,,$$

so that

$$\frac{\partial V}{\partial x} = -L\frac{\partial I}{\partial t}\,. \tag{5.60}$$

Combining equations (5.59) and (5.60), we find that V and I both satisfy the wave equation (5.1),

$$LC\frac{\partial^2 V}{\partial t^2} = \frac{\partial^2 V}{\partial x^2}\,, \qquad LC\frac{\partial^2 I}{\partial t^2} = \frac{\partial^2 I}{\partial x^2} \tag{5.61}$$

with wave speed

$$c = (LC)^{-1/2}\,. \tag{5.62}$$

Typical wave speeds of signals in conductors are two-thirds the speed of light, many orders of magnitude faster than the average speed of the electrons (see problems 1 and 14).

[2]The sign of δV is important. It does not represent the push of Faraday emf itself, but rather the drop in line voltage necessary to compensate for it. The induced emf is like the emf in a transient battery: it pushes negative charge toward the negative terminal and positive charge toward the positive one until balanced by the oppositely directed push from the voltage difference between the terminals.

Solutions of (5.1) are linear combinations of the plane waves (5.15), and the physical solutions are the real parts of such linear combinations. For analysis, it is useful to concentrate on solutions at a single angular frequency ω:

$$V(x,t) = \left\langle \left(v_+ e^{ikx} + v_- e^{-ikx}\right) e^{-i\omega t} \right\rangle_{\Re} . \tag{5.63}$$

The two terms represent forward and backward running waves with the complex amplitudes v_+ and v_-, respectively. To simplify the expressions, we generally suppress the time dependence and the brackets $\langle ... \rangle_{\Re}$. Thus we write

$$V(x) = \left(v_+ e^{ikx} + v_- e^{-ikx}\right) =: V_+(x) + V_-(x) . \tag{5.64}$$

The wave is generated at $x = 0$ and is initially forward running. The backward-running wave arises from reflections at the terminating end of the line at $x = l$. From equation (5.60), the corresponding current is

$$I(x) = \sqrt{\frac{C}{L}} \left(v_+ e^{ikx} - v_- e^{-ikx}\right) =: I_+(x) + I_-(x) . \tag{5.65}$$

For sinusoidal input, the *signal* measured at position x by an oscilloscope is a rapidly oscillating voltage of amplitude $|V(x)|$.

5.7 Impedance and Reflections

The *impedance* is defined by the complex ratio of the voltage to the current:

$$Z(x) = V(x)/I(x) . \tag{5.66}$$

The *characteristic impedance* of the line is the constant impedance for waves without reflections:

$$Z_0 = \frac{V_+(x)}{I_+(x)} = \sqrt{\frac{L}{C}} = -\frac{V_-(x)}{I_-(x)} . \tag{5.67}$$

Note that for a lossless line, Z_0 is real, that is, a purely resistive impedance. If the line is terminated at $x = l$ by a resistive load of the *same impedance*, then $Z(l) = Z_0$ and relation (5.66) is satisfied by the incident wave by itself with no reflected component. The impedance of the line is then said to be *matched* by the load and there will be no reflection. However, any other termination will require the addition of reflected waves in order to satisfy (5.66). Let's relate the load impedance to the reflection.

The reflection coefficient is the ratio of the reflected to the incident wave:

$$\Gamma(x) = \frac{V_-(x)}{V_+(x)} = \frac{v_-}{v_+}e^{-2ikx} = -\frac{I_-(x)}{I_+(x)}. \tag{5.68}$$

Any part of the wave that is not reflected is transmitted or absorbed (or a combination). Thus, $\Gamma = 0$ means no reflection and hence total transmission and/or absorption, whereas $|\Gamma| = 1$ implies total reflection and no transmission. Like the impedance, it is generally a complex number that depends on ω. With the definition (5.68), we can re-express the impedance (5.66) by

$$Z(x) = Z_0 \left[\frac{1+\Gamma(x)}{1-\Gamma(x)}\right]. \tag{5.69}$$

The value of $Z(0)$ is the *input impedance* of the line. It generally depends on the output impedance. We can also solve for the reflection coefficient in terms of the impedance at x:

$$\Gamma(x) = \frac{Z(x) - Z_0}{Z(x) + Z_0}. \tag{5.70}$$

Reflections occur wherever there is a sudden change of impedance. We can control the impedance at the ends of a transmission line by the load we place there. Note that the reflection coefficient at $x = l$ vanishes if and only if the output impedance matches the characteristic impedance of the line. In particular, if the termination of the line is shorted, $Z(l) = 0$ and $\Gamma(l) = -1$, whereas if the end is open, $Z(l) = \infty$ and $\Gamma(l) = 1$. Thus, an open line reflects signals totally with no change in sign, whereas a shorted line reflects totally but with a change in sign. Other resistive terminations of the line will give less than total reflection, and the incident and reflected signals will have the same sign if and only if the load resistance is greater than the characteristic impedance Z_0. A purely reactive impedance (termination by a coil or capacitor) will give total reflection with a frequency-dependent change of phase. However, resistive terminations give reflection coefficients $\Gamma(l)$ which are independent of frequency and therefore also valid for pulses, which may be viewed as linear combinations of sinusoidal signals of different frequencies.

In the case of sinusoidal excitation, the total signal

$$V(x) = v_+e^{ikx}\left[1 + \Gamma(x)\right] \tag{5.71}$$

varies in magnitude as kx is varied from a minimum of $|v_+|\,[1-|\Gamma|]$ to a maximum of $|v_+|\,[1+|\Gamma|]$. Generally

$$\Gamma(x) = \Gamma(l)\exp\left[2ik\,(l-x)\right] . \tag{5.72}$$

If $Z(l)$ is resistive, $|\Gamma| = \Gamma(l)$. For a more general load impedance $Z(l)$, the reflection coefficient (5.70) is complex: $\Gamma(l) = |\Gamma|\exp(i\phi)$. The signal at $x=0$ then is

$$\begin{aligned} |V(0)| &= |v_+\,[1+\Gamma(0)]| \\ &= |v_+|\left\{1+|\Gamma|^2+2\,|\Gamma|\cos(\phi+2kl)\right\}^{1/2} . \end{aligned}$$

5.8 Problems

1. Make an order-of-magnitude estimate of the average **velocity of electrons** in a copper wire of 1 mm diameter when it carries one ampere of current. You may assume that each copper atom contributes one electron to the conduction band. [Answer: roughly 0.1 mm/s, and thus much, much less than the wave or signal velocity in the wire, which is about 0.7 c.]

2. Consider the **electron flow** in a copper wire. Represent the wire by a solid cylinder and show that the Lorentz force on charge carriers from the constant magnetic field of a direct current flow pushes the carriers toward the center of the wire, whereas Faraday induction in a high-frequency flow forces most of the current to the outer surface of the conductor. [Hint: consider the emf around a current loop inside the wire, one side of which lies on the axis of the cylinder, one side on the surface directed along the length of the cylinder, and these two lengths joined by short radial segments.]

3. Find the Fourier series for a string with boundary conditions $y=0$ at $x=0$ and $x=L$, with an initial **triangular-wave shape**

$$y(x,0) = \begin{cases} xy_0/x_0\,, & 0<x<x_0 \\ (L-x)\,y_0/\,(L-x_0)\,, & x_0<x<L \end{cases} , \tag{5.73}$$

with a peak of y_0 at x_0, and $\partial y/\partial t = 0$ at $t=0$.

4. Describe the **Helmholtz motion** of a bowed string of length L, given by

$$y(x,t) = b\sum_{n=1}^{\infty}\frac{1}{n^2}\sin(k_n x)\sin(k_n ct) . \tag{5.74}$$

Show that the shape at any instant in time has the triangular shape of the wave in problem 3. Determine how the shape changes in time.[3]

5. **Uncertainty relation.** Find the x, t dependence of a wave packet (5.27) with a Gaussian amplitude distribution:

$$a(k) = \exp\left[-b^2 (k - k_0)^2 / 2\right]$$

where b is a constant length and b^{-1} is a measure (the root-mean square deviation from the mean) of the width of the distribution $a(k)$:

$$\Delta k = \left[\frac{\int dk\, (k - k_0)^2 a(k)}{\int dk\, a(k)}\right]^{1/2} = b^{-1}. \tag{5.75}$$

Show that the wave packet $V(x, 0)$ also has a Gaussian shape but that its width Δx is b rather than b^{-1}. The result $\Delta x\, \Delta k \geq 1$ is a form of the uncertainty relation for waves. Argue that therefore a sharply defined wave vector k implies a poorly defined position and *vice versa*.

6. Show that the **Gaussian wave packet** with amplitude function $a(k) = \exp(-b^2 k^2 / 2)$ splits into two packets traveling in opposite directions. Prove that the real part of its shape as a function of time is

$$\Re\{V(x,t)\} = \sqrt{\frac{\pi}{2}} b^{-1} \left\{\exp\left[-\frac{(x+ct)^2}{2b^2}\right] + \exp\left[-\frac{(x-ct)^2}{2b^2}\right]\right\}. \tag{5.76}$$

7. **Wave-packet spreading.** In section 5.5.2 we saw that wave-packet components at wave vector k experienced a group velocity $v_g = d\omega/dk$. Assume that the dispersion is well-described by a second-order expansion of $\omega(k)$, namely

$$\omega(k) \simeq \omega(k_0) + d\omega/dk|_{k_0} (k - k_0) + \frac{1}{2} d^2\omega/dk^2|_{k_0} (k - k_0)^2. \tag{5.77}$$

Use the results of the previous problem to show that the spreading of the wave packet is given approximately by the difference in group velocities

$$\Delta v_g = \frac{1}{b} d^2\omega/dk^2|_{k_0}, \tag{5.78}$$

where b is the width of the packet.

[3] See, for example, Fletcher and Rossing (1991), pp. 45–46 and 238–241.

8. **Finite wave train.** Let $V(x,0)$ be the finite wave train

$$V(x,0) = \begin{cases} e^{ik_0x}, & |x| < x_0 \\ 0, & |x| > x_0 \end{cases} . \tag{5.79}$$

Find the amplitude function $a(k)$ and show that the result is consistent with the uncertainty relation $\Delta x \, \Delta k \geq 1$.

9. **Beats.** Consider the superposition of two waves of slightly different frequencies:

$$y(t) = \cos \omega_1 t + \cos \omega_2 t . \tag{5.80}$$

(a) Re-express this sum in terms of the difference frequency $\Delta\omega = \omega_2 - \omega_1$ and the average frequency $\bar{\omega} = (\omega_1 + \omega_2)/2$, and show that it is a rapidly varying wave of frequency $\bar{\omega}$ modulated by a slowly varying envelope of "beats":

$$2\cos\left(\frac{1}{2}\Delta\omega t\right)\cos\bar{\omega}t . \tag{5.81}$$

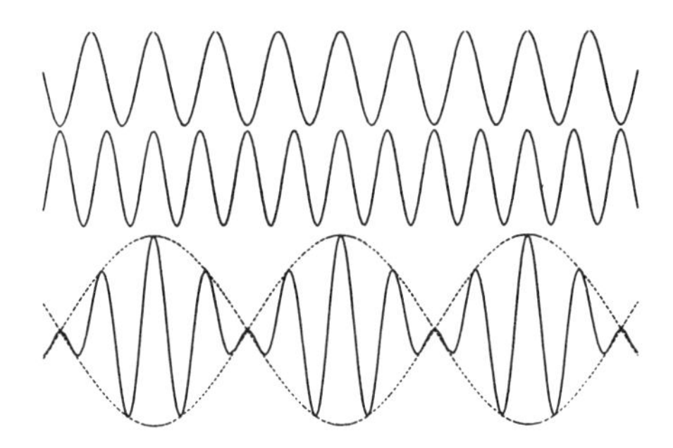

Fig. 5.5. The superposition of sinusoidal waves of slightly different frequencies (shown at the top) produces a beating envelope of waves at the average frequency (bottom).

(b) Explain why the beat frequency is usually identified as $\Delta\omega$ rather than $\frac{1}{2}\Delta\omega$. (Consider, for example, how the superposition would sound if the two sinusoid components were pure tones produced by different musical instruments. Note that the envelope curve in Figure 5.5 comprises two sinusoids.)

(c) Replace the time dependence $\omega_j t$ by a space and time dependence $k_j x - \omega_j t$ and show that the velocity of the envelope is $\Delta\omega / \Delta k$. Relate this to the group velocity of a wave packet.

10. **Dirac delta function.** Sketch the function $\Delta_X(K) = \sin(KX) / (\pi K)$ for a given X as a function of KX, for $-3\pi < KX < 3\pi$. Indicate the KX values at which $\Delta_X(K)$ vanishes and give the value of $\Delta_X(0)$. Show that $\lim_{X\to\infty} \{\Delta_X(K)\}$ does have the required properties of the Dirac delta function (see section 5.5.1).

11. **Solenoid current.** Use Maxwell's equation to find the current j that gives the electromagnetic field

$$\mathbf{F} = icB_0 \mathbf{e}_3 \theta(\rho_0 - \rho) , \qquad (5.82)$$

where $\theta(\rho_0 - \rho)$ is the Heaviside step function (see section 5.5.1) and ρ is the radial cylindrical coordinate (see Appendix). Relate the result to an ideal solenoid. [Ans.: $(B_0/\mu_0)\,\hat{\phi}$.]

12. **Derivatives of the delta function.** The singular nature of the delta "function" would seem to preclude any sensible discussion of its derivative.

(a) Use integration by parts to show nevertheless that the integral of a continuous bounded function $g(x)$ times the derivative $\delta'(x - x_0) \equiv d\delta(x - x_0) / dx$ of the delta function $\delta(x)$ gives

$$\int dx\, g(x)\; \delta'(x - x_0) = -g'(x_0) \qquad (5.83)$$

as long as the domain of integration includes a neighborhood of x_0.

(b) Find an expression for the integral of $g(x)$ times the second derivative of $\delta(x - x_0)$.

(c) Let $\exp(x_0 d/dx)$ be the differential operator obtained by power-series expansion:

$$\exp\left(x_0 \frac{d}{dx}\right) = 1 + x_0 \frac{d}{dx} + \frac{x_0^2}{2!}\frac{d^2}{dx^2} + \cdots \qquad (5.84)$$

and extend the argument of parts (a) and (b) above to show that

$$\exp\left(x_0 \frac{d}{dx}\right) \delta(x) = \delta(x + x_0) . \qquad (5.85)$$

13. Let $y(x)$ be a monotonic function of x. Change variables in the integral $\int dx\, g(x)\, \delta\left[y(x)\right]$ to prove the relation

$$\delta\left[y(x)\right] = \frac{\delta(x_0)}{|dy/dx|}, \tag{5.86}$$

where x_0 is defined as the root of y: $y(x_0) = 0$. (Note: if $y(x)$ is not monotonic, the inverse function $x(y)$ will have distinct branches possibly with distinct roots x_j of $y(x)$ that must be summed over.)

14. Estimate the **characteristics of a twin lead** with two wires of 1 mm diameter whose centers are separated by 1 cm. Approximate the charge on each wire, at a given position x along the twin lead (see Figure 5.4), by a uniform distribution around the circumference of its circular cross section.

    (a) Find the voltage V between the wires associated with equal but opposite static charges Q and $-Q$ per unit length on the two leads. Use this to determine the capacitance C per unit length.

    (b) Use Faraday induction to determine the emf along the two axial segments of a loop that runs for a unit length along the inner surfaces of the leads and is joined at the ends by short transverse segments. From this, determine the self-inductance L per unit length.

    (c) From L and C of parts (a) and (b), find the wave speed c and the impedance Z_0 of the line.

15. Repeat the calculation of $L, C, c,$ and Z_0 (see previous problem) for a **coaxial line** with a central conductor 1 mm in diameter and an outer cylindrical shell 0.5 mm in diameter. (Ignore the effects of any dielectric material in the cable for this problem.)

16. Use the condition (5.72) on the reflection coefficient to show that only discrete frequencies are allowed when the ends of a transmission line are either both open or both shorted. What are the frequencies? What are the allowed frequencies for a line that is open at one end and shorted at the other?

17. Show that total reflection occurs [$\Gamma(l)$ has a magnitude of unity] if the load impedance is purely reactive (purely capacitive or purely inductive). Find an expression for the voltage signal at $x = 0$ for sinusoidal waves of angular frequency ω.

18. Suppose a capacitor with capacitance C_1 is used to terminate a transmission line, so that

$$Z(l) = (-i\omega C_1)^{-1}$$

and that the frequency is adjusted to $\omega = 1/(Z_0 C_1)$. Find the complex reflection coefficient $\Gamma(x)$ as a function of $kx - kl$ and sketch the result in the complex plane.

19. The ends of four 50-ohm coaxial cables are spliced together to form an X junction. From the viewpoint of one of the cables, the junction represents a certain termination impedance.

    (a) How great is this impedance?

    (b) What happens to an electromagnetic pulse that is sent down one cable toward the junction?

    (c) How can resistors be incorporated into the junction in order to achieve complete impedance matching?

    (d) Suppose the impedances have been successfully matched. If a signal of amplitude V_+ is sent down one cable toward the junction, what signal appears in the other cables.

20. Assume that the frequency has been adjusted to make $kl = \pi/2$, The line is then said to be a *quarter-wave line*. Show that then $Z(0)\,Z(l) = Z_0^2$.

Chapter 6

Electromagnetic Waves in Vacuum

As we turn our attention from waves in one spatial dimension to waves in three-dimensional physical space, Clifford's geometric algebra resumes its guiding role in describing the symmetries and geometrical relations that mold the fundamental physics. Not only are the electric and magnetic fields of the wave unified in a single biparavector known as the electromagnetic field (the Faraday **F**) that acts in one or two planes in spacetime, but also the *projector*, a noninvertible real element of the algebra, proves to be a potent new computational tool that illuminates important properties of *directed plane waves*.

The directed plane waves referred to here are not necessarily monochromatic; they may be pulses that are well-localized in the propagation direction. Algebraically, they are all minimal left and right *ideals*, and this algebraic structure determines many of their physical properties, such as the fact that at any point in space, the electric and magnetic fields and the propagation direction of such waves form an orthogonal right-handed set of vectors. The algebraic structure, together with the spinorial form of the Lorentz-force equation introduced in Chapter 4, also allows rather simple derivations of the relativistic motion of charges in directed plane waves, with a couple of surprising results.

Unlike the common complex notation for monochromatic plane waves, the algebraic expressions are fully physical; there is no imaginary part to be discarded. When waves of opposite propagation direction are superimposed to form standing waves, new properties arise, such as electric and magnetic fields that are parallel at every point in spacetime. Here, projectors and what has become known as their "pacwoman" property allow simple derivations and can be used to isolate the wave components propagating in different directions.

The material in this chapter is important background for studies in the following chapters on waves in media and radiation from accelerating particles.

6.1 Maxwell's Equation in a Vacuum

Recall Maxwell's equation for the electromagnetic field $\mathbf{F}$ when the source term vanishes, $j = 0$:

$$\bar{\partial}\mathbf{F}(x) = 0\,. \tag{6.1}$$

In terms of the paravector potential A, the relation

$$\mathbf{F} = c\partial\bar{A} \tag{6.2}$$

gives the wave equation for A,

$$\partial\bar{\partial}A = \Box A = \left(c^{-2}\partial_t^2 - \nabla^2\right)A = 0\,, \tag{6.3}$$

where the Lorenz gauge condition $\left\langle\bar{\partial}A\right\rangle_S = 0$ has been imposed. Since the D'Alembertian $\Box$ is a scalar operator, the wave equation holds for each component of A. It also holds for each component of the electromagnetic field $\mathbf{F}$, as is seen by differentiating (6.1):

$$\Box\mathbf{F} = \partial\bar{\partial}\mathbf{F} = 0. \tag{6.4}$$

6.1.1 Directed Plane-Wave Solutions

What implications does the wave equation have for the paravector potential A or for the biparavector field $\mathbf{F}$? Suppose for the moment that A depends on position through the scalar product

$$s = \left\langle\bar{k}x\right\rangle_S = \langle k\bar{x}\rangle_S = \omega t - \mathbf{k}\cdot\mathbf{x}\,, \tag{6.5}$$

where $k = \omega/c + \mathbf{k}$ is a constant spacetime vector. The potential A may be said to be a *directed plane wave* since it is the same everywhere on the *hypersurface* $\langle k\bar{x}\rangle_S =$ const, and at any given time t, it is the same everywhere on the spatial surface $\mathbf{k}\cdot\mathbf{x} =$ const (see Figure 6.1).

During the interval dt, every constant surface moves a distance $d\mathbf{x}$ perpendicular to the surface and hence parallel to $\mathbf{k}$ by an amount that satisfies $\omega dt = \mathbf{k}\cdot d\mathbf{x}$. Thus, $d\mathbf{x} = \mathbf{k}^{-1}\omega\, dt$. The velocity $d\mathbf{x}/dt = \mathbf{k}^{-1}\omega$ is called the *wave (phase) velocity* of the wave.

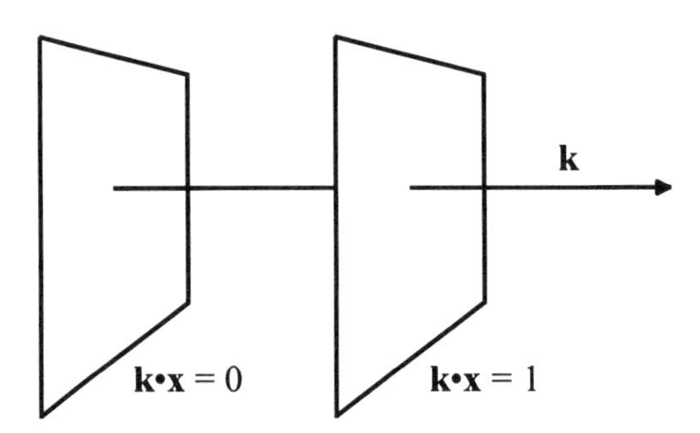

Fig. 6.1. The constant spatial planes of a directed plane wave propagate with velocity $\omega/\mathbf{k}$.

If ω is thought of as an angular velocity, then since the proper-time derivative

$$\dot{s} = c \langle k\bar{u} \rangle_S = \omega_{\text{rest}} \tag{6.6}$$

is that angular velocity in the rest frame, s is the angle swept out in the rest frame. In order to see whether $A(s)$ is a possible solution of the wave equation (6.3), we need to compute the field $\partial \bar{A}(s)$. By the chain rule,

$$\partial \bar{A}(s) = (\partial s) \bar{A}'(s) \ , \tag{6.7}$$

where the prime indicates differentiation with respect to the scalar argument:

$$A'(s) := \frac{d}{ds} A(s) \ . \tag{6.8}$$

Recall that in component notation, the scalar product s is[1]

$$s = \left\langle \bar{k} x \right\rangle_S = k^\lambda x^\mu \langle \bar{\mathbf{e}}_\lambda \mathbf{e}_\mu \rangle_S = k^\lambda x^\mu \eta_{\lambda\mu} = k_\mu x^\mu. \tag{6.9}$$

[1]Recall our use of the Einstein summation convention that repeated indices are summed over. Usually one of the indices is an upper ("contravariant") index while the other is a lower ("covariant") one. The result does not depend on which factor has the upper and which the lower index. The metric tensor $\eta_{\mu\nu} = \langle \mathbf{e}_\mu \bar{\mathbf{e}}_\nu \rangle_S$ can be used to lower indices: $k^\lambda \eta_{\lambda\mu} = k_\mu$, and similarly, $\eta^{\mu\nu} = \langle \mathbf{e}^\mu \bar{\mathbf{e}}^\nu \rangle$ raises indices: $\eta^{\mu\nu} x_\mu = x^\nu$. The symbol

$$\eta^\mu_\nu = \langle \mathbf{e}^\mu \bar{\mathbf{e}}_\nu \rangle_S = \langle \bar{\mathbf{e}}^\mu \mathbf{e}_\nu \rangle_S$$

Since the gradient is

$$\partial = \mathbf{e}_\mu \partial^\mu = \mathbf{e}^\mu \partial_\mu = \mathbf{e}^\mu \frac{\partial}{\partial x^\mu}\,, \tag{6.10}$$

the gradient of s is

$$\partial s = \mathbf{e}^\mu \partial_\mu \left(k_\nu x^\nu\right) = \mathbf{e}^\mu k_\mu = \mathbf{e}_\mu k^\mu = k\,. \tag{6.11}$$

Consequently, the field (6.2) is

$$\mathbf{F} = c\partial \bar{A}(s) = c\left(\partial s\right)\bar{A}'(s) = ck\bar{A}'(s)\,. \tag{6.12}$$

Similarly,

$$\partial\bar{\partial}A(s) = \Box A(s) = k\bar{k}A''(s)\;. \tag{6.13}$$

The wave equation (6.3) tells us that this vanishes, so that unless $A'' = 0$ at all positions, in which case A is either constant throughout spacetime or grows without limit with a term proportional to s, then k must be null:

$$k\bar{k} = \omega^2/c^2 - \mathbf{k}^2 = 0. \tag{6.14}$$

Consequently,

$$k = \frac{\omega}{c}\left(1 + \hat{\mathbf{k}}\right). \tag{6.15}$$

Any function A of $s = \langle \bar{k}x \rangle_S$, where k is any constant null spacetime vector, is seen to be a solution of the wave equation (6.3). More generally, since (6.3) is linear, any superposition of solutions with different values of k is also a solution.

Exercise 6.1. The wave equation (6.3) for A follows from Maxwell's equation if the Lorenz gauge condition is imposed. Show that the wave equation $\Box\mathbf{F} = 0$ implies that any non-constant, bounded directed plane-wave solution $\mathbf{F}(s)$ has a null k regardless of the choice of gauge.

---

simply replaces one index by another at the same position, for example $\eta^\nu_\mu x^\mu = x^\nu$ and is thus exactly the Kronecker delta

$$\eta^\mu_\nu = \delta^\mu_\nu\,.$$

6.1.2 Gauge Choices

As noted in chapter 3, many different choices of the vector potential A give rise to exactly the same electromagnetic field $\mathbf{F}$. In particular, $\mathbf{F}$ is invariant under the gauge transformation

$$A \to A + \partial\chi\,, \tag{6.16}$$

where χ is any scalar function. The Lorenz-gauge condition

$$\langle\partial\bar{A}\rangle_S = c^{-2}\partial_t\phi + \nabla\cdot\mathbf{A} = \langle k\bar{A}'(s)\rangle_S = 0 \tag{6.17}$$

limits the gauge freedom and relates the components of $A = \phi/c + \mathbf{A}$: integration of the condition (6.17) with respect to s gives

$$\left\langle\left(1+\hat{\mathbf{k}}\right)\bar{A}\right\rangle_S = \text{const.} \tag{6.18}$$

and hence

$$\phi = \phi_0 + \hat{\mathbf{k}}\cdot\mathbf{A}c\,, \tag{6.19}$$

where ϕ_0 is a constant. Consequently, $\hat{\mathbf{k}}\phi' = c\mathbf{A}'_\parallel$, where $\mathbf{A}_\parallel = \mathbf{A} - \mathbf{A}_\perp = \hat{\mathbf{k}}\hat{\mathbf{k}}\cdot\mathbf{A}$, and therefore

$$\begin{aligned}\mathbf{F} &= c\langle k\bar{A}'\rangle_V \\ &= \mathbf{k}\phi' - \omega\mathbf{A}' - c\mathbf{k}\mathbf{A}'_\perp \\ &= -ck\mathbf{A}'_\perp \\ &= \left(1+\hat{\mathbf{k}}\right)\mathbf{E}\,,\end{aligned} \tag{6.20}$$

where $\mathbf{E} = -\omega\mathbf{A}'_\perp(s)$. If the Coulomb gauge ($\nabla\cdot\mathbf{A} = 0$) is also imposed, then $\hat{\mathbf{k}}\cdot\mathbf{A}$ is constant and one can take simply $\phi = \hat{\mathbf{k}}\cdot\mathbf{A}c = 0$. The Coulomb gauge is thus compatible with the Lorenz-gauge condition in source-free space. However, the Lorenz-gauge condition is Lorentz invariant whereas the Coulomb-gauge condition is not.

Exercise 6.2. Prove that the Lorenz-gauge condition implies that the paravector potential $A(s)$ satisfies

$$\left(1+\hat{\mathbf{k}}\right)\bar{A}'(s) = -A'(s)\left(1-\hat{\mathbf{k}}\right). \tag{6.21}$$

Split this expression into scalar and vector parts to prove that it is equivalent to the condition $\hat{\mathbf{k}}\phi' = c\mathbf{A}'_{\parallel}$ given above.

It is significant that the field $\mathbf{F}$ associated with any directed plane wave has the form $k\bar{A}'(s)$ where k is a null paravector. The projector properties of k determine much of the behavior of $\mathbf{F}$. Before we look at specific examples of directed plane waves, we need to review some of these properties.

6.2 Projectors

Since k is not only null (6.14) but also real, it can be factored into

$$kc = \omega\left(1 + \hat{\mathbf{k}}\right) = 2\omega P_{\hat{\mathbf{k}}} \tag{6.22}$$

where $P_{\hat{\mathbf{k}}} = \frac{1}{2}\left(1 + \hat{\mathbf{k}}\right)$ is a *projector*, defined to be a *real null idempotent* element with the properties

$$P_{\hat{\mathbf{k}}}^2 = P_{\hat{\mathbf{k}}} = P_{\hat{\mathbf{k}}}^\dagger \text{ (idempotent and real)} \tag{6.23}$$

$$P_{\hat{\mathbf{k}}}\bar{P}_{\hat{\mathbf{k}}} = 0 = \bar{P}_{\hat{\mathbf{k}}}P_{\hat{\mathbf{k}}} \text{ (null).} \tag{6.24}$$

It follows that the bar conjugate $\bar{P}_{\hat{\mathbf{k}}}$ is also a projector. Furthermore, $P_{\hat{\mathbf{k}}} + \bar{P}_{\hat{\mathbf{k}}}$ is idempotent:

$$\begin{aligned}\left(P_{\hat{\mathbf{k}}} + \bar{P}_{\hat{\mathbf{k}}}\right)^2 &= P_{\hat{\mathbf{k}}}^2 + \bar{P}_{\hat{\mathbf{k}}}^2 + P_{\hat{\mathbf{k}}}\bar{P}_{\hat{\mathbf{k}}} + \bar{P}_{\hat{\mathbf{k}}}P_{\hat{\mathbf{k}}} \\ &= P_{\hat{\mathbf{k}}} + \bar{P}_{\hat{\mathbf{k}}}\,.\end{aligned} \tag{6.25}$$

Since $P_{\hat{\mathbf{k}}} + \bar{P}_{\hat{\mathbf{k}}}$ is a nonvanishing real scalar, it can only be unity:

$$P_{\hat{\mathbf{k}}} + \bar{P}_{\hat{\mathbf{k}}} = 1\,. \tag{6.26}$$

The projectors $P_{\hat{\mathbf{k}}}$ and $\bar{P}_{\hat{\mathbf{k}}}$ are said to be *complementary.*

Because of the null property of projectors, the product of a projector and its complement vanishes. One sometimes refers to the “orthogonality” of complementary projectors, and indeed such projectors often do represent orthogonal states. However, complementary projectors are not orthogonal spacetime vectors. Rather, each projector is null and hence orthogonal to itself.

In what geometrical sense do "projectors" *project* anything? For any null element P, the sandwich PxP of an arbitrary cliffor $x = x^0 + \mathbf{x}$ is just the projection of x along P in the sense

$$PxP = \left(Px + \bar{x}\bar{P}\right) P = 2 \langle Px \rangle_S P \,, \tag{6.27}$$

and if P is the projector $P = \frac{1}{2}(1 + \hat{\mathbf{n}})$, then $2 \langle Px \rangle_S = x^0 + \mathbf{x} \cdot \hat{\mathbf{n}}$. Similarly, since $\hat{\mathbf{n}} P = P$,

$$Px\bar{P} = Px(1 - P) = P\mathbf{x}_\perp = \mathbf{x}_\perp \bar{P} \,, \tag{6.28}$$

where

$$\mathbf{x}_\perp := x - \mathbf{x} \cdot \hat{\mathbf{n}} \, \hat{\mathbf{n}} \,. \tag{6.29}$$

Example 6.3. Let $P = \frac{1}{2}(1 + \mathbf{e}_3)$. The vector $\mathbf{r} \equiv x\mathbf{e}_1 + y\mathbf{e}_2 + z\mathbf{e}_3$ has components $z\mathbf{e}_3$ and $r_\pm (\mathbf{e}_1 \mp i\mathbf{e}_2)/2$ that arise from the decomposition

$$\mathbf{r} = \left(P + \bar{P}\right) \mathbf{r} \left(P + \bar{P}\right) = P\mathbf{r}P + \bar{P}\mathbf{r}P + P\mathbf{r}\bar{P} + \bar{P}\mathbf{r}\bar{P} \,. \tag{6.30}$$

In particular,

$$\begin{aligned} z\mathbf{e}_3 &= P\mathbf{r}P + \bar{P}\mathbf{r}\bar{P} \\ r_+ (\mathbf{e}_1 - i\mathbf{e}_2)/2 &= \bar{P}\mathbf{r}P \\ r_- (\mathbf{e}_1 + i\mathbf{e}_2)/2 &= P\mathbf{r}\bar{P} \,, \end{aligned} \tag{6.31}$$

where $r_\pm = x \pm iy$.

Projectors are examples of nonzero *zero-divisors*, elements that are not zero but which are nevertheless factors of zero. It is the existence of such elements which keeps the algebra from being a division algebra and which, together with noncommutivity, distinguishes the algebra from a mathematical field. In the field of complex numbers, for example, all elements commute, and if the product of two elements is zero, say $z_1 z_2 = 0$, then at least one of the factors, z_1 or z_2, must vanish. Clifford algebras other than the real and complex numbers and the quaternions are not division algebras. The existence of projectors adds richness to the algebra and allows "magic" manipulations, operations that are not possible in fields.

6.2.1 Pacwoman Property

One useful bit of such magic is the *pacwoman* ability of a projector to "gobble up" a unit vector with which it commutes: since $\hat{\mathbf{k}} = P_{\hat{\mathbf{k}}} - \bar{P}_{\hat{\mathbf{k}}}$, it follows from the null property (6.24) that

$$\begin{aligned} P_{\hat{\mathbf{k}}}\hat{\mathbf{k}} &= \hat{\mathbf{k}}P_{\hat{\mathbf{k}}} = P_{\hat{\mathbf{k}}} \\ \bar{P}_{\hat{\mathbf{k}}}\hat{\mathbf{k}} &= \hat{\mathbf{k}}\bar{P}_{\hat{\mathbf{k}}} = -\bar{P}_{\hat{\mathbf{k}}} . \end{aligned} \tag{6.32}$$

This bit of magic can be applied over and over again, so that if $f(\mathbf{k})$ is any function of the real vector $\mathbf{k} = |\mathbf{k}|\ \hat{\mathbf{k}}$ with a power series expansion, then

$$f(\mathbf{k})\, P_{\hat{\mathbf{k}}} = f(\,|\mathbf{k}|)\, P_{\hat{\mathbf{k}}} \tag{6.33}$$

and

$$f(\mathbf{k})\, \bar{P}_{\hat{\mathbf{k}}} = f(-\,|\mathbf{k}|)\, \bar{P}_{\hat{\mathbf{k}}} . \tag{6.34}$$

Of course, the pacwoman property can also be applied in reverse to insert or to "ungobble" factors of the unit vector $\hat{\mathbf{k}}$: if s and t are any scalars,

$$\begin{aligned} P_{\hat{\mathbf{k}}} f(s+t) &= P_{\hat{\mathbf{k}}} f\left(s\hat{\mathbf{k}} + t\right) \\ &= P_{\hat{\mathbf{k}}} f\left(s + t\hat{\mathbf{k}}\right) \\ &= P_{\hat{\mathbf{k}}} f\left(s\hat{\mathbf{k}} + t\hat{\mathbf{k}}\right) . \end{aligned} \tag{6.35}$$

Equations (6.33) and (6.34) may be considered *eigenvalue equations*, where $f(\pm\,|\mathbf{k}|)$ are the *eigenvalues* and the projectors $P_{\hat{\mathbf{k}}}$ and $\bar{P}_{\hat{\mathbf{k}}}$ are *eigenprojectors* of $f(\mathbf{k})$. Since $P_{\hat{\mathbf{k}}} + \bar{P}_{\hat{\mathbf{k}}} = 1$, the function $f(\mathbf{k})$ can be expanded in its eigenprojectors

$$f(\mathbf{k}) = f(\mathbf{k})\left(P_{\hat{\mathbf{k}}} + \bar{P}_{\hat{\mathbf{k}}}\right) = f(\,|\mathbf{k}|)\, P_{\hat{\mathbf{k}}} + f(-\,|\mathbf{k}|)\, \bar{P}_{\hat{\mathbf{k}}} . \tag{6.36}$$

Such an expansion is known as the *spectral decomposition* of $f(\mathbf{k})$.

Example 6.4. The rotation element $R = \exp(-i\boldsymbol{\theta}/2)$ has eigenprojectors $P = \frac{1}{2}\left(1 + \hat{\boldsymbol{\theta}}\right)$ and $\bar{P} = \frac{1}{2}\left(1 - \hat{\boldsymbol{\theta}}\right)$, and corresponding eigenvalues $\exp(\mp i\theta/2)$. Its spectral decomposition is thus

$$\begin{aligned} \exp(-i\boldsymbol{\theta}/2) &= e^{-i\theta/2}P + e^{i\theta/2}\bar{P} && (6.37) \\ &= \cos\theta/2 - i\hat{\boldsymbol{\theta}}\sin\theta/2 . && (6.38) \end{aligned}$$

More generally, any analytic function of a real vector can be decomposed into even and odd function parts:

$$f(\boldsymbol{\theta}) = f(\boldsymbol{\theta})\left(P + \bar{P}\right) = \frac{f(\theta) + f(-\theta)}{2} + \hat{\boldsymbol{\theta}}\frac{f(\theta) - f(-\theta)}{2}. \tag{6.39}$$

For reference, note that similar decompositions can be written down for more general functions of algebraic elements [see also problems (3–4) at the end of the chapter]. Consider a general element $z = z^0 + \mathbf{z}$ of $\mathcal{C}\ell_3$ where z^0 is a complex scalar and the complex vector part is written $\mathbf{z} = e^{i\phi}(\mathbf{a} + i\mathbf{b})$. Here, $\mathbf{a}$ and $\mathbf{b}$ are real vectors and the real angle ϕ is chosen so that $\mathbf{a} \cdot \mathbf{b} = 0$ and $\mathbf{a}^2 \geq \mathbf{b}^2$. The elements

$$\pi_{\pm} = \frac{1}{2}\left(1 + i\mathbf{b}\mathbf{a}^{-1} \pm \mathbf{a}^{-1}\sqrt{\mathbf{a}^2 - \mathbf{b}^2}\right) \tag{6.40}$$

are eigenprojectors of z with eigenvalues

$$\lambda_{\pm} = z^0 \pm e^{i\phi}\sqrt{\mathbf{a}^2 - \mathbf{b}^2}. \tag{6.41}$$

Any analytic function $f(z)$ can be projected onto $\pi_{\pm}$:

$$f(z)\,\pi_{\pm} = f(\lambda_{\pm})\,\pi_{\pm}\,, \tag{6.42}$$

but the projectors $\pi_{\pm}$ are complementary only in the limit $\mathbf{b} \to 0$. Note that in the limit $\mathbf{b}^2 \to \mathbf{a}^2$, the eigenvalues become degenerate and there is only one distinct projector.

Exercise 6.5. Show explicitly that the $\pi_{\pm}$ are projectors, and prove the eigenvalue relations

$$z\pi_{\pm} = \lambda_{\pm}\pi_{\pm}\,. \tag{6.43}$$

6.3 Directed Plane Waves and Flags

The paravector potential $A(s)$ discussed above, which depends on x only through the scalar $s = \left\langle \bar{k}x \right\rangle_S$, is the potential of a directed plane wave. When the Lorenz gauge condition (6.17) condition is imposed, the field $\mathbf{F}$ has the form (6.20):

$$\mathbf{F} = \left(1 + \hat{\mathbf{k}}\right)\mathbf{E}\,. \tag{6.44}$$

A number of important properties follow from the projector factor $P_{\hat{\mathbf{k}}} = \frac{1}{2}\left(1+\hat{\mathbf{k}}\right)$ in such waves.

Since $k = (\omega/c)\left(1+\hat{\mathbf{k}}\right) = (2\omega/c)\, P_{\hat{\mathbf{k}}}$ is null,

$$\bar{k}\mathbf{F} = (\omega/c - \mathbf{k})\,\mathbf{F} = 0\,. \tag{6.45}$$

The scalar part gives the orthogonality of the fields with $\mathbf{k}$

$$\mathbf{k}\cdot\mathbf{F} = (\mathbf{kF}+\mathbf{Fk})\,/2 = 0\,, \tag{6.46}$$

and the vector part gives

$$\mathbf{F} = \hat{\mathbf{k}}\mathbf{F} = i\hat{\mathbf{k}}\times\mathbf{F} \tag{6.47}$$

with real and imaginary parts $-\hat{\mathbf{k}}\times c\mathbf{B} = \mathbf{E}$, $\hat{\mathbf{k}}\mathbf{E} = ic\mathbf{B}$. This implies that the vectors $\{\mathbf{E},\mathbf{B},\mathbf{k}\}$ form a right-handed orthogonal vector basis of three-dimensional space. In terms of projectors $P_{\hat{\mathbf{k}}} = \frac{1}{2}\left(1+\hat{\mathbf{k}}\right)$ and $\bar{P}_{\hat{\mathbf{k}}} = 1 - P_{\hat{\mathbf{k}}}$,

$$\mathbf{F} = P_{\hat{\mathbf{k}}}\mathbf{F} = \mathbf{F}\bar{P}_{\hat{\mathbf{k}}} = P_{\hat{\mathbf{k}}}\mathbf{F}\bar{P}_{\hat{\mathbf{k}}}\,. \tag{6.48}$$

It follows that $\mathbf{F}$ is a null biparavector and hence simple:

$$\mathbf{F}^2 = 0\,, \tag{6.49}$$

which means that $\mathbf{E}$ and $c\mathbf{B}$ are of equal magnitude and perpendicular to each other. This result will be important when we solve for particle motion in the electromagnetic field.

The propagating plane-wave field $\mathbf{F} = \left(1+\hat{\mathbf{k}}\right)\mathbf{E} = \mathbf{E}\left(1-\hat{\mathbf{k}}\right)$ is sometimes called a *flag*. It is a biparavector lying in the *flag plane*, a semi-infinite plane containing all real linear combinations

$$a\left(1+\hat{\mathbf{k}}\right) + b\mathbf{E} \tag{6.50}$$

of the lightlike paravector $\left(1+\hat{\mathbf{k}}\right)$, known as the *flagpole*, and the orthogonal vector $\mathbf{E}$ in which $b \geq 0$. The flagpole $1+\hat{\mathbf{k}}$ lies on the light cone, but every other paravector in the flag plane is spacelike:

$$\left[a\left(1+\hat{\mathbf{k}}\right) + b\mathbf{E}\right]\left[a\left(1-\hat{\mathbf{k}}\right) - b\mathbf{E}\right] = -b^2\mathbf{E}^2 \leq 0\,. \tag{6.51}$$

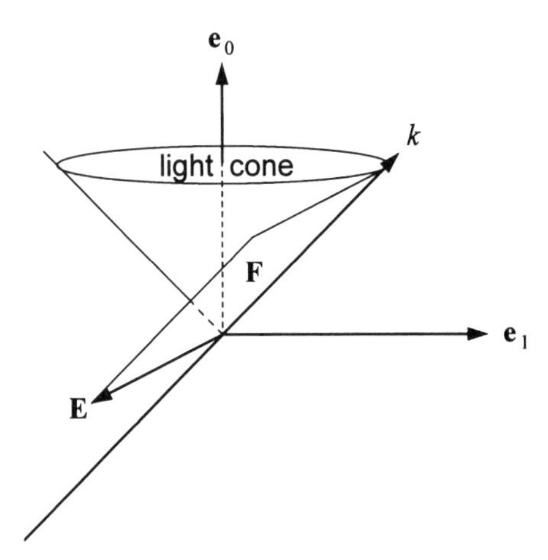

Fig. 6.2. The electromagnetic field of a directed plane wave is a flag tangent to the light cone. Its flagpole lies along the propagation paravector k.

In other words, every paravector in the plane except the flagpole lies outside of the light cone: the flag plane of $\mathbf{F}$ is *tangent to the light cone.* The flagpole is orthogonal to itself and to every paravector in the flag plane. Because of the pacwoman property, any flag can be rotated about its flagpole by multiplication by a scalar phase factor:

$$\begin{aligned} e^{-i\phi}\mathbf{F} &= \mathbf{e}^{-i\phi}\left(1+\hat{\mathbf{k}}\right)\mathbf{E} \\ &= e^{-i\phi\hat{\mathbf{k}}}\left(1+\hat{\mathbf{k}}\right)\mathbf{E} \\ &= e^{-i\phi\hat{\mathbf{k}}/\mathbf{2}}\left(1+\hat{\mathbf{k}}\right)\mathbf{E}\, e^{i\phi\hat{\mathbf{k}}/\mathbf{2}}. \end{aligned} \tag{6.52}$$

In particular, a phase factor of $-i$ rotates the flag by 90° while multiplication by -1 corresponds to a 180° rotation.[2] The complex phase factor performs a *duality rotation,* and the factor $-i$ rotates $c\mathbf{B} \to \mathbf{E}$ and $\mathbf{E} \to -c\mathbf{B}$. We see that the duality rotation of a directed plane wave is equivalent to a spatial rotation about the propagation direction $\hat{\mathbf{k}}$ of the wave.

[2]This description is due to Penrose and Rindler (1984) and was developed to give a geometrical meaning to null (ideal, reduced) spinors. The null spinor associated with the directed plane wave $\mathbf{F}$ may be identified as $R\left(1+\mathrm{e}_3\right)/2$, where $R\mathrm{e}_3R^\dagger = \hat{\mathbf{k}}$ and $R\mathrm{e}_1R^\dagger$ is directed along $\mathbf{E}$.

Since $\mathbf{F} = \left(1 + \hat{\mathbf{k}}\right) \mathbf{E} = \left(1 + \hat{\mathbf{k}}\right) ic\mathbf{B}$, the energy density $\mathcal{E}$ and Poynting vector $\mathbf{S}$ in the lab are given by

$$\begin{aligned} \mathcal{E} + \mathbf{S}/c &= \frac{\varepsilon_0}{2}\mathbf{F}\mathbf{F}^\dagger = \varepsilon_0 \mathbf{E}^2 \left(1 + \hat{\mathbf{k}}\right) \\ &= \varepsilon_0 \mathbf{E}_0^2 \left|f(s)\right|^2 \left(1 + \hat{\mathbf{k}}\right) . \end{aligned} \tag{6.53}$$

In particular, $\mathcal{E} + \mathbf{S}/c$ is constant in spacetime for circularly polarized directed plane waves, for which, as discussed below, $|f(s)| = |\exp(\pm is)| = 1$.

6.4 Monochromatic Plane Waves

Consider the particularly simple case in which A is a vector of constant magnitude whose direction at x is rotated by the angle s. The Lorenz gauge condition (6.19) $\phi = \phi_0 + \hat{\mathbf{k}} \cdot \mathbf{A}c$ can be valid only if the axis of rotation is $\pm\hat{\mathbf{k}}$. Thus with the suggested Coulomb-gauge choice $\phi = \hat{\mathbf{k}} \cdot \mathbf{A}c = 0$, we put

$$A(s) = \mathbf{a}e^{\pm i\hat{\mathbf{k}}s} = e^{\mp i\hat{\mathbf{k}}s}\mathbf{a}\,, \tag{6.54}$$

where $A(0) = \mathbf{a}$ is a constant vector perpendicular to $\hat{\mathbf{k}}$.

Because the vector potential $\mathbf{A}(s)$ rotates in a circle, the waves (6.54) are said to be *circularly polarized.* The sign of the rotation angle gives the *helicity* of the wave. Thus the wave $\mathbf{a}e^{i\hat{\mathbf{k}}s}$ is said to have positive helicity whereas the wave $\mathbf{a}e^{-i\hat{\mathbf{k}}s}$ has negative helicity. The waves (6.54) are *monochromatic* because they rotate at a single angular frequency (color). The corresponding electromagnetic field can be written as the simple biparavector

$$\mathbf{F}(s) = c\left\langle \partial \bar{A}\right\rangle_V = \pm ikc\hat{\mathbf{k}}A(s)\ , \tag{6.55}$$

which, since $k\hat{\mathbf{k}} = k$, can be simplified to

$$\mathbf{F}(s) = \pm ikce^{\mp i\hat{\mathbf{k}}s}\mathbf{a} = \pm ikc\mathbf{a}e^{\mp is}. \tag{6.56}$$

This relates a spatial rotation of $\mathbf{F}(0) = \mathbf{E}(0) + ic\mathbf{B}(0) = \pm ikc\mathbf{a}$ to a rotation in the complex plane. The latter mixes $\mathbf{F}$ with its dual $-i\mathbf{F}$ and is therefore called a duality (or sometimes, "dyality") rotation (see also Section 3.1).

Linear combinations of the form

$$\mathbf{F}(s) = ikc\hat{\mathbf{a}}\left(a_+ e^{-is} + a_- e^{is}\right) \tag{6.57}$$

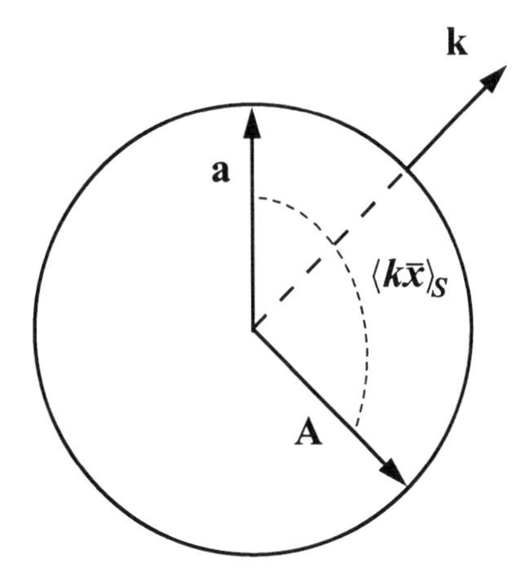

Fig. 6.3. The vector potential of a plane wave is a real rotating vector or a linear combination thereof. Here, **k** is directed into the page, and the wave is circularly polarized with positive helicity.

of the two solutions for a given direction $\hat{\mathbf{k}}$ and frequency ω represent monochromatic directed plane waves of arbitrary polarization. If one of a_+ or a_- is zero, the wave is circularly polarized. If $a_+ = a_-$, the wave is plane polarized. Other values of $a_\pm$ give elliptical polarization. All such cases can be expressed as the single lightlike spacetime plane

$$\begin{aligned} \mathbf{F}(s) &= \left(1+\hat{\mathbf{k}}\right)\mathbf{E}(s) \\ &= \left(1+\hat{\mathbf{k}}\right)\mathbf{E}_0 f(s)\,, \end{aligned} \tag{6.58}$$

where $f(s)$ is the (possibly complex) dimensionless scalar function[3]

$$f(s) = \frac{a_+ e^{-is} + a_- e^{is}}{\sqrt{2\left(a_+^2 + a_-^2\right)}} \tag{6.59}$$

and

$$\mathbf{E}_0 = \omega\sqrt{2}\left(a_+^2 + a_-^2\right)^{1/2} \hat{\mathbf{a}} \times \hat{\mathbf{k}}\,. \tag{6.60}$$

[3]The normalization is arbitrary. The choice here is made for convenience when defining the coherency density in the next chapter.

If $\mathbf{F}$ is linearly polarized, $f(s) = \cos s$, and if it is circularly polarized, $f(s) = \exp(\pm is)/\sqrt{2}$.

Exercise 6.6. Verify that (6.58) with the given relations for $f(s)$ and $\mathbf{E}_0$ do yield the linear combination (6.57).

One may be tempted to conclude from (6.58) that $\mathbf{E}(s) = \mathbf{E}_0 f(s)$, but since $\mathbf{E}(s)$ is real, this is true only if $f(s)$ is real. The problem is that the factor $\left(1+\hat{\mathbf{k}}\right)$ is not invertible, and consequently a relation of the form $\left(1+\hat{\mathbf{k}}\right) A = \left(1+\hat{\mathbf{k}}\right) B$ does not generally imply that $A = B$. However, by taking the real part of $\left(1+\hat{\mathbf{k}}\right)\mathbf{E}(s) = \left(1+\hat{\mathbf{k}}\right)\mathbf{E}_0 f(s)$, we find in general

$$\mathbf{E}(s) = \left\langle \left(1+\hat{\mathbf{k}}\right)\mathbf{E}_0 f(s)\right\rangle_{\Re} \tag{6.61}$$

$$= \mathbf{E}_0 \left\langle f(s)\right\rangle_{\Re} + \hat{\mathbf{k}}\mathbf{E}_0 \left\langle f(s)\right\rangle_{\Im} . \tag{6.62}$$

In particular, if $f(s) = \exp(-is)$,

$$\mathbf{E}(s) = \mathbf{E}_0 \cos s + \hat{\mathbf{k}} \times \mathbf{E}_0 \sin s \tag{6.63}$$

$$= \mathbf{E}_0 \exp\left(is\hat{\mathbf{k}}\right) . \tag{6.64}$$

This shows that the sense of the rotation of the positive-helicity wave with increasing $s = \omega t - \mathbf{k}\cdot\mathbf{x}$ is from $\mathbf{E}_0$ to $\hat{\mathbf{k}} \times \mathbf{E}_0$.

Exercise 6.7. Show that (6.64) also follows directly from (6.61) by the inverse pacwoman property of projectors: $\left(1+\hat{\mathbf{k}}\right) f(s) = \left(1+\hat{\mathbf{k}}\right) f\left(s\hat{\mathbf{k}}\right)$. Hint: note that $\mathbf{k}$ and $\mathbf{E}_0$ anticommute and that $\exp\left(-is\hat{\mathbf{k}}\right)\mathbf{E}_0 = \mathbf{E}_0 \cos s + \hat{\mathbf{k}} \times \mathbf{E}_0 \sin s$ is real.

Polarization is discussed further in the following chapter.

The form (6.58) is seen to be a solution of the source-free Maxwell equation for any scalar function $f(s)$. For example, consider the pulse with

$$f(s) = \exp\left(-\beta s^2/2\right) . \tag{6.65}$$

The field (6.58) still describes a fully polarized directed plane wave, but it is no longer monochromatic. Indeed, since

$$\exp\left(-\beta s^2/2\right) = \sqrt{\frac{2}{\pi\beta}} \int_0^\infty dy \exp\left(-\frac{y^2}{2\beta}\right)\cos(ys) , \tag{6.66}$$

the pulse may be viewed as a superposition of plane-polarized directed plane waves of frequencies $y\omega$.

Any directed plane wave can be written in the form (6.58). It will be plane-polarized if and only if the complex phase of $f(s)$ is constant, and it will be monochromatic at frequency $y\omega$ if and only if $f(s)$ is harmonic, with $f''(s) = -y^2 f(s)$.

6.5 Motion of Charges in Plane Waves

We consider here the motion of a charge e in a directed plane wave [see equation (6.12)]

$$\mathbf{F} = ck\bar{A}'(s) \tag{6.67}$$

with $s = \langle k\bar{x}\rangle_S = \omega t - \mathbf{k}\cdot\mathbf{x}$, where k is a null paravector and the Lorenz gauge condition (6.17) is assumed:

$$\langle \partial \bar{A}\rangle_S = \langle k\bar{A}'(s)\rangle_S = 0\,. \tag{6.68}$$

We hasten to add that no further assumptions on the paravector potential A are required. It may be a monochromatic plane wave, but it could just as well be a wave packet or a string of pulses.

We want to solve the spinorial Lorentz-force equation

$$\dot{\Lambda} = \frac{e}{2mc}\mathbf{F}\Lambda \tag{6.69}$$

for the eigenspinor Λ of the charge. At first blush, the problem appears intractable because the position of the charge, which we need in order to evaluate $\mathbf{F}$ there, requires the solution of the equation. However, since $\bar{k}\mathbf{F} = 0$,

$$\bar{k}\dot{\Lambda} = 0 = \dot{\Lambda}^\dagger \bar{k}\,, \tag{6.70}$$

and this implies that $k_{\text{rest}} = \bar{\Lambda}k\bar{\Lambda}^\dagger$, the propagation paravector in the rest frame of the charge, *is constant*:

$$\frac{d}{d\tau}\bar{k}_{\text{rest}} = \dot{\Lambda}^\dagger \bar{k}\Lambda + \Lambda^\dagger \bar{k}\dot{\Lambda} = 0\,. \tag{6.71}$$

Consequently, the rest-frame "frequency"

$$\begin{aligned} \dot{s} &= c\left\langle \bar{k}u\right\rangle_S = c\left\langle \Lambda^\dagger \bar{k}\Lambda\right\rangle_S \\ &= \omega_{\text{rest}} = c\left\langle k_{\text{rest}}\right\rangle_S \qquad (6.72)\\ &= \gamma\omega\left(1-\hat{\mathbf{k}}\cdot\mathbf{v}/c\right) \qquad (6.73)\end{aligned}$$

is also constant. If $\mathbf{F}$ is a monochromatic plane wave with propagation vector k, then $\dot{s}$ is its frequency as seen from the instantaneous rest frame of the charge.

This seems quite surprising: If $\mathbf{F}$ is a monochromatic wave, the Doppler-shifted frequency in the frame of the charge is constant, even while that charge is being accelerated by the wave. The acceleration must be arranged so that the part of the Doppler shift linear in the velocity exactly cancels that of the higher-order terms. As the velocity of the charge increases, its direction must become increasingly aligned with $\mathbf{k}$:

$$\hat{\mathbf{k}}\cdot\frac{\mathbf{v}}{c} = 1 - \frac{\dot{s}}{\gamma\omega}. \qquad (6.74)$$

In spacetime terms, k_{rest} is invariant because it lies in the invariant plane orthogonal to the field $\mathbf{F}$, whereas the spacetime rotation induced by the field lies in the plane $\mathbf{F}$ itself.[4] The result is very fortunate since it permits exact solutions to the Lorentz-force equation.

In terms of the eigenspinor Λ, it means that if Λ is split into complementary "irreducible" parts,[5]

$$\Lambda = P\Lambda + \bar{P}\Lambda \qquad (6.75)$$

where the projector P is $P = \frac{1}{2}\left(1+\hat{\mathbf{k}}\right)$, then only the part $P\Lambda$ changes under the action of the directed plane wave; $\bar{P}\Lambda$ is constant.

[4]Because k is a null paravector, it is spacetime-orthogonal to itself. As discussed in Chapter 3, it is both *in* the spacetime plane of $\mathbf{F} = ck\bar{A}'$ (*i.e.*, the spacetime plane of the incremental Lorentz transformation) and *orthogonal to it.*

[5]The term "irreducible" arises from the theory of group representations. A 2×2-matrix representation of the eigenspinor Λ for arbitrary motion provides an irreducible and faithful matrix representation of the group SL(2,C). However, as elements upon which the operators of the group act, the eigenspinors are also said to *carry* a 4×4-matrix representation of SL(2,C) and that representation is *reducible*: It can be reduced to a uniform block-diagonal form by the application of complementary projectors.

The spinorial equation for the Lorentz force can be expressed in terms of the scalar angle s :

$$\begin{aligned} \frac{d}{ds}\Lambda &= \frac{1}{\dot{s}}\frac{d}{d\tau}\Lambda = \frac{ek\bar{A}'(s)}{2m\omega_{\text{rest}}}\Lambda \\ &= -\frac{e}{2m\omega_{\text{rest}}}A'(s)\,\bar{k}\Lambda\,. \end{aligned} \tag{6.76}$$

Because both $\bar{k}\Lambda = \bar{\Lambda}^\dagger \bar{k}_{\text{rest}}$ and ω_{rest} are constant, integration gives

$$\begin{aligned} \Lambda(s) &= \Lambda(s_0) - \frac{e}{2m\omega_{\text{rest}}}\int_{s_0}^{s} ds'\,A'(s')\,\bar{k}\Lambda(s_0) \\ &= \Lambda(s_0) - \frac{e\Delta A}{2m\omega_{\text{rest}}}\bar{k}\Lambda(s_0) \qquad (6.77)\\ &= \Lambda(s_0) - \frac{e\Delta A}{2m\omega_{\text{rest}}}\bar{\Lambda}^\dagger(s_0)\,\bar{k}_{\text{rest}} \\ &= \Lambda(s_0) - \frac{e}{mc}\Delta A\,\bar{\Lambda}^\dagger(s_0)\,\bar{P}_{\text{rest}}\,, \qquad (6.78) \end{aligned}$$

where $P_{\text{rest}} = \frac{1}{2}\left(1 + \hat{\mathbf{k}}_{\text{rest}}\right)$. This is a beautifully simple result to what at first appeared to be an intractable problem. It predicts a change in the eigenspinor proportional to the change in paravector potential $\Delta A = A(s) - A(s_0)$. The proper velocity is

$$\begin{aligned} u(s) = \Lambda(s)\,\Lambda^\dagger(s) &= u(s_0) - \frac{e}{m\omega_{\text{rest}}}\left\langle \Delta A\,\bar{k}u(s_0)\right\rangle_{\Re} \\ &\quad + \left(\frac{e}{2m\omega_{\text{rest}}}\right)^2 \Delta A\,\bar{k}u(s_0)\,\bar{k}\,\Delta A\,. \end{aligned} \tag{6.79}$$

Since $k\bar{k} = 0$,

$$\begin{aligned} \bar{k}u\bar{k} &= \left(\bar{k}u + \bar{u}k\right)\bar{k} \\ &= 2\left\langle \bar{k}u\right\rangle_S \bar{k} \\ &= 2\omega_{\text{rest}}\bar{k}/c\,. \end{aligned} \tag{6.80}$$

With help from the Lorenz-gauge condition (6.17) we can consequently write

$$\begin{aligned} u(s) &= u(s_0) - \frac{e}{m\omega_{\text{rest}}}\left\langle \left\langle \Delta A\,\bar{k}\right\rangle_V u(s_0)\right\rangle_{\Re} \\ &\quad - \left(\frac{e^2 k}{2m^2\omega_{\text{rest}}c}\right)\left(\Delta A\overline{\Delta A}\right)\,. \end{aligned} \tag{6.81}$$

Exercise 6.8. Show that the Lorenz-gauge condition (6.68) implies $\langle \Delta A\, \bar{k} \rangle_S = 0$ and that therefore

$$\Delta A^0 = \Delta \mathbf{A} \cdot \hat{\mathbf{k}}\,. \tag{6.82}$$

Apply this result to demonstrate that

$$\Delta A \bar{k} = -k \Delta \bar{A} = k \Delta \mathbf{A}_\perp \tag{6.83}$$

and

$$\Delta A \bar{k} \Delta A = -k \left(\Delta \bar{A} \Delta A \right) = \left(\Delta \mathbf{A}_\perp \right)^2 k\,. \tag{6.84}$$

Exercise 6.9. Show that both the first and third terms on the RHS of equation (6.81) are orthogonal to the second term. [Hint: recall Theorem 3.1.] Use this result to prove $u(s)\, \bar{u}(s) = u(s_0)\, \bar{u}(s_0)\,.$

Exercise 6.10. Assume the ΔA oscillates rapidly with an average value of 0. Show that if p_{av} is the momentum of the charge averaged over these rapid oscillations of ΔA, then

$$p_{av} \bar{p}_{av} = m^2 c^2 + \left(e \Delta \mathbf{A}_\perp \right)^2. \tag{6.85}$$

The result (6.81) can be massaged in various ways. In particular, by noting that

$$\left\langle \bar{k} u(s_0) \right\rangle_S = \frac{\omega_{\text{rest}}}{c}\,, \tag{6.86}$$

we can find the change in the *canonical momentum* $p + eA = mcu - eA$:

$$\Delta (p + eA) = \frac{k}{2m\omega_{\text{rest}}} \left\langle e \Delta \bar{A} \left[2p(s_0) - e \Delta A \right] \right\rangle_S\,, \tag{6.87}$$

where $\Delta(p + eA) = p(s) + eA(s) - p(s_0) - eA(s_0)$. One can verify that the components of p along k and in the direction $\hat{\mathbf{k}} \times \Delta \mathbf{A}$ are invariant:

$$\begin{aligned} \left\langle p(s)\, \bar{k} \right\rangle_S &= \left\langle p(s_0)\, \bar{k} \right\rangle_S \\ \left\langle p(s)\, \hat{\mathbf{k}} \times \Delta \mathbf{A} \right\rangle_S &= \left\langle p(s_0)\, \hat{\mathbf{k}} \times \Delta \mathbf{A} \right\rangle_S\,. \end{aligned} \tag{6.88}$$

Thus, only the components of p along $\bar{k}$ (*i.e.*, $\langle pk \rangle_S$) and in the direction of $\Delta \mathbf{A}_\perp$ can change during the interaction, and they are related by the *mass-shell condition* $p\bar{p} = m^2 c^2$.

Exercise 6.11. Combine the invariants and the mass-shell condition to show

$$\Delta E = c\Delta \mathbf{p}\cdot\hat{\mathbf{k}} = \left(\frac{\omega}{\omega_{\text{rest}}}\right)\frac{\Delta(\mathbf{p}\cdot\hat{\mathbf{a}})^2}{2m}, \tag{6.89}$$

where $\Delta p = p(s) - p(s_0)$ and $\hat{\mathbf{a}}$ is a unit vector in the direction of $\Delta\mathbf{A}_\perp$. Note that this is a fully relativistic relation. [Hint: it may be useful to show the following relation:

$$\begin{aligned}\Delta(\mathbf{p}\cdot\hat{\mathbf{a}})^2 &= (\mathbf{p}(s)\cdot\hat{\mathbf{a}})^2 - (\mathbf{p}(s_0)\cdot\hat{\mathbf{a}})^2 \\ &= (\Delta\mathbf{p}\cdot\hat{\mathbf{a}})^2 + 2\mathbf{p}_0\cdot\hat{\mathbf{a}}\,(\Delta\mathbf{p}\cdot\hat{\mathbf{a}})\ .]\end{aligned} \tag{6.90}$$

Exercise 6.12. Use the result (6.87) to prove for any direction $\hat{\mathbf{b}}$ perpendicular to $\mathbf{k}$ that

$$\Delta\left(\mathbf{p}\cdot\hat{\mathbf{b}} + e\mathbf{A}\cdot\hat{\mathbf{b}}\right) = 0. \tag{6.91}$$

It is important to recognize that these results hold for any plane wave $A(s)$, including a pulse or string of pulses. The plane wave does not need to be even approximately monochromatic (see, for example, problem 6). Furthermore, since only the field at the position of the charge enters the Lorentz-force equation, the results of this section should be useful whenever the field at the position of the charge is well-approximated by a plane wave.

Note that the change in the momentum of the charge vanishes whenever there is no net change in the paravector potential A. Thus, the maximum momentum gain that can be achieved with an oscillating plane wave is limited to what can be extracted in a half cycle in the frame of the charge.

As an example, consider the plane-polarized monochromatic wave with paravector potential

$$A(s) = \mathbf{a}\sin s\,, \tag{6.92}$$

with $\mathbf{a}\cdot\hat{\mathbf{k}} = 0$. If the velocity of the charge at s_0 has no component in the direction of $\mathbf{a}$, then

$$p(s) - p(s_0) = -e\mathbf{a}\left(\sin s - \sin s_0\right) + \frac{e^2\mathbf{a}^2 k}{2m\omega_{\text{rest}}}\left(\sin s - \sin s_0\right)^2, \tag{6.93}$$

and its change in energy is

$$mc^2\left[\gamma(s) - \gamma(s_0)\right] = \frac{e^2\omega\mathbf{a}^2}{2m\omega_{\text{rest}}}\left(\sin s - \sin s_0\right)^2. \tag{6.94}$$

Exercise 6.13. Verify the last two results.

The energy gained can be quite large for a charge traveling with the wave, since then the Doppler-shifted frequency (the wave frequency as seen in the frame of the charge) $\omega_{\text{rest}} \ll \omega$:

$$\frac{\omega}{\omega_{\text{rest}}} = \left(\frac{c+v_0}{c-v_0}\right)^{1/2} = \gamma(s_0) + \sqrt{\gamma^2(s_0) - 1}, \tag{6.95}$$

where $v_0 = \mathbf{v}(s_0) \cdot \hat{\mathbf{k}}$ is the initial particle speed and $\gamma(s_0)\, mc^2$ is its initial energy. The charge can continue to gain energy for relatively long proper times up to $\pi\omega_{\text{rest}}^{-1}$ as it "rides" the wave. The effect is known as "surfing."

For large values of the dimensionless field strength,

$$\frac{e\,|\mathbf{a}|}{mc}\frac{\omega}{\omega_{\text{rest}}} \gg 1\,, \tag{6.96}$$

most of the momentum gain is seen to be in the propagation direction $\hat{\mathbf{k}}$. Of course the force on slowly moving charges, $v \ll c$, is in the direction of the electric field $\mathbf{E}$, perpendicular to $\hat{\mathbf{k}}$, but at relativistic speeds, the magnetic field $\mathbf{B}$ is also important and combines with the electric field to produce an acceleration component in the direction $\mathbf{E} \times \mathbf{B}$, *i.e.* the direction of $\hat{\mathbf{k}}$.

Exercise 6.14. Let the initial velocity of the charge (at $s = s_0$) be perpendicular to $\Delta\mathbf{A}_\perp$, the vector component of ΔA perpendicular to $\hat{\mathbf{k}}$. Use (6.87) to prove

$$\Delta p(s) + e\Delta\mathbf{A}_\perp(s) = \frac{e^2}{2m\omega_{rest}}\left[\Delta\mathbf{A}_\perp(s)\right]^2 k\,. \tag{6.97}$$

The last term in (6.97) is a light-like spacetime momentum called the *ponderomotive momentum.* Its scalar part is the ponderomotive energy.

The treatment given here may be extended to include a constant field in the propagation direction of the plane wave (see problem 11).

6.6 Monochromatic Standing Waves

Standing waves are created when directed waves of opposite $\mathbf{k}$ are superimposed. Unlike the directed monochromatic waves, standing waves are monochromatic only in frames that have no longitudinal motion (along $\pm\mathbf{k}$) with respect to the lab frame. In other frames, the two oppositely directed components do not share the same frequency. In standing waves, the electromagnetic field $\mathbf{F}$ may be compound, and in particular, its electric and magnetic components can be parallel.

If the directed plane wave is written in the form of (6.58)

$$\left(1+\hat{\mathbf{k}}\right)\mathbf{E}_0 f(s) \tag{6.98}$$

where

$$f(s)=\frac{a_+e^{-is}+a_-e^{is}}{\sqrt{2\left(a_+^2+a_-^2\right)}} \tag{6.99}$$

is a complex scalar function of $s=\langle k\bar{x}\rangle_S=\omega t-\mathbf{k}\cdot\mathbf{x}$, then we can investigate linear combinations such as[6]

$$\mathbf{F}(x)=\left(1+\hat{\mathbf{k}}\right)\mathbf{E}_0 f(s)-\left(1-\hat{\mathbf{k}}\right)\mathbf{E}_0 f'(s')\,, \tag{6.100}$$

where $s'=\omega t+\mathbf{k}\cdot\mathbf{x}$. From the orthogonality of $\mathbf{k}$ and $\mathbf{E}_0$, the Lorentz invariant $\mathbf{F}^2$ and the spacetime momentum density $\frac{1}{2}\varepsilon_0\mathbf{F}\mathbf{F}^\dagger=\mathcal{E}+\mathbf{S}/c$ are evaluated to be

$$\mathbf{F}^2=-4\mathbf{E}_0^2 f(s)\,f'(s') \tag{6.101}$$

$$\frac{1}{2}\varepsilon_0\mathbf{F}\mathbf{F}^\dagger=\varepsilon_0\mathbf{E}_0^2\left[\left(1+\hat{\mathbf{k}}\right)|f(s)|^2+\left(1-\hat{\mathbf{k}}\right)|f'(s')|^2\right]. \tag{6.102}$$

Thus, the spacetime momentum density of the superposition is simply a sum of the momentum densities of the two oppositely directed waves whereas the Lorentz invariant $\mathbf{F}^2$ vanishes unless both waves are nonzero.

Exercise 6.15. Show that for the plane-polarized standing wave with $f(s)=\cos s$ and $f'(s')=\cos s'$,the time averaged value of $\mathbf{F}^2$ is $-2\mathbf{E}_0^2\cos 2\mathbf{k}\cdot\mathbf{x}$ and the Poynting vector is

$$\mathbf{S}=\varepsilon_0 c\mathbf{E}_0^2\hat{\mathbf{k}}\sin 2\omega t\sin 2\mathbf{k}\cdot\mathbf{x}\,. \tag{6.103}$$

---

[6] The relative phase is chosen here so that when the scalar functions $f(s)$ and $f'(s)$ are equal, the electric field has a node at $\mathbf{k}\cdot\mathbf{x}=0$, as is appropriate for a plane wave incident on a conducting plane there. However, the relative phase changes rapidly with position so that all other choices of relative phase can be obtained by translation.

The electric and magnetic fields are parallel if and only if $\left\langle \mathbf{F}\mathbf{F}^\dagger \right\rangle_V = 0$. Since the scalar functions $f(s)$ and $f'(s')$ are independent functions of x, this condition implies

$$|f(s)| = |f'(s')| = const.$$

In other words, the electric and magnetic fields are parallel if and only if the two oppositely directed waves are circularly polarized. Depending on whether the waves have the same or opposite helicities, $\mathbf{F}^2$ is a function only of time or only of position, respectively [see problem (8)]. In both cases, $|f| = 1/\sqrt{2}$ and the energy density is constant and uniform:

$$\frac{1}{2}\varepsilon_0 \mathbf{F}\mathbf{F}^\dagger = \mathcal{E} = \varepsilon_0 \mathbf{E}_0^2 \,. \tag{6.104}$$

With the help of the inverse pacwoman property of $1 \pm \hat{\mathbf{k}}$, we can find alternative expressions for $\mathbf{F}$. Note

$$\begin{aligned} \left(1+\hat{\mathbf{k}}\right)\mathbf{E}_0 f(s) &= \mathbf{E}_0\left(1-\hat{\mathbf{k}}\right) f(\omega t - \mathbf{k}\cdot\mathbf{x}) \\ &= \left(1+\hat{\mathbf{k}}\right)\mathbf{E}_0 f\left(\omega t + \hat{\mathbf{k}}\mathbf{k}\cdot\mathbf{x}\right) \\ &= \left(1+\hat{\mathbf{k}}\right)\mathbf{E}_0 f\left(-\hat{\mathbf{k}}\omega t - \mathbf{k}\cdot\mathbf{x}\right), \end{aligned} \tag{6.105}$$

and analogously for the second term,

$$\begin{aligned} \left(1-\hat{\mathbf{k}}\right)\mathbf{E}_0 f'(s') &= \left(1-\hat{\mathbf{k}}\right)\mathbf{E}_0 f'\left(\omega t + \hat{\mathbf{k}}\mathbf{k}\cdot\mathbf{x}\right) \\ &= \left(1-\hat{\mathbf{k}}\right)\mathbf{E}_0 f'\left(\hat{\mathbf{k}}\omega t + \mathbf{k}\cdot\mathbf{x}\right). \end{aligned} \tag{6.106}$$

Taken together, these relations also allow us to write

$$\mathbf{F}(x) = 2\hat{\mathbf{k}}\mathbf{E}_0 f\left(\omega t + \hat{\mathbf{k}}\mathbf{k}\cdot\mathbf{x}\right) \tag{6.107}$$

if the helicities are equal, $f(s) = f'(s) = \exp(\pm is)/\sqrt{2}$, and

$$\mathbf{F}(x) = 2\hat{\mathbf{k}}\mathbf{E}_0 f'\left(\hat{\mathbf{k}}\omega t + \mathbf{k}\cdot\mathbf{x}\right) \tag{6.108}$$

if the helicities are opposite, $f(-s) = f'(s) = \exp(\pm is)/\sqrt{2}$.

Exercise 6.16. Show that the projectors $\frac{1}{2}\left(1 \pm \hat{\mathbf{k}}\right)$ project out directed plane waves propagating in directions $\pm\hat{\mathbf{k}}$ out of any standing wave of the form (6.100). Verify explicitly that

$$\frac{1}{2}\left(1 \pm \hat{\mathbf{k}}\right) 2\hat{\mathbf{k}}\mathbf{E}_0 f\left(\omega t + \hat{\mathbf{k}}\mathbf{k}\cdot\mathbf{x}\right) = \pm\left(1 \pm \hat{\mathbf{k}}\right)\mathbf{E}_0 f(\omega t \mp \mathbf{k}\cdot\mathbf{x}) . \tag{6.109}$$

Exercise 6.17. Show that if the relative phase of the waves is changed by π, so that in place of the field (6.100) we have

$$\mathbf{F}(x) = \left(1 + \hat{\mathbf{k}}\right)\mathbf{E}_0 f(s) + \left(1 - \hat{\mathbf{k}}\right)\mathbf{E}_0 f'(s') , \tag{6.110}$$

the standing waves become

$$\mathbf{F}(x) = 2\mathbf{E}_0 f\left(\omega t + \hat{\mathbf{k}}\mathbf{k}\cdot\mathbf{x}\right) \tag{6.111}$$

if the helicities are equal, $f(s) = f'(s) = \exp(\pm is)/\sqrt{2}$, and

$$\mathbf{F}(x) = 2\mathbf{E}_0 f'\left(\hat{\mathbf{k}}\omega t + \mathbf{k}\cdot\mathbf{x}\right) \tag{6.112}$$

if the helicities are opposite, $f(-s) = f'(s) = \exp(\pm is)/\sqrt{2}$.

Duality and spatial rotations about $\hat{\mathbf{k}}$ are distinct for these waves. Since $\mathbf{E}$ and $\mathbf{B}$ are parallel, the duality rotation $\exp(\pm i\phi)$ interchanges $\mathbf{E}$ and $c\mathbf{B}$ without performing any rotation. In one of the standing waves, the parallel fields form a fixed spiral which at one instant is purely electric throughout space, but a quarter of a cycle later becomes purely magnetic, whereas in the other standing wave, both fields exist together at any instant in time and all lie in a single plane, but that plane rotates in time. (See problem 8.) One may note the appropriate symmetries of the standing-wave spacetime dependencies under the reversal $\mathbf{k} \to -\mathbf{k}$:

$$\left(\omega t + \hat{\mathbf{k}}\mathbf{k}\cdot\mathbf{x}\right) \quad \to \quad \left(\omega t + \hat{\mathbf{k}}\mathbf{k}\cdot\mathbf{x}\right) \tag{6.113}$$

$$\left(\hat{\mathbf{k}}\omega t + \mathbf{k}\cdot\mathbf{x}\right) \quad \to \quad -\left(\hat{\mathbf{k}}\omega t + \mathbf{k}\cdot\mathbf{x}\right) . \tag{6.114}$$

6.7 Problems

1. **Spectral decomposition.** Consider an element of $C\ell_3$ that can be expressed by $x = \alpha + \beta\hat{\mathbf{x}}$, where α and β are complex numbers and $\hat{\mathbf{x}}$ is a real unit vector.

(a) Find the two scalar eigenvalues $\lambda_\pm$ and the corresponding eigenprojectors $P, \bar{P}$ of x, satisfying the eigenvalue equations

$$xP = \lambda_+ P,\ x\bar{P} = \lambda_- \bar{P}\,. \tag{6.115}$$

(b) Use these to show that x has the spectral decomposition

$$x = \lambda_+ P + \lambda_- \bar{P}\,. \tag{6.116}$$

(c) Show that if x is unimodular, $\lambda_+\lambda_- = 1$; if x is real, $\lambda_+ = \lambda_+^*$ and $\lambda_- = \lambda_-^*$; and if x is unitary, $|\lambda_+| = |\lambda_-| = 1$.

2. Show that any boost $B = \exp(\mathbf{w}/2)$ has a spectral decomposition of the form $B = \lambda P + \lambda^{-1}\bar{P}$. Find the two possible values of λ and the corresponding projector P in terms of $\mathbf{w}$, and show that n applications of the boost gives a composite transformation with the decomposition

$$B^n = \lambda^n P + \lambda^{-n}\bar{P}\,. \tag{6.117}$$

Relate λ to γ and $|\mathbf{u}|$ by putting $B^2 = u$, and use the result to derive the relation

$$B = \frac{u+1}{\sqrt{2(\gamma+1)}}\,. \tag{6.118}$$

[Hint: To derive the last expression, try expressing everything in terms of $\sqrt{\gamma \pm 1}$ and $\hat{\mathbf{u}}$.]

3. Use the null property of projectors to show that any analytic function $f(x)$ (assume the existence of a power-series expansion) of an element x with the spectral decomposition $x = \lambda_+ P + \lambda_- \bar{P}$ can be expanded

$$f(x) = f(\lambda_+)\,P + f(\lambda_-)\,\bar{P}\,. \tag{6.119}$$

In particular, find the spectral decomposition of $\exp(sP)$, where s is a scalar and P is a projector. [The relation (6.119) is sometimes used to *define* nonanalytic functions of cliffors, such as the square root and log functions.]

4. **Idempotent decomposition.** Let $z = z^0 + \mathbf{z}$ with $\mathbf{z} = \mathbf{x} + i\mathbf{y}$ be a general element of $C\ell_3$, where z^0 is a complex scalar and $\mathbf{x}$ and $\mathbf{y}$ are real vectors with $\mathbf{x}^2 \neq \mathbf{y}^2$. In analogy with the discussion of drift frames in Chapter 4,

there exists a proper "drift velocity" $u_D = \gamma_D + \mathbf{u}_D$ with $\mathbf{u}_D \cdot \mathbf{x} = \mathbf{u}_D \cdot \mathbf{y} = 0$ such that

$$z_D \equiv \bar{u}_D^{1/2} z u_D^{1/2} = z^0 + e^{i\delta}\mathbf{d}\,, \tag{6.120}$$

where δ is a real scalar and $\mathbf{d}$ is a real vector.

(a) Derive the relations

$$\begin{aligned} u_D &= \mathbf{z}e^{-i\delta}\mathbf{d}^{-1} \\ e^{2i\delta} &= \mathbf{z}^2/\mathbf{d}^2 \\ u_D^2 &= \frac{\mathbf{z}\mathbf{z}^\dagger}{|\mathbf{z}^2|}\,. \end{aligned} \tag{6.121}$$

(b) Show that for any analytic function f,

$$f(z) = u_D^{1/2} f(z_D)\, \bar{u}_D^{1/2}. \tag{6.122}$$

(c) Use the spectral decomposition of z_D,

$$z_D = \lambda_+ P + \lambda_- \bar{P} \tag{6.123}$$

with $P = \left(1 + \hat{\mathbf{d}}\right)/2$ and $\lambda_\pm = z^0 \pm |\mathbf{d}|\, e^{i\delta}$, to find a spectral decomposition of $f(z_D)$.

(d) Combine results of parts (b) and (c) to derive the *idempotent decomposition*

$$f(z) = f(\lambda_+)\mathcal{I} + f(\lambda_-)\bar{\mathcal{I}}\,, \tag{6.124}$$

where $\mathcal{I}$ and $\bar{\mathcal{I}}$ are complementary, null, primitive idempotents.[7]

(e) Show that the idempotent $\mathcal{I}$ can be written as the product of $\gamma_D\hat{\mathbf{d}}$ with a projector $\frac{1}{2}(1+\mathbf{e})$, where $\mathbf{e}$ is the real unit vector

$$\mathbf{e} = \gamma_D^{-1}\left(\hat{\mathbf{d}} - \mathbf{u}_D\right), \tag{6.125}$$

whereas $\bar{\mathcal{I}} = -\frac{1}{2}\gamma_D\hat{\mathbf{d}}\left(1 - \hat{\mathbf{d}}\hat{\mathbf{e}}\hat{\mathbf{d}}\right)$.[8]

[7] An idempotent element is *primitive* if it cannot be written as the sum of two other idempotents. For example, the unit scalar is idempotent but not primitive because it is the sum $1 = P + \bar{P}$ of two other idempotents.

[8] As a consequence, the minimum left ideals $\mathcal{C}\ell_3\mathcal{I}$ and $\mathcal{C}\ell_3\bar{\mathcal{I}}$ generated by the idempotents $\mathcal{I}$ and $\bar{\mathcal{I}}$ are not generally orthogonal, in contrast to the ideals generated by projectors.

5. Find the world line of a charge e released from rest at the origin in a circularly polarized plane wave

$$\mathbf{F}(\tau) = \left(1 + \hat{\mathbf{k}}\right) \mathbf{E}(0) \exp \left\langle i\bar{k}x \right\rangle_S .$$

6. Consider a pulse of radiation given by the paravector potential

$$A(s) = \mathbf{a} \tanh s \,, \tag{6.126}$$

where $s = \omega t - \mathbf{k} \cdot \mathbf{x}$ and $\mathbf{a}$ is perpendicular to $\mathbf{k}$. Find and sketch the electric and magnetic fields of the pulse. Find the proper velocity of charge, initially (at $\tau = -\infty$) at rest at the origin, after the pulse has passed.

7. Find the maximum energy gain by an electron in the field of a laser that can concentrate 100 J of energy in a cubic mm of space

   (a) if the electron is initially at rest;

   (b) if the electron is injected down the laser beam with a speed of $0.9\,c$.

8. Study the **standing waves** formed from the superposition of oppositely directed circularly polarized plane waves of the same amplitude and frequency. Write out the electric and magnetic fields separately (remember, they should be real vectors!) and verify that they are parallel at every point in spacetime. Interpret the time and space structure of the standing wave in the two cases of equal and opposite helicities. Show that in one case the fields everywhere at any instant in time point in the same direction, whereas in the other case, the fields lie along a fixed spatial spiral.

9. Consider **plane-polarized standing waves** formed from oppositely directed plane waves (6.100) with $f(s) = f'(s) = \cos s$. Find the separate electric and magnetic fields. Show that both have fixed nodal points, but that the magnetic and electric nodes are displaced from each other by a quarter of a wavelength.

10. **Reflection at a mirror.**

    (a) Show that the field (6.57) can also be written

$$\mathbf{F}(x) = kc\left(\mathbf{a}\cos\left\langle k\bar{x}\right\rangle_S + \mathbf{b}\sin\left\langle k\bar{x}\right\rangle_S\right) ,$$

   where $\mathbf{a} = \hat{\mathbf{a}} \times \hat{\mathbf{k}}\,(a_+ + a_-)$ and $\mathbf{b} = \hat{\mathbf{a}}\,(a_+ - a_-)$.

(b) Write down the reflected wave if a good conductor is oriented perpendicular to $\mathbf{k}$ at $\mathbf{x} = 0$. Ensure that the chosen coefficient vectors (corresponding to $\mathbf{a}$ and $\mathbf{b}$ above) make the electric field of the reflected wave opposite to that of the incident wave at $x = 0$.

(c) Show that the standing wave resulting from the sum of the incident and reflected waves has electric and magnetic parts given by

$$\begin{aligned} \mathbf{E} &= 2\omega\left(\mathbf{a}\sin\omega t - \mathbf{b}\cos\omega t\right)\sin\mathbf{k}\cdot\mathbf{x} \\ \mathbf{B} &= 2\mathbf{k}\times\left(\mathbf{a}\cos\omega t + \mathbf{b}\sin\omega t\right)\cos\mathbf{k}\cdot\mathbf{x}\,. \end{aligned}$$

(d) Describe the special cases of linear polarization (take $\mathbf{b} = 0$) and circular polarization (take $a_- = 0$). In particular, where are the nodes and what is the relative orientation of the electric and magnetic fields at any spacetime position for the two cases?

11. Extend the treatment of a charge in a directed plane wave by adding a constant electric field $\mathbf{E} = E_0\hat{\mathbf{k}}$ in the propagation direction of the plane wave. Use the projector $P = \frac{1}{2}\left(1 + \hat{\mathbf{k}}\right)$ and its bar conjugate to split the eigenspinor equation.

    (a) Find the equations of motion for $P\Lambda$ and $\bar{P}\Lambda$. Solve for $\bar{P}\Lambda(\tau)$.

    (b) Show that $\dot{s}(\tau)$ can be found if $\bar{P}\Lambda(\tau)$ is known. Integrate to obtain $s(\tau)$ in the form

$$s(\tau) = \frac{\omega_{\mathrm{rest}}(0)}{\alpha}\left(1 - e^{-\alpha\tau}\right) \tag{6.127}$$

    and identify the constant α.

    (c) Now solve for $Pe^{-\alpha\tau/2}\Lambda(\tau)$ and combine with part (a) to obtain

$$\Lambda(\tau) = \left\{e^{-\alpha\tau/2}\bar{P} + e^{\alpha\tau/2}P\left[1 + \bar{\xi}\left(\tau\right)\right]\right\}\Lambda(0)\,, \tag{6.128}$$

    where

$$\xi(\tau) = \frac{e\omega}{mc\omega_{\mathrm{rest}}(0)}\left[A(s) - A(0)\right]\,. \tag{6.129}$$

    (d) Show that the result can be further massaged into the form

$$\Lambda(\tau) = e^{\alpha\tau\hat{\mathbf{k}}/2}\left[1 - \xi(\tau)\,\bar{P}\right]\Lambda(0)\,. \tag{6.130}$$

(e) Verify that the solution $\Lambda(\tau)$ from part (d) has the correct form in the limit $\xi \to 0$ as well as in the limit $E_0 \to 0$.

(f) Show that if the charge starts from rest at $\tau = 0$, its proper velocity at τ is

$$u(\tau) = e^{\alpha\tau\hat{\mathbf{k}}} - \frac{1}{2}\left(\xi - \hat{\mathbf{k}}\xi\hat{\mathbf{k}}\right) - \frac{1}{2}\xi\bar{\xi}e^{\alpha\tau}\left(1 + \hat{\mathbf{k}}\right). \qquad (6.131)$$

(g) Suppose the plane wave is rapidly oscillating such that the average value $\langle\xi\rangle_{av}$ of ξ vanishes. Show that the proper velocity of the charge, averaged over the rapid oscillations, is then

$$\langle u(\tau)\rangle_{av} = e^{\alpha\tau\hat{\mathbf{k}}}\left[1 - \frac{1}{2}\langle\xi\bar{\xi}\rangle_{av}\left(1 + \hat{\mathbf{k}}\right)\right], \qquad (6.132)$$

and that the value of

$$p_{av}\bar{p}_{av} = m^2c^2\langle u(\tau)\rangle_{av}\langle\bar{u}(\tau)\rangle_{av} \qquad (6.133)$$

is the same as found in exercise (6.5) when $E_0 = 0$ and $u(0) = 1$.

12. Find the eigenspinor for the motion of a charge e in a circularly-polarized monochromatic plane wave plus a constant axial magnetic field. Describe the resonance phenomenon that can occur when the proper cyclotron frequency matches the Doppler-shifted wave frequency in the frame of the charge.

Chapter 7

Polarization

Any transverse vibration can be polarized. Polarization generally describes the distribution of vibrational energy among two or more degenerate normal modes. (Modes are said to be degenerate when their natural frequencies are equal.) A simple example is a pendulum on a string: it can oscillate back and forth or from side to side, or in any superposition of these two perpendicular degrees of freedom. If only one of these oscillatory modes is excited, the pendulum oscillates along a line and can be said to be linearly polarized. Any linear superposition of the two linear modes in which the phases are equal or differ by 180° also produces linearly polarized vibrations, but along other directions.

If the two linear modes are combined with equal amplitudes but with phases that differ by a quarter of a cycle, that is by 90°, a circularly polarized mode will result. There are two independent circular modes, one circling clockwise, the other counter-clockwise. Linear combinations of the two circular modes can be found to reproduce any other possible motion of the pendulum.

Mathematically, the oscillations of the pendulum on a string are described by a vector in a two-dimensional complex linear space. The space is spanned by any two linearly independent vectors in the space. In particular, any vector in the space can be expressed as a complex linear combination of the orthonormal basis vectors $\{\mathbf{e}_1, \mathbf{e}_2\}$, or by any unitary transformation of this basis, such as the complex circular-vector basis $\{\boldsymbol{\epsilon}_+, \boldsymbol{\epsilon}_-\}$, where

$$\boldsymbol{\epsilon}_\pm = (\mathbf{e}_1 \pm i\mathbf{e}_2)/\sqrt{2} = (1 \pm \mathbf{e}_3)\,\mathbf{e}_1/\sqrt{2}\,. \tag{7.1}$$

The circular basis elements are nilpotent, $\boldsymbol{\epsilon}_+^2 = \boldsymbol{\epsilon}_-^2 = 0$, but satisfy the normalization condition

$$\boldsymbol{\epsilon}_+ \cdot \boldsymbol{\epsilon}_+^\dagger = \boldsymbol{\epsilon}_- \cdot \boldsymbol{\epsilon}_-^\dagger = 1\,. \tag{7.2}$$

Exercise 7.1. Show more generally that

$$\epsilon_+\epsilon_+^\dagger = 1 + \mathbf{e}_3 \tag{7.3}$$
$$\epsilon_-\epsilon_-^\dagger = 1 - \mathbf{e}_3\,. \tag{7.4}$$

If the oscillations of the pendulum are spread as a disturbance in space along the direction normal to the plane of oscillation, they become a directed transverse wave. The wave velocity is the rate at which the disturbance spreads, and the wave in this case has the same polarization properties as the vibrations of the pendulum.

The waves generated by simple-harmonic motions of an oscillator, spreading at a constant rate in space, are monochromatic. Monochromatic directed plane waves are invariably fully polarized; like the pendulum motion, they may be linearly polarized, circularly polarized, or somewhere in between, namely elliptically polarized, and as with the pendulum, any type of polarization can be created from a superposition of orthogonal polarizations. Attempts to rapidly change the polarization of a monochromatic wave induce new frequency components and destroy the monochromaticity.

However, monochromatic radiation is an idealized fiction. It requires infinite waves of constant amplitude, conditions that are approximated—but not quite met—by continuous single-mode lasers. Real radiation always contains a finite bandwidth of different frequencies, and the components at different frequencies can have different polarizations. The radiation is then partially polarized.

Polarization is efficiently described in the Pauli algebra, and the full power of the Mueller-matrix method and of Stokes parameters for treating partially polarized light is contained in the algebraic treatment of the coherency density. The projectors in this treatment take on a new role: they represent ideal polarization filters.

7.1 Polarization of Monochromatic Waves

Consider a directed electromagnetic plane wave with positive helicity: at a fixed position $\mathbf{x}$ in space, the field $\mathbf{F} = \mathbf{E} + ic\mathbf{B}$ rotates around $\hat{\mathbf{k}}$ in a right-handed sense, that is, in the sense that you would turn a right-handed screw in order to advance it along $\hat{\mathbf{k}}$, but at any given time, a "snapshot" of the wave would show a left-handed spiral in space, that is a spiral that rotates to the left as one progresses in the propagation direction $\hat{\mathbf{k}}$. The wave of positive helicity is therefore sometimes said to be circularly polarized in a "right-handed" sense, and sometimes in a "left-handed" sense. To avoid misunderstanding, we use the helicity (which most authors agree on) to describe the sense of circular polarization.

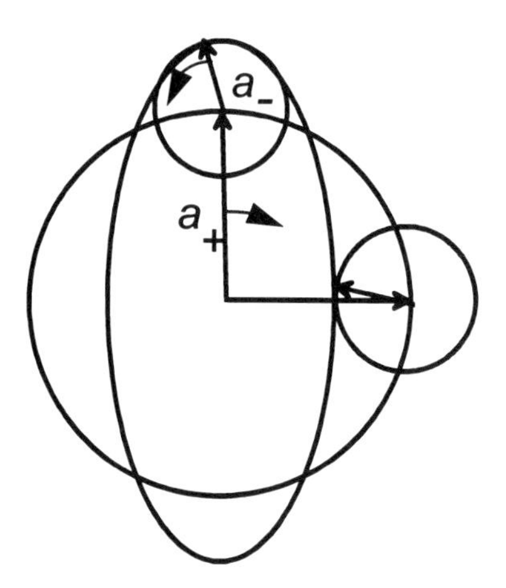

Fig. 7.1. Elliptical polarization is the sum of counter-rotating circular polarizations.

As seen in the previous chapter, linear combinations of two waves of the same k but opposite helicity generally give waves of elliptical polarization. The result can be written

$$\mathbf{F}(x) = \left(1+\hat{\mathbf{k}}\right)\mathbf{E} \tag{7.5}$$

$$= \left(1+\hat{\mathbf{k}}\right)\mathbf{E}_0 f(s)\,, \tag{7.6}$$

with $s = \langle k\bar{x}\rangle_S$, where

$$f(s) = \frac{a_+ e^{-is} + a_- e^{is}}{\sqrt{2\left(a_+^2 + a_-^2\right)}} \tag{7.7}$$

$$\mathbf{E}_0 = \omega\sqrt{2\left(a_+^2 + a_-^2\right)}\,\hat{\mathbf{a}}\times\hat{\mathbf{k}}\,. \tag{7.8}$$

It is often convenient to refer the major axis direction $\hat{\mathbf{a}}\times\hat{\mathbf{k}}$ to a fixed spatial direction $\widehat{\boldsymbol{\xi}}$ in the plane $i\mathbf{k}$. If $\hat{\mathbf{a}}\times\hat{\mathbf{k}}$ is related to $\widehat{\boldsymbol{\xi}}$ by a rotation about $\hat{\mathbf{k}}$ by the angle δ, then $\hat{\mathbf{a}}\times\hat{\mathbf{k}} = \exp\left(-i\hat{\mathbf{k}}\delta\right)\widehat{\boldsymbol{\xi}}$ and

$$\mathbf{F}(x) = \left(1+\hat{\mathbf{k}}\right)\widehat{\boldsymbol{\xi}}\omega\alpha_k(x)\,, \tag{7.9}$$

where $\alpha_k(x)$ is the complex scalar function

$$\alpha_k(x) = e^{-i\delta}\left(a_+ e^{-is} + a_- e^{is}\right). \tag{7.10}$$

The polarization of a wave is usually defined in terms of the electric field $\mathbf{E} = \langle \mathbf{F} \rangle_{\Re}$ which can be expressed in terms of a real amplitude E_0 and a complex polarization vector $\boldsymbol{\epsilon}$ as

$$\mathbf{E} = \langle \mathbf{F} \rangle_{\Re} = E_0 \left\langle \boldsymbol{\epsilon} e^{-is} \right\rangle_{\Re}. \tag{7.11}$$

The polarization vector is normalized so that

$$\boldsymbol{\epsilon} \cdot \boldsymbol{\epsilon}^{\dagger} = 1. \tag{7.12}$$

Exercise 7.2. Show

$$E_0 \boldsymbol{\epsilon} = kc\widehat{\boldsymbol{\xi}} a_+ e^{-i\delta} + \widehat{\boldsymbol{\xi}} kc a_- e^{i\delta}. \tag{7.13}$$

If $\boldsymbol{\epsilon}$ is a real unit vector times a complex scalar, the polarization is linear and collinear with $\boldsymbol{\epsilon}$. On the other hand, if $\boldsymbol{\epsilon}$ has the form of a complex scalar times

$$kc\widehat{\boldsymbol{\xi}}/\omega = \left(1 + \hat{\mathbf{k}}\right)\widehat{\boldsymbol{\xi}} = \left(\widehat{\boldsymbol{\xi}} + i\widehat{\boldsymbol{\eta}}\right) := \boldsymbol{\epsilon}_+ \sqrt{2} \tag{7.14}$$

or times

$$\widehat{\boldsymbol{\xi}} kc/\omega = \widehat{\boldsymbol{\xi}}\left(1 + \hat{\mathbf{k}}\right) = \left(\widehat{\boldsymbol{\xi}} - i\widehat{\boldsymbol{\eta}}\right) := \boldsymbol{\epsilon}_- \sqrt{2} \tag{7.15}$$

where $\widehat{\boldsymbol{\eta}} = \hat{\mathbf{k}} \times \widehat{\boldsymbol{\xi}}$, the wave is circularly polarized. Note that $\boldsymbol{\epsilon}_- = \boldsymbol{\epsilon}_+^{\dagger}$. To study polarization more generally, it is convenient to expand $E_0 \boldsymbol{\epsilon}$ in a polarization basis.

7.2 Poincaré Spinors

The complex field coefficient (7.13) for a given wave vector k and direction $\widehat{\boldsymbol{\xi}}$ is described in the *helicity basis* (*circular-polarization basis)* $\{\boldsymbol{\epsilon}_+, \boldsymbol{\epsilon}_-\}$ by a complex 2-component *Poincaré spinor*

$$\Phi := \sqrt{2}\omega \begin{pmatrix} a_+ e^{-i\delta} \\ a_- e^{+i\delta} \end{pmatrix}. \tag{7.16}$$

Thus,

$$\mathbf{E} = \left\langle E_0 \boldsymbol{\epsilon} e^{-is} \right\rangle_{\Re} = \left\langle \left(\boldsymbol{\epsilon}_+, \boldsymbol{\epsilon}_- \right) \Phi \, e^{-is} \right\rangle_{\Re} . \tag{7.17}$$

The equivalent expression in the Cartesian basis (*linear polarization basis*) $\left\{ \widehat{\boldsymbol{\xi}}, \widehat{\boldsymbol{\eta}} \right\}$

$$E_0 \boldsymbol{\epsilon} = \left(\widehat{\boldsymbol{\xi}}, \widehat{\boldsymbol{\eta}} \right) \Phi_J \tag{7.18}$$

is related to (7.17) by the unitary transformation

$$U_J = \frac{1}{\sqrt{2}} \begin{pmatrix} 1 & 1 \\ i & -i \end{pmatrix} \tag{7.19}$$

through

$$\Phi_J = U_J \Phi, \quad \left(\widehat{\boldsymbol{\xi}}, \widehat{\boldsymbol{\eta}} \right) = \left(\boldsymbol{\epsilon}_+, \boldsymbol{\epsilon}_- \right) U_J^{\dagger} . \tag{7.20}$$

The electromagnetic spinor Φ_J in the linear-polarization basis is also known as the *Jones vector*.[1] Any unitary transformation of Φ can be used to describe the polarization of a monochromatic directed wave.

7.3 Stokes Parameters

Experiments on electromagnetic waves usually measure various intensities rather than the fields directly. A polarization measurement, for example, will typically find the intensities of radiation passed through several different polarization filters. The measurements are conveniently related to the fields through the *coherency density*

$$\rho := \frac{1}{\Omega} \Phi \Phi^{\dagger} = \rho^{\mu} \boldsymbol{\sigma}_{\mu} , \tag{7.21}$$

where Ω is a normalization constant and in the helicity representation, the $\boldsymbol{\sigma}_{\mu}$ are the 2×2 Pauli spin matrices:

$$\boldsymbol{\sigma}_0 = \begin{pmatrix} 1 & 0 \\ 0 & 1 \end{pmatrix}, \; \boldsymbol{\sigma}_1 = \begin{pmatrix} 0 & 1 \\ 1 & 0 \end{pmatrix}, \; \boldsymbol{\sigma}_2 = \begin{pmatrix} 0 & -i \\ i & 0 \end{pmatrix}, \; \boldsymbol{\sigma}_3 = \begin{pmatrix} 1 & 0 \\ 0 & -1 \end{pmatrix} .$$

[1] Named after R. Clark Jones of the Polaroid Corporation, who published a series of articles on his method in the *Journal of the Optical Society of America* in the 1940's.

Exercise 7.3. Demonstrate that any 2×2 matrix can be expressed as a complex linear superposition of the four Pauli spin matrices.

Exercise 7.4. Show that the Pauli spin matrices are matrix representations of the unit paravectors $\mathbf{e}_\mu$. That is, show that they obey the same multiplicative relations among themselves. Also show that $\mathrm{tr}\{\boldsymbol{\sigma}_\mu\} = 2\langle \mathbf{e}_\mu \rangle_S$, and more generally that the trace of any 2×2 matrix equals twice the scalar part of the paravector that it represents.

The coefficients ρ^μ are known as the *Stokes parameters.* The scalar part ρ^0 of ρ is proportional to the time-averaged energy density and the ratio of the vector part $\boldsymbol{\rho}$ to ρ^0 gives the polarization. With the normalization $\Omega = \varepsilon_0^{-1}$, the energy density and Poynting's vector, averaged over oscillations at the angular frequency 2ω, is

$$\langle \mathcal{E} + \mathbf{S}/c \rangle_{t-av} = \rho^0 \left(1 + \hat{\mathbf{k}}\right) . \tag{7.22}$$

A direct calculation in the helicity representation gives

$$\begin{aligned} \Omega\rho^0 &= \omega^2 \left(a_+^2 + a_-^2\right), \quad & \Omega\rho^1 &= \omega^2 \left(2a_+a_- \cos\phi\right) \\ \Omega\rho^2 &= \omega^2 \left(2a_+a_- \sin\phi\right), \quad & \Omega\rho^3 &= \omega^2 \left(a_+^2 - a_-^2\right) , \end{aligned} \tag{7.23}$$

where $\phi = 2\delta$ is the azimuthal angle of $\boldsymbol{\rho}$. It is seen from (7.23) that in monochromatic directed plane waves, the vector part $\boldsymbol{\rho}$ has a length equal to the scalar part ρ^0. Furthermore, $\boldsymbol{\rho}$ lies along the $\boldsymbol{\sigma}_3$ axis if the wave is circularly polarized and in the $\boldsymbol{\sigma}_1\boldsymbol{\sigma}_2$ plane if it is linearly polarized. Elliptically polarized waves generally have $\boldsymbol{\rho}$ with components in both the $\boldsymbol{\sigma}_1\boldsymbol{\sigma}_2$ plane and on the $\boldsymbol{\sigma}_3$ axis.

The coherency density is a real element of the Pauli algebra generated by the three-dimensional space spanned by the basis $\{\boldsymbol{\sigma}_1, \boldsymbol{\sigma}_2, \boldsymbol{\sigma}_3\}$ (see Fig. 7.2.) This space, called the *Stokes subspace*, is not the same as physical space. Not only do the directions in Stokes subspace correspond to different types of polarization, rotations in the $i\mathbf{k}$ plane in physical space correspond to rotations of twice the angle in Stokes subspace. Thus, the transformation that rotates the fields about $\hat{\mathbf{k}}$ by α is

$$\Phi \to e^{-i\sigma_3\alpha}\Phi ,$$

which rotates ρ in Stokes subspace by 2α:

$$\rho \to e^{-i\sigma_3\alpha}\rho e^{i\sigma_3\alpha} . \tag{7.24}$$

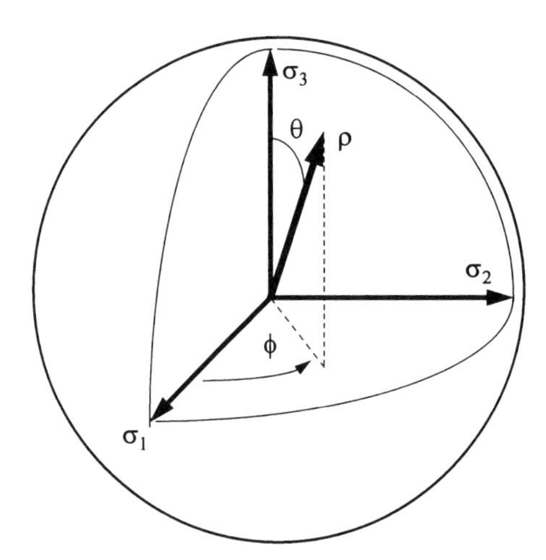

Fig. 7.2. The vector part of the coherency density ρ in units of the scalar part ρ^0 lies on or within the unit Poincaré sphere in Stokes subspace. Its length equals the degree of polarization, and its direction gives the type of polarization. For simplicity, $\rho^0 = 1$ in this figure.

The same factor of 2 relates polarization directions to the azimuthal angles of $\widehat{\boldsymbol{\rho}}$. Thus, elliptical polarization with its major axis $\hat{\mathbf{a}} \times \hat{\mathbf{k}}$ in the physical-space direction

$$\hat{\mathbf{a}} \times \hat{\mathbf{k}} = \widehat{\boldsymbol{\xi}} \cos \phi/2 + \widehat{\boldsymbol{\eta}} \sin \phi/2 \tag{7.25}$$

has a coherency density with the Stokes subspace direction $\widehat{\boldsymbol{\rho}}$ given by

$$\widehat{\boldsymbol{\rho}} = (\boldsymbol{\sigma}_1 \cos \phi + \boldsymbol{\sigma}_2 \sin \phi) \sin \theta + \boldsymbol{\sigma}_3 \cos \theta \,, \tag{7.26}$$

where $\tan \theta/2 = a_-/a_+$ and $-\pi \leq \phi \leq \pi$.

Exercise 7.5. Write down the 2×2 matrix for ρ and use this to verify the last relation. Note that $\rho^\mu = \frac{1}{2} tr \left\{ \rho \boldsymbol{\sigma}_\mu \right\}$, and recall that $\phi = 2\delta$.

The length of the ratio $\widehat{\boldsymbol{\rho}}/\rho^0$ gives the degree of polarization (see Fig. 7.2). Its maximum length is 1, for which case the radiation is fully polarized and $\widehat{\boldsymbol{\rho}}/\rho^0$ lies on the unit (or *Poincaré*) sphere. This is the situation for monochromatic waves.

For an incoherently mixed beam (see below) whose polarization is less than 100%, $\widehat{\boldsymbol{\rho}}/\rho^0$ has a length of less than 1 and lies within the Poincaré sphere. Unpolarized radiation has $\widehat{\boldsymbol{\rho}} = 0$.

7.4 Polarizers and Phase Shifters

The action of ideal polarizers and phase shifters on the wave can be modeled mathematically by transformations on the Poincaré spinor Φ of the form $\Phi \to T\Phi$. Polarizers are represented by projectors

$$P_{\mathbf{n}} = \frac{1}{2}(1 + \mathbf{n}) , \tag{7.27}$$

where $\mathbf{n}$ is a real unit vector in Stokes subspace that specifies the type of polarization, and the action of the polarizer is indicated by the transformation

$$\Phi \to P_{\mathbf{n}}\Phi . \tag{7.28}$$

For example, a circular polarizer allowing only waves of positive helicity corresponds to the projector

$$P_{\sigma_3} = \frac{1}{2}(1 + \boldsymbol{\sigma}_3) = \begin{pmatrix} 1 & 0 \\ 0 & 0 \end{pmatrix}, \tag{7.29}$$

which when applied to Φ in the helicity representation (7.16) eliminates the contribution a_- of negative helicity without affecting the positive-helicity part. Thus, the effect of such a polarizer on a wave represented by Φ is to transform

$$\Phi := \sqrt{2}\omega \begin{pmatrix} a_+ e^{-i\delta} \\ a_- e^{i\delta} \end{pmatrix} \to P_{\sigma_3}\Phi := \sqrt{2}\omega \begin{pmatrix} a_+ e^{-i\delta} \\ 0 \end{pmatrix}. \tag{7.30}$$

The polarizer represented by the complementary projector $\bar{P}_{\sigma_3}$ similarly eliminates the upper component of Φ. More generally, since $\bar{P}_{\mathbf{n}}P_{\mathbf{n}} = P_{-\mathbf{n}}P_{\mathbf{n}} = 0$, opposite directions in Stokes subspace correspond to orthogonal polarizations.

A wave is phase-shifted by an angle α through multiplication by the phase factor $\exp(i\alpha)$. An overall phase shift in the wave is hardly noticeable since the total phase is in any case changing very rapidly, but the effect of giving different polarization parts different shifts can be important. If the wave is split into orthogonal polarization components ($\pm\mathbf{n}$) and the two components are given a relative shift of α, the result is equivalent to rotating $\boldsymbol{\rho}$ by α about $\hat{\mathbf{n}}$ in Stokes subspace:

$$P_{\mathbf{n}}e^{i\alpha/2} + \bar{P}_{\mathbf{n}}e^{-i\alpha/2} = e^{i\mathbf{n}\alpha/2}. \tag{7.31}$$

If $\mathbf{n} = \boldsymbol{\sigma}_3$, this operator represents the effects of passing the waves through a medium with different indices of refraction for circularly polarized light of different helicities, as in the Faraday effect or in optically active organic solutions, and the result is a rotation of the plane of linear polarization by $\alpha/2$ about $\mathbf{k}$. If $\mathbf{n}$ lies in the $\boldsymbol{\sigma}_1\boldsymbol{\sigma}_2$ plane, the transformation (7.31) represents a birefringent medium with polarization types $\mathbf{n}$ and $-\mathbf{n}$ corresponding to the slow and fast axes, respectively. In a quarter-wave plate, for example, $\alpha = \pi/2$.

7.5 Coherent and Incoherent Superpositions

A superposition of two waves of the same frequency is *coherent* because their relative phase is fixed. Mathematically, one adds spinors in such cases:

$$\Phi = \Phi_1 + \Phi_2 . \tag{7.32}$$

where the subscripts refer to the two waves, not to spinor components. Coherent superpositions of monochromatic waves are always fully polarized and can be represented by a single Poincaré spinor or Jones vector. The corresponding coherency density is

$$\rho = \frac{1}{\Omega}\Phi\Phi^\dagger = \rho_1 + \rho_2 + \frac{1}{\Omega}\left(\Phi_1\Phi_2^\dagger + \Phi_2\Phi_1^\dagger\right) \tag{7.33}$$

and can always be shown to be a null element, which can be written

$$\rho = \rho^0\left(1 + \widehat{\boldsymbol{\rho}}\right) . \tag{7.34}$$

However, as emphasized at the beginning of the chapter, real beams of waves are never fully monochromatic. Whether or not the frequency spread is significant depends on the product of the frequency spread with the time over which the measurement is averaged. A beam of waves may be said to be effectively monochromatic when this product is much smaller than unity. On the other hand, two waves of different frequencies have a continually changing relative phase, and when their product $\Phi_1\Phi_2^\dagger$ is averaged over a period $2\pi/\left|\omega_2 - \omega_1\right|$ equal to the inverse frequency difference, the cross terms vanish. The waves then combine *incoherently*, and one can add their coherency densities rather than their Poincaré spinors:

$$\rho = \rho_1 + \rho_2 . \tag{7.35}$$

In such an incoherent superposition, the polarization can vary from 0 to 100%.

Any transformation T of spinors, $\Phi \rightarrow T\Phi$, will transform the coherency density by

$$\rho \rightarrow T\rho T^{\dagger}, \tag{7.36}$$

but in addition, many other transformations that do not preserve the polarization can be applied. For example, depolarization of a fraction f of the radiation is modeled by

$$\rho \rightarrow (1-f)\,\rho + f\,\langle \rho \rangle_S\,, \tag{7.37}$$

and detection itself takes the form

$$\rho \rightarrow \langle \rho D \rangle_S\,, \tag{7.38}$$

where the detection operator may equal $D = 1$ in an ideal case, but more generally it can have different efficiencies $D_{\pm}$ for opposite polarization types:

$$D = D_{+}P_{\mathbf{n}} + D_{-}P_{-\mathbf{n}}\,. \tag{7.39}$$

A number of other transformations are possible.[2]

7.6 Problems

1. Calculate the electric field $\mathbf{E}$, the Poincaré spinor Φ, and the Stokes parameters for the following monochromatic cases:

   (a) $(a_{+}, a_{-}, \delta) = (1, 0, \delta)\,,$

   (b) $(a_{+}, a_{-}, \delta) = (0, 1, \delta)\,,$

   (c) $(a_{+}, a_{-}, \delta) = (1, 1, 0)\,,$

   (d) $(a_{+}, a_{-}, \delta) = (1, 1, \pi/2)\,.$

   Briefly describe the polarization for each case.

[2] See W. E. Baylis, J. Bonenfant, J. Derbyshire, and J. Huschilt, *Am. J. Phys.* **61**, 534 (1993).

2. When a beam of light is interrupted by a pair of crossed polarizers, its intensity is extinguished. However, if you insert a third linear polarizer between the crossed ones and rotate it so that its plane of polarization bisects the other two, some light will be transmitted. If the initial beam is unpolarized, find—using the formalism of the coherency density—the fraction of intensity that gets through the set of three filters. You may assume ideal optical elements.

3. Show that, if a directed plane-wave beam of radiation is passed through two ideal polarizers of polarization types $\mathbf{n}$ and $\mathbf{m}$, the transmitted intensity is proportional to $\frac{1}{2}(1 + \mathbf{n} \cdot \mathbf{m})$. Now let $\mathbf{n}$ and $\mathbf{m}$ represent linear polarization types and find the dependence of the transmitted intensity on the angle between the polarization directions in the laboratory.

4. A quarter-wave plate shifts the phase of radiation linearly polarized along the *slow axis* by a quarter of a cycle relative to radiation orthogonally polarized along the *fast axis*. Show mathematically that if the radiation incident on a quarter-wave plate is polarized along a direction that bisects the fast and slow axes, then the transmitted radiation is circularly polarized. How can you control the helicity of the resultant circular polarization?

5. Show with the coherency density how a quarter-wave plate and a linear polarizer can by configured to give elliptical polarization with the Stokes unit vector $\hat{\boldsymbol{\rho}} = (\mathbf{e}_1 + \mathbf{e}_3)/\sqrt{2}$.

6. Show that the polarization of an incident beam can be completely determined from the intensity transmitted through a quarter-wave plate followed by a fixed linear polarizer, plotted as a function of the angle between the fast axis of the $\lambda/4$-plate and the direction of linear polarization.

Chapter 8

Waves in Media

Maxwell's equation $\bar{\partial}\mathbf{F} = (c\varepsilon_0)^{-1}\bar{j} = c\mu_0\bar{j}$ relates the electromagnetic field $\mathbf{F} = \mathbf{E} + ic\mathbf{B}$ to its source, the current $\bar{j} = \rho c - \mathbf{j}$. It can be conveniently applied to isolated currents in a vacuum. However, most physics measurements occur in a medium where there are usually many bound electric charges and magnetic dipoles that can contribute to the current. There are in fact so many (typically between 10^{20} and 10^{30} per cubic meter) small charges and dipoles that it would be hopeless to keep track of them individually. We prefer to treat them as part of the background continuum and to consider only the *free* charges in j.

In this chapter, we look at the modifications needed in Maxwell's equation to describe electromagnetic phenomena in polarizable and conducting media. We concentrate mainly on media that are homogeneous, isotropic, and linear. Plane-wave solutions are discussed and related to the vacuum solutions. In the following chapter, the matching conditions at media boundaries are found and applied to several simple cases.

8.1 Maxwell's Macroscopic Equations

Maxwell's microscopic (or vacuum) equations can be applied to currents in a medium if the contributions to the source from the *bound polarization densities* of the medium are separated from the current density of *free* charges j:

$$j_{total} = j + j_{bound} \,. \tag{8.1}$$

The bound currents can be related to the electromagnetic *polarization* $c\mathbf{P} + i\mathbf{M}$ by the spacetime gradient

$$j_{bound} = \langle \partial \left(c\mathbf{P} + i\mathbf{M} \right) \rangle_{\Re} \, . \tag{8.2}$$

Here, the electric polarization $\mathbf{P}$ is the electric dipole moment per unit volume, and similarly, $\mathbf{M}$ is the magnetic dipole moment per unit volume. We have made explicit the fact that j_{bound} is real, which it must be since j_{total} and j are real (corresponding to the experimental result that there are no magnetic monopoles). Maxwell's equation thus becomes

$$\begin{aligned} \bar{\partial}\mathbf{F} &= \left(c\varepsilon_0 \right)^{-1} \left[\bar{j} + \left\langle \overline{\partial \left(c\mathbf{P} + i\mathbf{M} \right)} \right\rangle_{\Re} \right] \\ &= \left(c\varepsilon_0 \right)^{-1} \left[\bar{j} + \left\langle -c\bar{\partial}\mathbf{P} + i\bar{\partial}\mathbf{M} \right\rangle_{\Re} \right] \end{aligned} \tag{8.3}$$

where we used the reality of $\partial, \mathbf{P}$, and $\mathbf{M}$ to equate

$$\begin{aligned} \left\langle \overline{\partial \left(c\mathbf{P} + i\mathbf{M} \right)} \right\rangle_{\Re} &= \left\langle \overline{\partial \left(c\mathbf{P} + i\mathbf{M} \right)}^{\dagger} \right\rangle_{\Re} = \langle \bar{\partial}^{\dagger} c \bar{\mathbf{P}}^{\dagger} - i \bar{\partial}^{\dagger} \bar{\mathbf{M}}^{\dagger} \rangle_{\Re} \\ &= \left\langle -c\bar{\partial}\mathbf{P} + i\bar{\partial}\mathbf{M} \right\rangle_{\Re} \end{aligned} \tag{8.4}$$

Taking the polarization terms to the left-hand side we find Maxwell's macroscopic equations

$$\begin{aligned} \langle \bar{\partial}\mathbf{F}_{mac} \rangle_{\Re} &= \bar{j}/c \\ \langle \bar{\partial}\mathbf{F} \rangle_{\Im} &= 0 \end{aligned} \tag{8.5}$$

where the macroscopic field is $\mathbf{F}_{mac} = \mathbf{D} + i\mathbf{H}/c$ with

$$\begin{aligned} \mathbf{D} &= \varepsilon_0 \mathbf{E} + \mathbf{P} \\ \mathbf{H} &= \mathbf{B}/\mu_0 - \mathbf{M} \end{aligned} \tag{8.6}$$

In Gaussian units, the fields $\mathbf{D}, \mathbf{H}$ have the same dimensions as $\mathbf{E}, \mathbf{B}$, but in SI units, it is customary to use fields $\mathbf{D}, \mathbf{H}$ with the same dimensions as the polarizations $\mathbf{P}, \mathbf{M}$, and these are obtained from the dimensions of $\mathbf{E}, \mathbf{B}$ by multiplying $\mathbf{E}$ by ε_0 or, equivalently, by dividing $\mathbf{B}$ by $\mu_0 = c^{-2}\varepsilon_0^{-1}$.

It is worthwhile to recall that the "true" field is $\mathbf{F} = \mathbf{E} + ic\mathbf{B}$, in the sense that the force on a moving charge is determined by $\mathbf{F}$. The effect of the medium on the macroscopic field $\mathbf{F}_{mac}$ is different for the electric and magnetic parts mainly because of the difference in the sources: there exist electric charges on which $\mathbf{E}$ field lines can start or end, but there are no magnetic monopoles. All magnetic fields are created by currents, that is by moving charges. Consequently field lines of $\mathbf{B}$ are continuous; they do not stop or start anywhere. In dielectric media in the absence of *free* charges, it is $\mathbf{D}$ that is continuous and divergence-free. Thus it is $\mathbf{B}$ and $\mathbf{D}$ that are most similar in the equations they obey. Similarly, $\mathbf{E}$ and $\mathbf{H}$ obey similar equations, but it is $\mathbf{F} = \mathbf{E} + ic\mathbf{B}$ that is most fundamental.

8.2 Linear, Isotropic media

In a linear isotropic medium, the polarizations are linear in the fields:

$$\mathbf{P} = \chi_e \epsilon_0 \mathbf{E}, \ \mathbf{M} = \chi_m \mathbf{H}, \tag{8.7}$$

where χ_e and χ_m are the *electric and magnetic susceptibilities*. In anisotropic media, the susceptibilities are tensors (or dyads) instead of scalars so that the polarizations may point in a different direction than the field. Furthermore, at high field strengths, as produced for example with high-power lasers, most media have nonlinear (e.g., cubic) contributions which can produce higher harmonics of the incident frequency. Here, however, we concentrate on the simpler case of linear, isotropic media, for which one can define a scalar *dielectric constant* (or *electric permittivity*) ϵ and a *magnetic permeability* μ by[1]

$$\varepsilon/\varepsilon_0 = 1 + \chi_e, \ \mu/\mu_0 = 1 + \chi_m \tag{8.8}$$

so that the macroscopic fields can be written

$$\mathbf{D} = \varepsilon \mathbf{E}, \ \mathbf{H} = \mathbf{B}/\mu\,. \tag{8.9}$$

Note that $\epsilon \geq \epsilon_0$ for all dielectric materials, whereas $\mu < \mu_0$ for diamagnetic materials and $\mu > \mu_0$ for paramagnetic materials. Ferromagnetic materials have large, positive susceptibilities, and they are not linear in the field. The value of the magnetization $\mathbf{M}$ at a given field $\mathbf{H}$ for ferromagnets depends on the *magnetic history* of the material, and the dependence is called a *hysteresis curve*. Of course it is the *memory* of magnetizable media that makes magnetic recording possible. The value of ϵ/ϵ_0 can be strongly dependent on the frequency. For example, water has a static dielectric constant of about 80, but at optical frequencies it is much smaller, $\epsilon \approx 1.8\epsilon_0$.

---

[1] In Gaussian units, $\varepsilon = 1 + 4\pi\chi_e, \ \mu = 1 + 4\pi\chi_m$.

In the next section, we study electromagnetic waves in nonconducting media. We will need to derive a wave equation from Maxwell's equations (8.5). For this derivation, it is convenient to recast Maxwell's macroscopic equations in terms of the field

$$\mathbf{F} := \mathbf{E} + ic\mathbf{B}\,, \tag{8.10}$$

the current density

$$j := c\rho + \mathbf{j}\,, \tag{8.11}$$

and the gradient operator

$$\bar{\partial} := c^{-1}\partial_t + \nabla\,, \tag{8.12}$$

where c, which will turn out to be the speed of waves in the medium, is c_0, which we now use for the speed of light in vacuum, divided by n, the *index of refraction*,

$$c^{-1} = n/c_0 := \sqrt{\epsilon\mu}\,. \tag{8.13}$$

The above definitions of $\mathbf{F}$ and j may be thought of as a *rescaling* of our original definitions to take account of the different speed of light in the medium. We used the familiar symbols because they reduce to the original forms in a vacuum, where $n \to 1$. In a linear, isotropic medium, Maxwell's equations (8.5) can now be written

$$\bar{\partial}\mathbf{F} = (c\varepsilon)^{-1}\bar{j}\,, \tag{8.14}$$

and the energy density

$$\mathcal{E} = \frac{\varepsilon}{2}\left(\mathbf{E}^2 + c^2\mathbf{B}^2\right) = \frac{1}{2}\left(\mathbf{E}\cdot\mathbf{D} + \mathbf{H}\cdot\mathbf{B}\right) \tag{8.15}$$

and Poynting vector

$$\mathbf{S} = \varepsilon\mathbf{E}\times\mathbf{B}c^2 = \mathbf{E}\times\mathbf{H} \tag{8.16}$$

are given by

$$\frac{\varepsilon}{2}\mathbf{F}\mathbf{F}^\dagger = \mathcal{E} + \mathbf{S}/c\,. \tag{8.17}$$

8.3 Wave Equation in Dielectric Media

A wave equation is obtained for the potential

$$A = c^{-1}\phi + \mathbf{A} \tag{8.18}$$

if we impose the gauge condition

$$\langle \partial \bar{A} \rangle_S = c^{-2}\partial_t \phi + \nabla \cdot \mathbf{A} = 0\,. \tag{8.19}$$

The field (8.10) is then

$$\mathbf{F} = c\partial \bar{A}\,, \tag{8.20}$$

so that Maxwell's equation (8.14) with no free sources gives the wave equation for the vector potential

$$\bar{\partial}\mathbf{F} = c\bar{\partial}\partial\bar{A} = \left[c^{-1}\partial_t^2 - c\nabla^2\right]\bar{A} = 0\,, \tag{8.21}$$

and a further differentiation yields a similar equation for the field $\mathbf{F}$

$$\partial\bar{\partial}\mathbf{F} = 0\,. \tag{8.22}$$

8.4 Directed Plane Waves

Consider solutions to (8.22) of the form

$$\mathbf{F} = \mathbf{F}(s)\,,\; s = \langle \bar{k}x \rangle_S = \omega t - \mathbf{k}\cdot\mathbf{x}\,. \tag{8.23}$$

Since we assume $\mathbf{F}'' \neq 0$, the wave equation (8.22) implies

$$\left(\frac{\omega}{c}\right)^2 - \mathbf{k}^2 = 0\,. \tag{8.24}$$

The wave velocity is thus

$$\omega \mathbf{k}^{-1} = c\hat{\mathbf{k}}\,. \tag{8.25}$$

Since the dielectric constant ϵ and hence the index of refraction $n = c_0\sqrt{\epsilon\mu}$ can depend on the frequency ω, there will in general be *dispersion*, and wave packets will tend to spread. As shown in chapter 5, the group velocity of a wave packet of fixed propagation direction $\hat{\mathbf{k}}$ is

$$\left(\frac{d\mathbf{k}}{d\omega}\right)^{-1} = \left(\hat{\mathbf{k}}\frac{d}{d\omega}n\omega\right)^{-1} = c\hat{\mathbf{k}}\left(1+\frac{\omega}{n}\frac{dn}{d\omega}\right)^{-1}.$$

For a field $\mathbf{F}$ whose average value is zero, the wave equation (8.21) implies

$$(\omega - c\mathbf{k})\,\mathbf{F} = \omega\left(1-\hat{\mathbf{k}}\right)\mathbf{F} = 0\,, \tag{8.26}$$

and thus

$$\begin{aligned}\hat{\mathbf{k}}\mathbf{F} = \mathbf{F} = -\mathbf{F}\hat{\mathbf{k}} &= \mathbf{E} + ic\mathbf{B}\\ &= \left(1+\hat{\mathbf{k}}\right)\mathbf{E} = \left(1+\hat{\mathbf{k}}\right)ic\mathbf{B}\\ &= \mathbf{E}\left(1-\hat{\mathbf{k}}\right) = ic\mathbf{B}\left(1-\hat{\mathbf{k}}\right).\end{aligned} \tag{8.27}$$

The directed plane wave satisfies

$$\begin{aligned}\mathbf{F}^2 &= 0\\ \mathcal{E} + \mathbf{S}/c = \tfrac{\varepsilon}{2}\mathbf{F}\mathbf{F}^{\dagger} &= \varepsilon\mathbf{E}^2\left(1+\hat{\mathbf{k}}\right).\end{aligned} \tag{8.28}$$

Circularly polarized monochromatic plane waves can be written as in Chapter 6 as real vectors $\mathbf{A}$ rotating in spacetime:

$$\mathbf{A} = \mathbf{a}e^{i\kappa s\hat{\mathbf{k}}}, \tag{8.29}$$

where $\kappa = \pm 1$ is the helicity, and linear superpositions may also be taken as in Chapter 6. The electromagnetic field corresponding to (8.29) is

$$\mathbf{F} = c\partial\bar{A} = i\kappa\omega\left(1+\hat{\mathbf{k}}\right)\mathbf{a}e^{-i\kappa s} = \mathbf{F}(0)\,e^{-i\kappa s}. \tag{8.30}$$

As far as the propagation of electromagnetic fields is concerned, all results of the previous chapter can be taken over directly. The only change is a reduction in the wave speed of the radiation from c_0 to $c = c_0/n$. If we were to measure distances in units of time, according to which the distance is the "length" of time required for radiation to travel that distance, then the effect of a uniform, isotropic, linear, nonconducting medium is simply to contract our length scales by the factor c. Essentially all the results of the previous chapter apply, and in particular the entire polarization treatment with Stokes' parameters can be adopted without change.

One difference should be mentioned, however. Since the speed of light in a medium can be less than the limiting velocity of light in a vacuum, it is now possible for a charge to travel faster than its field. Such charges create a shock front analogous to the sonic boom of a plane or meteor traveling faster than the speed of sound. The pulse of light ("photonic boom") emitted by a superluminal (faster than light) charge is called *Čerenkov radiation*[2] (see Figure 8.1).

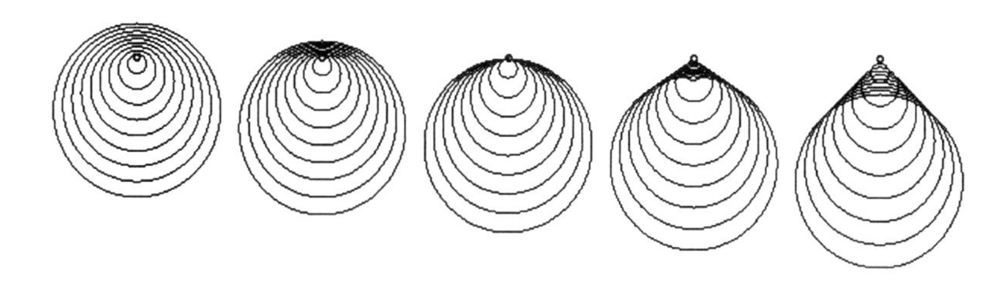

Fig. 8.1. If a source moves faster than the field it generates, wave fronts pile up, giving a shock wave. This gives sonic booms in acoustics and Čerenkov radiation for charges traveling faster than the speed of light in a medium. In the five plots above, from left to right, the source is moving upward at speeds $0.6, 0.8, 1.0, 1.2$, and 1.4 times c.

Exercise 8.1. Show that the Čerenkov radiation of a charge moving at constant velocity **v** forms a cone that makes an angle θ to **v** given by

$$\sin\theta = c/v\,, \tag{8.31}$$

where $c = nc_0$ is the speed of light in the medium. Find θ for charges moving at c_0, the speed of light in vacuum, in water whose index of refraction is $n = 4/3$. [Ans. 48.6°]

---

[2]This name is often transliterated in English writing as "Cherenkov."

8.5 Waves in Conducting Media

8.5.1 Ohm's Law

If the conductivity of the medium is nonzero, a current term

$$\mathbf{j} = \sigma \mathbf{E}\,, \tag{8.32}$$

where σ is the *conductivity* of the medium in units of ohm^{-1}/m, must be added to the source. This is a form of *Ohm's law,* a phenomenological expression that results after an average over electronic collisions that occur on the atomic scale. In accordance with the Lorentz-force equation, the electric field **E** accelerates the charges, but in most media, well before the speeds become relativistic, collisions scatter the charges, change their velocities, and transform some of their energy into heat. The net result of the field and the collisions is approximated by Ohm's law (8.32). Fortunately it is a good approximation in most conducting solids, liquids, and plasmas.

An immediate consequence of any significant conductivity is that no appreciable charge density ρ is built up. By combining the conductivity relation (8.32) for a uniform conductivity σ with the continuity equation and Coulomb's law, we obtain

$$\nabla \cdot \mathbf{j} = \sigma \nabla \cdot \mathbf{E} = \sigma \varepsilon^{-1} \rho = -\frac{\partial}{\partial t}\rho\,, \tag{8.33}$$

whose solution

$$\rho(t) = \rho(0) \exp(-\sigma t/\varepsilon) \tag{8.34}$$

implies a decay time ε/σ for any non-zero current density ρ. In copper, for example, $\varepsilon/\sigma = 1.6 \times 10^{-19}$s. This value is too small for the assumptions behind the phenomenological approximation to be valid, since electronic collisions in a solid at room temperature occur on a time scale of roughly 10^{-14}s, but it does show that even in relatively poor conductors such as distilled water (where well below optical frequencies $\varepsilon/\sigma \simeq 3 \times 10^{-6}$s), it is difficult for any appreciable charge density to build up. We are therefore usually justified in ignoring ρ in comparison to $\mathbf{j}$ in conducting media.

8.5.2 Differential Equation for Conductors

With the source term $\bar{j} \simeq -\mathbf{j} \simeq -\sigma\mathbf{E}$, Maxwell's macroscopic equation (8.14) becomes

$$\bar{\partial}\mathbf{F} = -\left(\varepsilon c\right)^{-1}\sigma\mathbf{E}\,. \tag{8.35}$$

Note that the right-hand side of this expression is a real vector and hence minus the hermitian conjugate of its spatial reversal. The same must also be true of the left-hand side:

$$\bar{\partial}\mathbf{F} = \partial\mathbf{F}^{\dagger}. \tag{8.36}$$

Adding $\partial\mathbf{F}$ to both sides gives

$$\left(\partial + \bar{\partial}\right)\mathbf{F} = \partial\left(\mathbf{F} + \mathbf{F}^{\dagger}\right), \tag{8.37}$$

and thus

$$c^{-1}\partial_t\mathbf{F} = \partial\mathbf{E}\,. \tag{8.38}$$

8.5.3 Physical and Unphysical Solutions

Consider now a directed plane wave of the form

$$\mathbf{F} = \mathbf{F}(s)\,,\ s = \left\langle\bar{k}x\right\rangle_s = \omega t - \mathbf{k}\cdot\mathbf{x}\,. \tag{8.39}$$

In other words, its only dependence on the space-time position $x = ct + \mathbf{x}$ is through the scalar s. Maxwell's equation (8.35) gives

$$\left(\omega - c\mathbf{k}\right)\mathbf{F}' = -\varepsilon^{-1}\sigma\mathbf{E}\,, \tag{8.40}$$

the scalar part of which is

$$\mathbf{k}\cdot\mathbf{F}' = 0\,.$$

Integration then shows that $\mathbf{k}\cdot\mathbf{F}$ is constant, and as long as it vanishes somewhere (which it does), it vanishes everywhere:

$$\mathbf{k}\cdot\mathbf{F} = 0\,. \tag{8.41}$$

The fields of a directed plane wave are therefore everywhere perpendicular to the propagation direction. The condition (8.38) gives

$$\omega \mathbf{F}' = \omega\left(\mathbf{E}' + ic\mathbf{B}'\right) = \left(\omega + c\mathbf{k}\right)\mathbf{E}',$$

which, when integrated over s, shows that the electric and magnetic fields are also perpendicular to each other:

$$c\mathbf{B} = \frac{c}{\omega}\mathbf{k} \times \mathbf{E}\,. \tag{8.42}$$

The integration constant must vanish if the average value of $\mathbf{F}$ over s is zero. We assume this to be the case. The total field can thus be written

$$\mathbf{F} = \mathbf{E} + ic\mathbf{B} = \left(1 + \frac{c}{\omega}\mathbf{k}\right)\mathbf{E} = \mathbf{E}\left(1 - \frac{c}{\omega}\mathbf{k}\right) \tag{8.43}$$

in analogy with the nonconducting case (8.27).

We have thus shown that the vectors $\mathbf{E}, \mathbf{B}, \mathbf{k}$ of any directed plane wave form a right-handed orthogonal system of vectors in a conducting medium, just as they do in vacuum. However, we have not yet shown that any directed plane waves are *physical solutions* to Maxwell's equation (8.35) in a conducting medium. In fact, because of unavoidable reflections and absorption in the conducting media, none are! Nevertheless, *nonphysical* directed plane waves are solutions, and linear superpositions of such solutions exist that are physical. We shall investigate the solutions further with the help of a wave equation.

8.5.4 Wave/Diffusion Equation

A useful macroscopic equation for conducting media is obtained by substituting the result (8.38) into the gradient of (8.35):

$$\partial\bar{\partial}\mathbf{F} = -\mu\sigma\partial_t\mathbf{F}\,. \tag{8.44}$$

This is the generalization of the wave equation (8.22). For slowly varying fields in good conductors, where $\sigma/\varepsilon \gg \omega$, one can often ignore $c^{-2}\partial_t^2\mathbf{F}$ compared to $\mu\sigma\partial_t\mathbf{F}$. The equation (8.44) then takes the form of a *diffusion equation:*

$$\nabla^2\mathbf{F} = \mu\sigma\partial_t\mathbf{F}\,, \tag{8.45}$$

which describes the diffusion of fields through the conducting medium. More generally, the wave/diffusion equation (8.44) is a linear equation with real scalar operators, but it is *not* invariant under time reversal $t \to -t$. Since it is linear in $\mathbf{F}$, any superposition of solutions is also a solution. Since the differential operators on both sides are real and scalar-valued, the equation must hold for each real and imaginary vector component of $\mathbf{F}$, and if $\mathbf{F}$ is a solution, so is $\mathbf{F}^\dagger$. Any solution can be expressed as a linear combination of monochromatic directed plane waves of the form:

$$\mathbf{F}(s) = \mathbf{F}(0)\, e^{-i\kappa s}, \; s := \omega t - \mathbf{k}\cdot\mathbf{x}, \tag{8.46}$$

where $\kappa = \pm 1$ is the helicity. Substitution of this wave (8.46) into the wave equation (8.44) yields a complex propagation vector $\mathbf{k}$:

$$\mathbf{k}^2 = \left(\frac{\omega}{c}\right)^2 (1 + i\kappa\zeta), \tag{8.47}$$

where the dimensionless conductivity is $\zeta := \sigma/(\varepsilon\omega)$. To find the real and imaginary parts of $\mathbf{k}$, we put

$$\mathbf{k} = \frac{\omega}{c}\hat{\mathbf{k}}\,(\alpha + i\kappa\beta), \tag{8.48}$$

substitute this into (8.47), and solve for real positive values of the dimensionless parameters $\alpha, \; \beta$:

$$\alpha^2 - \beta^2 = 1, \; 2\alpha\beta = \zeta.$$

The solutions are

$$\alpha = \frac{1}{\sqrt{2}}\left(\sqrt{1+\zeta^2} + 1\right)^{1/2} \tag{8.49}$$

$$\beta = \frac{1}{\sqrt{2}}\left(\sqrt{1+\zeta^2} - 1\right)^{1/2}. \tag{8.50}$$

8.5.5 Penetration Depth

Because the imaginary part of $\mathbf{k}$ (8.48) changes sign with the helicity of the wave, waves of both helicities are attenuated exponentially as they pass through the conducting medium in the direction of $\hat{\mathbf{k}}$:

$$\mathbf{F}(s) = \mathbf{F}(0)\, e^{-\hat{\mathbf{k}}\cdot\mathbf{x}/\delta} e^{i\omega\kappa\left(\hat{\mathbf{k}}\cdot\mathbf{x}\alpha/c - t\right)}, \tag{8.51}$$

and their *penetration depth* ("skin depth") into the medium is

$$\delta \equiv \frac{c}{\omega\beta}, \tag{8.52}$$

which is much less than the wavelength $\lambda = 2\pi c/\omega$ for good conductors. From the definition (8.52) we see the physical significance of β: it is the reduced wavelength $\lambda/2\pi$ in units of the skin depth:

$$\beta = \frac{\lambda}{2\pi\delta}. \tag{8.53}$$

All linear combinations of the two helicities are similarly attenuated in the direction of propagation. Such rapidly damped waves are known as *evanescent waves.* They are not directed plane waves because their x dependence contains a damping term that cannot be expressed as a function of $s = \omega t - \mathbf{k}\cdot\mathbf{x}$. The exponential decay is largely the result of reflection on the incident wave, although some absorption by resistive currents also occurs (see below).

The condition (8.42) means that no directed plane wave with complex $\mathbf{k}$ can be physical. With real $\mathbf{E}$ and complex $\mathbf{k}$ (8.48), the condition gives a complex magnetic field, which would be unphysical:

$$c\mathbf{B} = \hat{\mathbf{k}} \times \mathbf{E}\,(\alpha + i\kappa\beta) := c\mathbf{b}e^{i\kappa\phi}, \tag{8.54}$$

where $c\mathbf{b}$ is the real vector

$$\begin{aligned} c\mathbf{b} &= \left(\alpha^2 + \beta^2\right)^{1/2} \hat{\mathbf{k}} \times \mathbf{E}, \\ \tan\phi &= \beta/\alpha\,. \end{aligned}$$

To build a linearly polarized physical solution from the unphysical ones, we can combine waves of opposite helicity and equal amplitude:

$$\begin{aligned} \mathbf{F} &= \frac{e^{-\hat{\mathbf{k}}\cdot\mathbf{x}/\delta}}{2}\left[\left(\mathbf{E}_0 + ic\mathbf{B}_0 e^{i\phi}\right)e^{-is_{\Re}} + \left(\mathbf{E}_0 + ic\mathbf{B}_0 e^{-i\phi}\right)e^{is_{\Re}}\right] \\ &= e^{-\hat{\mathbf{k}}\cdot\mathbf{x}/\delta}\left[\mathbf{E}_0\cos(s_{\Re}) + ic\mathbf{B}_0\cos(s_{\Re} - \phi)\right], \end{aligned} \tag{8.55}$$

where a possible overall phase angle has been absorbed into ωt, and

$$\begin{aligned} s_{\Re} &= \left\langle \bar{k}x \right\rangle_{\Re S} = \omega t - \alpha\tfrac{\omega}{c}\hat{\mathbf{k}}\cdot\mathbf{x} \\ c\mathbf{B}_0 &= \left(\alpha^2 + \beta^2\right)^{1/2}\hat{\mathbf{k}} \times \mathbf{E}_0\,. \end{aligned}$$

In the limit of low conductivity, $\zeta \ll 1,\ \beta \ll \alpha$,

$$\alpha \to 1,\ \beta \to 0,\ \phi \to 0,$$

and the fields $\mathbf{E}$ and $c\mathbf{B}$ are nearly equal in magnitude and phase, as in a vacuum. However, in the limit of high conductivity (for most conductors, even at optical frequencies)

$$\alpha \approx \sqrt{\frac{\zeta}{2}} \approx \beta \gg 1, \ \phi \to \pi/4 ,$$

and $c\mathbf{B}$ lags about an eighth of a cycle behind $\mathbf{E}$ and is much larger in magnitude. Linear combinations of the plane-polarized solutions (8.55) with different phase shifts (*i.e.*, angles added to $s_{\Re}$) can give elliptically and circularly polarized waves (see problem 7).

8.5.6 Reflection

The solution (8.55) can be split into the incident wave propagating forward and reflected wave propagating backward:

$$\mathbf{F} = \left(P + \bar{P}\right)\mathbf{F} \, , \ \mathbf{P} = \frac{1}{2}\left(1 + \hat{\mathbf{k}}\right) . \tag{8.56}$$

As an incident wave penetrates into the conductor, it is partially reflected and partially absorbed; these combine to give the exponential decay. The better the conductor, the greater the reflection. The incident and reflected waves are found from $P\mathbf{F}$ and $\bar{P}\mathbf{F}$, respectively:

$$\mathbf{F}_{in}\left(\mathbf{x} = 0\right) = P\mathbf{E}_0 \left[\cos s_{\Re} + \sqrt{\alpha^2 + \beta^2}\cos(s_{\Re} - \phi)\right] \tag{8.57}$$

$$\mathbf{F}_{refl}\left(\mathbf{x} = 0\right) = \bar{P}\mathbf{E}_0 \left[\cos s_{\Re} - \sqrt{\alpha^2 + \beta^2}\cos(s_{\Re} - \phi)\right] , \tag{8.58}$$

where the projector has gobbled a unit vector $\hat{\mathbf{k}}$. The reflection coefficient for the intensity is the ratio of the time-averaged Poynting vectors:

$$R = \frac{\left\langle \mathbf{F}_{refl}\mathbf{F}^{\dagger}_{refl} \right\rangle_{\mathrm{V,t-av}}}{\left\langle \mathbf{F}_{in}\mathbf{F}^{\dagger}_{in} \right\rangle_{\mathrm{V,t-av}}} = \frac{\alpha - 1}{\alpha + 1} . \tag{8.59}$$

Exercise 8.2. Verify the second equality above and show that R vanishes in the limit of low conductivity, where $\alpha \to 1$, and approaches unity for good conductors. [Hint: show $\sqrt{\alpha^2 + \beta^2}\cos\phi = \alpha$.]

Note that R is the reflection coefficient *within* the conductor. The reflection coefficient at the interface of a conductor in a nonconducting medium depends also on the relative impedance of the media, as discussed in the following chapter. A thin conductor whose thickness is not much larger than the skin depth will generally transmit part of the wave. The portion of the energy that is not reflected or transmitted is absorbed (see problem 10.)

8.6 Problems

1. Verify equation (8.17), given the expressions for $\mathbf{F}$ (8.10), $\mathcal{E}$ (8.15), and $\mathbf{S}$ (8.16).

2. The index of refraction of water near room temperature is 4/3, while that of glass is about 1.5. For each, find c, the speed of light in both dimensionless units and in m/s, and give it also in meters per ns. (You may take the speed of light in vacuum to be 3×10^8m/s.)

3. Verify equation (8.28) for the energy-momentum density of a directed, mono chromatic plane wave. The Poynting vector $\mathbf{S}$ gives the energy flow per unit area. Show that for the directed plane wave, it has just the value expected for the energy density $\mathcal{E}$ moving at the speed c.

4. Sketch the "skin depth" as a function of the frequency in appropriate units. Find the limiting forms in the limit of high and low frequencies. Show that at high frequencies, the "skin depth" becomes independent of frequency. What do "high" and "low" mean in this context? Give numerical estimates of 'high" and "low" frequencies in the case of copper. In copper, $\sigma/\varepsilon = 6.5 \times 10^{18}\text{s}^{-1}$, $\mu = \mu_0, \epsilon \approx \varepsilon_0$.

5. Calculate the "skin depth" δ in copper at (linear) frequencies of 60 Hz and 2 MHz.

6. Sketch the electric and magnetic fields (not necessarily on the same scale) (8.55) at time $t = 0$ of a 1 GHz wave as it penetrates copper. Indicate the distance scale for your sketch.

7. Find circularly polarized wave solutions in a conducting medium by taking a linear combination of plane-polarized waves of the form (8.55), but with orthogonal fields and phase differences of $\pi/2$. Find the angle between the

electric and magnetic fields for waves of both positive and negative helicity in the limit of high conductivity. [Ans.: Replace $\cos(\cdots)$ in (8.55) by $\exp\left[\pm i\hat{\mathbf{k}}\,(\cdots)\right]$. Angles are 45 and 135 degrees for positive and negative helicity, respectively.]

8. Differentiate the field (8.55) in a conducting medium to show explicitly that $\varepsilon c\bar{\partial}\mathbf{F} = -\sigma\mathbf{E}$.

9. Assume a wave solution in a conducting medium of the form

$$\mathbf{F} = e^{-\boldsymbol{\kappa}\cdot\mathbf{x}}\left[\mathbf{E}_0\cos s + ic\mathbf{B}_0\cos(s-\phi)\right]\,, \tag{8.60}$$

where $s = \langle k\bar{x}\rangle_S = \omega t - \mathbf{k}\cdot\mathbf{x}$ and $\mathbf{k}, \boldsymbol{\kappa}, \mathbf{E}_0, \mathbf{B}_0$ are constant vectors and ϕ is a constant phase angle. Show that the homogeneous form (8.44) of Maxwell's equation for the medium implies $\boldsymbol{\kappa}^2 + k\bar{k} = 0$ and that $2\mathbf{k}\cdot\boldsymbol{\kappa} = \omega\mu\sigma$. Show further that the first-order form (8.35) restricts the solution to the result (8.55). [Hint: equate coefficients of $e^{-\boldsymbol{\kappa}\cdot\mathbf{x}}\cos s$ and $e^{-\boldsymbol{\kappa}\cdot\mathbf{x}}\sin s$.]

10. Use the reflection coefficient (8.59) to determine the absorption of radiation in a thick conductor as a function of conductivity and frequency.

Chapter 9

Waves at Boundaries

Much interesting physics occurs at the boundaries between regions. In the last chapter, we learned how to find plane-wave solutions in uniform, isotropic, linear media. Here we want to learn how to relate the plane-wave solutions in one region with those in another. The relation is expressed in terms of *matching conditions*, which we derive from Maxwell's equation. For dielectric media, the matching condition tells us how waves reflect and refract at boundaries.

Matching conditions at conducting surfaces will determine the reflection and constrain the field at the conductor. Any electric field in a good conductor would be quickly canceled by mobile charges. This, as we will see below, implies total normal reflection of waves at the conducting surface and no waves with **E** parallel to the surface. This in turn constrains the type of wave that can exist inside a simply connected hollow conductor. The *simply connected* adjective means here that there is no conductor inside the conducting shell. We will find that the familiar plane wave of free space, one with electric and magnetic fields perpendicular to the direction of propagation (the so-called *transverse electric and magnetic fields,* the TEM modes), are not possible in a hollow conductor. Instead, one can have TE (transverse electric) or TM (transverse magnetic) waves. On the other hand, TEM waves *are* possible if there is a conducting core, as in a coaxial cable. All this will lead us to a brief discussion of wave guides.

9.1 Nonconducting Boundaries

At boundaries separating different media, we cannot use Maxwell's equation for homogeneous media, since there the media is *not* homogeneous. Instead, we can determine matching conditions between regions by referring back to the more general macroscopic equations. *If there are no free charges*, $j = 0$ and Maxwell's macroscopic equations are

$$\left\langle \bar{\partial} \left(\mathbf{E} + ic\mathbf{B} \right) \right\rangle_{\Im} = \left\langle \bar{\partial} \left(c\mathbf{B} - i\mathbf{E} \right) \right\rangle_{\Re} = 0 = \left\langle \bar{\partial} \left(\mathbf{D} + i\mathbf{H}/c \right) \right\rangle_{\Re} . \tag{9.1}$$

Writing out the real and imaginary scalar and vector parts explicitly, we have

$$\begin{aligned} \nabla \cdot \mathbf{D} = 0 = \nabla \cdot \mathbf{B} \\ \partial_t \mathbf{B} + \nabla \times \mathbf{E} = 0 = \partial_t \mathbf{D} - \nabla \times \mathbf{H} . \end{aligned} \tag{9.2}$$

Note first the parallel structure between $\mathbf{B}$ and $\mathbf{D}$ on the one hand, and $\mathbf{E}$ and $-\mathbf{H}$ on the other. The fact that both $\mathbf{B}$ and $\mathbf{D}$ are *divergence free* (in the absence of free charges) means that the $\mathbf{B}$ and $\mathbf{D}$ field lines are *continuous*; they don't start or stop anywhere.

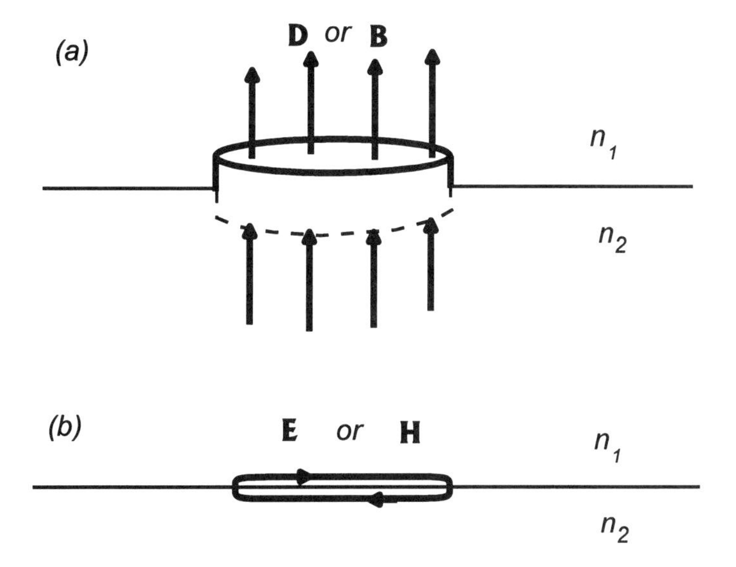

Fig. 9.1. Diagrams to show (a) that the normal components of $\mathbf{D}$ and $\mathbf{B}$ and (b) the tangential components of $\mathbf{E}$ and $\mathbf{H}$ are continuous at a dielectric boundary.

9.1.1 Continuity Conditions

First consider a small patch S of surface separating two different media. Surround the patch by a "pill box": an infinitesimal volume V enclosed between two surfaces S_1 and S_2, each equal in size to S and infinitesimally displaced from S on either side of the interface. Gauss's law together with the divergence equation above shows

$$\begin{aligned} \int_V d^3x\, \nabla \cdot \mathbf{D} &= \int_{S_1} d^2x\, \mathbf{n} \cdot \mathbf{D} - \int_{S_2} d^2x\, \mathbf{n} \cdot \mathbf{D} \\ &= 0\,, \end{aligned} \tag{9.3}$$

and similarly for $\mathbf{B}$, where $\mathbf{n}$ is the unit upward normal to the interface. Thus, the number of field lines $\mathbf{D}$ or $\mathbf{B}$ passing through one surface must also pass through the other. Since the size and shape of the surface S is arbitrary, the normal components $\mathbf{n} \cdot \mathbf{D}$ (or $\mathbf{n} \cdot \mathbf{B}$) on the two sides of the surface must be equal [see Fig. 9.1, part (a)].

Next consider a path that starts at a point in the interface, runs next to the interface in one medium, and then cuts to the other medium and returns to the starting point. The two legs l_1 and l_2 are parallel to each other and separated from the interface by only an infinitesimal distance. The distance around the path is not small, but the area $\mathbf{A}$ enclosed and hence the flux linked by the enclosed area is infinitesimal. Integration of $\nabla \times \mathbf{E}$ and $\nabla \times \mathbf{H}$ over $\mathbf{A}$ therefore vanishes, and this implies (by Stokes theorem and the arbitrary length of l_1 and l_2) that the tangential components of the fields $\mathbf{E}, \mathbf{H}$ are continuous [see Fig. 9.1, part (b)]:

$$\begin{aligned} \int_{\mathbf{A}} d\boldsymbol{\sigma} \cdot \nabla \times \mathbf{E} &= \int_{l_1} d\mathbf{l} \cdot \mathbf{E} - \int_{l_2} d\mathbf{l} \cdot \mathbf{E} \\ &= 0\,, \end{aligned} \tag{9.4}$$

and similarly for $\mathbf{H}$.

The matching conditions may be summarized by the continuity of

$$\begin{gathered} \mathbf{D} \cdot \mathbf{n},\ \mathbf{B} \cdot \mathbf{n} \\ \mathbf{E} \times \mathbf{n},\ \mathbf{H} \times \mathbf{n} \end{gathered} \tag{9.5}$$

in the spatial surface

$$\mathbf{x} = b\mathbf{n} + \mathbf{n} \times \mathbf{r}\,, \tag{9.6}$$

where b is the perpendicular distance from the origin to the surface, and $\mathbf{r}$ is any real vector. Note that in the presence of a surface charge $\mathbf{D}\cdot\mathbf{n}$ is not continuous, and if a surface current is present, $\mathbf{H}\times\mathbf{n}$ is different on the two sides of the interface.

Exercise 9.1. Use Maxwell's macroscopic equations to show that where a free surface charge Σ (with units of charge/area) exists, the discontinuity in $\mathbf{D}\cdot\mathbf{n}$ is given by

$$\mathbf{D}_1\cdot\mathbf{n}-\mathbf{D}_2\cdot\mathbf{n}=\Sigma\,. \tag{9.7}$$

Exercise 9.2. From Maxwell's macroscopic equations, show that where a free surface current $\mathbf{K}$ (with units of current/length) exists, the discontinuity in $\mathbf{n}\times\mathbf{H}$ is

$$\mathbf{n}\times(\mathbf{H}_1-\mathbf{H}_2)=\mathbf{K}\,. \tag{9.8}$$

Note that if $\hat{\mathbf{l}}=\hat{\mathbf{A}}\times\mathbf{n}$ is a unit vector in the direction of $d\mathbf{l}$ on the upper boundary of the infinitesimal area $\mathbf{A}$, then

$$\mathbf{H}\cdot\hat{\mathbf{l}}=\mathbf{H}\cdot\left(\hat{\mathbf{A}}\times\mathbf{n}\right)=(\mathbf{n}\times\mathbf{H})\cdot\hat{\mathbf{A}}$$

and scalar relations of the form $\mathbf{x}\cdot\hat{\mathbf{A}}=\mathbf{y}\cdot\hat{\mathbf{A}}$ which are true for three linearly independent real vectors $\hat{\mathbf{A}}$ imply $\mathbf{x}=\mathbf{y}$.

9.1.2 Snell's Law and Total Internal Reflection

Consider linearly polarized monochromatic plane waves in the two media. We assume nonconducting media with conductivity $\sigma=0$. An incident wave (0) in medium 1 will create both a transmitted wave (2) in medium 2 and a reflected wave (1) in medium 1 (see Fig. 9.2):

$$\begin{aligned}\mathbf{F}=\mathbf{F}_0\cos(\omega_0t-\mathbf{k}_0\cdot\mathbf{x}-\phi_0)+\mathbf{F}_1\cos(\omega_1t-\mathbf{k}_1\cdot\mathbf{x}-\phi_1)\,,\ \mathbf{x}\cdot\mathbf{n}<b\\ \mathbf{F}=\mathbf{F}_2\cos(\omega_2t-\mathbf{k}_2\cdot\mathbf{x}-\phi_2)\,,\ \mathbf{x}\cdot\mathbf{n}>b\,.\end{aligned} \tag{9.9}$$

From chapter 8, the coefficients are constant complex vectors of the form

$$\mathbf{F}_j=\mathbf{E}_j+ic_j\mathbf{B}_j=\left(1+\hat{\mathbf{k}}_j\right)\mathbf{E}_j\,, \tag{9.10}$$

where c_j is the speed of light in the appropriate medium, and the phase angles ϕ_j are constant scalars. The matching conditions can hold for all time t and positions

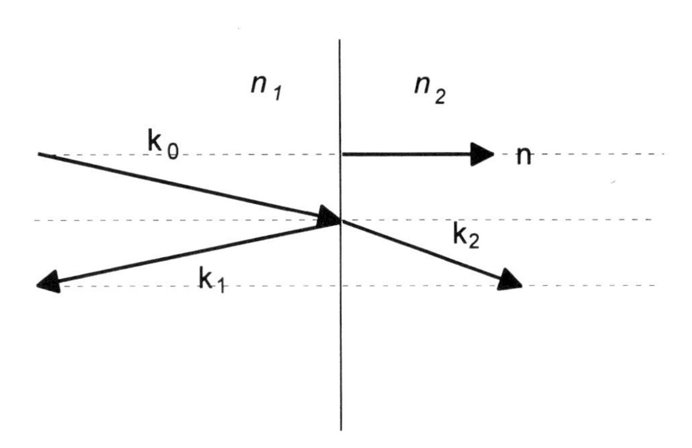

Fig. 9.2. Wave vectors at the boundary of two media.

$\mathbf{x} = \mathbf{x} \cdot \mathbf{n}\,\mathbf{n} + \mathbf{n} \times \mathbf{r}$ on the interface only if the angle arguments $\omega_j t - \mathbf{k}_j \cdot \mathbf{x} - \phi_j$ are equal for all t and $\mathbf{r}$. As a result,

$$\begin{aligned} \omega_0 = \omega_1 = \omega_2 \equiv \omega \\ \mathbf{k}_0 \times \mathbf{n} = \mathbf{k}_1 \times \mathbf{n} = \mathbf{k}_2 \times \mathbf{n}\,. \end{aligned} \tag{9.11}$$

In other words, the projections of the propagation vectors onto the interface are all equal in both length and direction. All three propagation vectors lie in a plane that also contains the normal $\mathbf{n}$ to the surface. Now using c_0 for the speed of light in a vacuum (see previous chapter), we have

$$\mathbf{k}_j = \frac{\omega}{c_j}\hat{\mathbf{k}}_j = \frac{\omega n_j}{c_0}\hat{\mathbf{k}}_j\,, \tag{9.12}$$

where $n_j = c_0/c_j\,,\, j = 1, 2$, is the index of refraction, so that the length condition of (9.11) gives *Snell's law*:

$$n_1 \sin\theta_1 = n_2 \sin\theta_2 = n_1 \sin\theta_0 \tag{9.13}$$

with θ_j the angle between $\mathbf{k}_j$ and the normal $\mathbf{n}$ to the interface, and $0 \le \theta_j \le \pi/2$. Snell's law shows that the angle of incidence θ_0 is equal to the angle of reflection θ_1.

9.1.3 Total Internal Reflection

If $n_1 > n_2$, Snell's law predicts $\sin\theta_2 = (n_1/n_2)\sin\theta_1 > \sin\theta_1$. For angles θ_1 larger than a *critical angle* θ_c where $\sin\theta_c = n_2/n_1$, the value $(n_1/n_2)\sin\theta_1$ predicted for $\sin\theta_2$ exceeds 1. In this case, that part of the beam transmitted through the second medium drops to zero, and one has *total internal reflection.* The continuity conditions together with the right-handed relationship of the vectors $\mathbf{E}, \mathbf{B}$, and $\mathbf{k}$ ensure that the field in medium 2 does not drop immediately to zero. Instead, there is a decaying *evanescent* wave there.

The basic property of the evanescent wave follows from the matching conditions (9.11) for $\mathbf{k} \times \mathbf{n}$ and ω together with the null property of k, namely $k\bar{k} = (\omega/c)^2 - \mathbf{k}^2 = 0$. The null property means that the length of $\mathbf{k}$ is $n\omega/c_0$ and is thus proportional to the index of refraction. In medium 2, with its smaller n, the total length of $\mathbf{k}$ is less than in medium 1, and at angles $\theta_1 > \theta_c$, one even has $|\mathbf{k}_2| = \sqrt{(\mathbf{k}_2 \times \mathbf{n})^2 + (\mathbf{k}_2 \cdot \mathbf{n})^2} < |\mathbf{k}_2 \times \mathbf{n}| = |\mathbf{k}_1 \times \mathbf{n}|$. This is possible if and only if $(\mathbf{k}_2 \cdot \mathbf{n})^2$ is negative, so that

$$\begin{aligned} \mathbf{k}_2 \cdot \mathbf{n} &= \pm i\sqrt{(\mathbf{k}_2 \times \mathbf{n})^2 - \mathbf{k}_2^2} \\ &= \pm i\sqrt{(\mathbf{k}_1 \times \mathbf{n})^2 - \left(\frac{n_2}{n_1}\right)^2 \mathbf{k}_1^2} \\ &= \pm i\kappa \end{aligned} \tag{9.14}$$

is purely imaginary, with

$$\kappa = |\mathbf{k}_1| \sqrt{\sin^2\theta_1 - \sin^2\theta_c}\,. \tag{9.15}$$

The solutions $\mathbf{F}_2$ to the wave equation in medium 2 are exponentially damped in the direction normal to the interface, with the form

$$\begin{aligned} \mathbf{F}_2 &= e^{-\mathbf{n}\cdot\mathbf{x}\,\kappa + \hat{\mathbf{b}} s_{\Re}} \mathbf{f}_0 \qquad (9.16) \\ &= e^{-\mathbf{n}\cdot\mathbf{x}\,\kappa} \left(\cos s_{\Re} + \hat{\mathbf{b}} \sin s_{\Re}\right) \mathbf{f}_0\,, \end{aligned}$$

where $\hat{\mathbf{b}}$ is the unit biparavector

$$\hat{\mathbf{b}} = -\frac{k_{\Re}\mathbf{n}}{\kappa}\,,\ \hat{\mathbf{b}}^2 = -1\,, \tag{9.17}$$

$$s_{\Re} = \omega t - \mathbf{k}_{\Re} \cdot \mathbf{x}\,, \tag{9.18}$$

and $\mathbf{f}_0$ is a constant vector parallel to $\mathbf{k}_{\Re}$,which lies in the interface and in the plane of $\mathbf{k}_0$ and $\mathbf{k}_1$ and has the magnitude $|\mathbf{k}_{\Re}| = |\mathbf{k}_1 \times \mathbf{n}|$.

Exercise 9.3. Show by direct differentiation that $\mathbf{F}_2$ as given in (9.16) is a solution to Maxwell's equation in a homogeneous medium with no free currents. (See also problem 5.)

An important example of total internal reflection is the transmission of light in a fibre-optic cable. Even after many oblique reflections, very little energy is lost from the light signal. The main losses are due to impurities in the fibre that scatter the light so that it strikes the interior wall at angles close to the normal and escapes. Another example is the reflecting prism. There, medium 1 is the prism material, usually glass or quartz, and medium 2 is air. The existence of the evanescent wave is demonstrated by placing a second piece of prism material close to, but not touching, the reflecting surface of the prism. If some of the evanescent wave reaches the second prism piece across the air gap, it will propagate as a plane wave in the material, but with a reduced amplitude as given by the damping factor (9.15). By modulating the gap distance, one can modulate the transmitted amplitude.

9.1.4 Matching Conditions

With the phase angles equal in (9.9), the continuity conditions (9.5) imply matching equalities among the constant vector coefficients:

$$\begin{aligned}(\mathbf{D}_0+\mathbf{D}_1 - \mathbf{D}_2) \cdot \mathbf{n} &=0 \\ (\mathbf{E}_0+\mathbf{E}_1 - \mathbf{E}_2) \times \mathbf{n} &=0 \,,\end{aligned} \tag{9.19}$$

and analogous relations for $\mathbf{B}$ and $\mathbf{H}$:

$$\begin{aligned}(\mathbf{B}_0+\mathbf{B}_1 - \mathbf{B}_2) \cdot \mathbf{n} &=0 \\ (\mathbf{H}_0+\mathbf{H}_1 - \mathbf{H}_2) \times \mathbf{n} &=0 \,.\end{aligned} \tag{9.20}$$

Using $\mathbf{D} = \varepsilon\mathbf{E}$, $\mathbf{H} = \mathbf{B}/\mu$, and $ic\mathbf{B} = \hat{\mathbf{k}}\mathbf{E}$, conditions appropriate to nonconducting media, we can express the continuity conditions in terms of matching conditions for the electric field $\mathbf{E}$:

$$\begin{aligned}
\left[\varepsilon_1\left(\mathbf{E}_0+\mathbf{E}_1\right)-\varepsilon_2\mathbf{E}_2\right]\cdot\mathbf{n}&=0\\
\left(\mathbf{E}_0+\mathbf{E}_1-\mathbf{E}_2\right)\times\mathbf{n}&=0\\
\left[n_1\left(\hat{\mathbf{k}}_0\mathbf{E}_0+\hat{\mathbf{k}}_1\mathbf{E}_1\right)-n_2\hat{\mathbf{k}}_2\mathbf{E}_2\right]\cdot\mathbf{n}&=0\\
\left[Z_1^{-1}\left(\hat{\mathbf{k}}_0\mathbf{E}_0+\hat{\mathbf{k}}_1\mathbf{E}_1\right)-Z_2^{-1}\hat{\mathbf{k}}_2\mathbf{E}_2\right]\times\mathbf{n}&=0\,,
\end{aligned} \tag{9.21}$$

where $Z=\sqrt{\mu/\varepsilon}=(\varepsilon c)^{-1}=\mu c=\mu c_0/n$ is the *impedance* of the medium.

9.2 Normal Incidence

Consider first the simple case of normal incidence: $\mathbf{n}\times\mathbf{k}_0=0$. Then from (9.11), $\mathbf{n}\times\mathbf{k}_1=\mathbf{n}\times\mathbf{k}_2=0$, and from (9.12),

$$\mathbf{k}_0=-\mathbf{k}_1=\frac{n_1}{n_2}\mathbf{k}_2\,. \tag{9.22}$$

The first and third of the four matching conditions (9.21) are trivially fulfilled since the fields are all normal to $\mathbf{n}$, and the other two conditions give

$$\mathbf{E}_0+\mathbf{E}_1=\mathbf{E}_2 \tag{9.23}$$

$$\mathbf{E}_0-\mathbf{E}_1=\frac{Z_1}{Z_2}\mathbf{E}_2\,. \tag{9.24}$$

These are easily solved to give

$$\mathbf{E}_1=\left(\frac{Z_2-Z_1}{Z_1+Z_2}\right)\mathbf{E}_0,\ \mathbf{E}_2=\left(\frac{2Z_2}{Z_1+Z_2}\right)\mathbf{E}_0\,. \tag{9.25}$$

Note that the direction of the reflected electric field is reversed if $Z_2<Z_1$ (for example, reflection in air at the surface of water) but not if $Z_1<Z_2$. The energy flow is given by the Poynting vector

$$\mathbf{S}=\frac{c\varepsilon}{2}\left\langle\mathbf{F}\mathbf{F}^{\dagger}\right\rangle_V=\mathbf{E}\times\mathbf{H}=Z^{-1}\mathbf{E}^2\hat{\mathbf{k}}\,. \tag{9.26}$$

The ratio of the transmitted energy flow to the incident flow is the *transmission coefficient* T, whereas that of the reflected flow to the incident flow is the *reflection coefficient* R:

$$T=\frac{|\mathbf{S}_2|}{|\mathbf{S}_0|}=\frac{Z_1\mathbf{E}_2^2}{Z_2\mathbf{E}_0^2}=\frac{4Z_1Z_2}{(Z_1+Z_2)^2}, \tag{9.27}$$

$$R = \frac{|\mathbf{S}_1|}{|\mathbf{S}_0|} = \frac{\mathbf{E}_1^2}{\mathbf{E}_0^2} = \frac{(Z_2 - Z_1)^2}{(Z_1 + Z_2)^2}.$$

Note that the reflection coefficient is minimized by matching the impedances. For a given change in impedance, the net reflection can be minimized by making the change in several smaller steps (see problem 1.) These are examples of the general rule, also experienced for transmission lines, that sudden changes of impedance lead to large reflections. One can easily verify that

$$T + R = 1\,, \tag{9.28}$$

which is an expression of the conservation of energy: what doesn't get through gets reflected.

9.3 Fresnel Equations

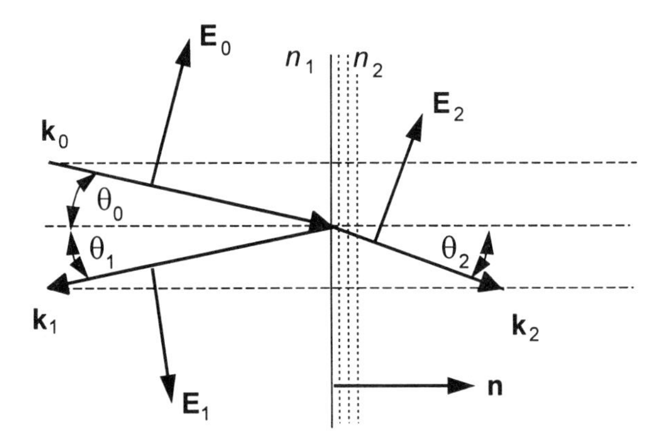

Fig. 9.3. Vectors for the case $\mathbf{B} \cdot \mathbf{n} = 0$. Here, $\mathbf{B}$ is out of the page and $\mathbf{E}$ (and hence the polarization) is in the plane of incidence.

For the more general case of oblique incidence of waves on an interface, it is useful to consider two independent polarizations, one with $\mathbf{B} \cdot \mathbf{n} = 0$ and one with $\mathbf{E} \cdot \mathbf{n} = 0$. The behavior of more general polarizations can be found from superpositions of these two cases. Consider the first case, in which the radiation is polarized with its electric field in the plane of the propagation vectors (the *plane of incidence*) and for which

$$\begin{aligned} \mathbf{E}_j \cdot \mathbf{n} &= E_j \sin\theta_j \\ \mathbf{B}_j \cdot \mathbf{n} &= 0 . \end{aligned} \tag{9.29}$$

The matching conditions (9.21) then give

$$\begin{aligned} \varepsilon_1 (E_0 + E_1) \sin\theta_1 &= \varepsilon_2 E_2 \sin\theta_2 \\ (E_0 - E_1) \cos\theta_1 &= E_2 \cos\theta_2 \\ (E_1 + E_0) / Z_1 &= E_2 / Z_2 \end{aligned} \tag{9.30}$$

Since $\varepsilon Z = n/c_0$, the first and third of these reproduce Snell's law (9.13) and can be combined with the second condition to give

$$E_1 = E_0 \frac{Z_1 \cos\theta_1 - Z_2 \cos\theta_2}{Z_1 \cos\theta_1 + Z_2 \cos\theta_2} \tag{9.31}$$

$$E_2 = E_0 \frac{2 Z_2 \cos\theta_1}{Z_1 \cos\theta_1 + Z_2 \cos\theta_2} . \tag{9.32}$$

These relations reduce to those for normal incidence when $\theta_1 = \theta_2 = 0$. Note that the reflected wave vanishes ($E_1 = 0$) if $Z_1 \cos\theta_1 = Z_2 \cos\theta_2$, in other words if

$$\mu_1 \sin\theta_1 \cos\theta_1 = \mu_2 \sin\theta_2 \cos\theta_2 . \tag{9.33}$$

In the common case that $\mu_1 = \mu_2$, this means that the angles are equal or complementary: $\theta_1 + \theta_2 = \pi/2$. The angles are equal only if $\theta_0 = 0$ or if $n_1 = n_2$. When they are complementary, the $\mathbf{E}_2$ field of the refracted wave would have to lie parallel to the propagation direction $\mathbf{k}_1$ of the reflected wave. However, it is precisely the oscillations of the charges at the interface in medium 2 along $\mathbf{E}_2$ that generate the reflected wave, and there is no propagation in the direction of the oscillation. The angle of incidence θ_B at which the reflected wave polarized in the plane of incidence disappears is known as the *Brewster angle*:

$$\frac{n_2}{n_1} = \frac{\sin\theta_1}{\sin\theta_2} = \tan\theta_B . \tag{9.34}$$

Any radiation reflected at this angle must be linearly polarized in the plane of the interface.

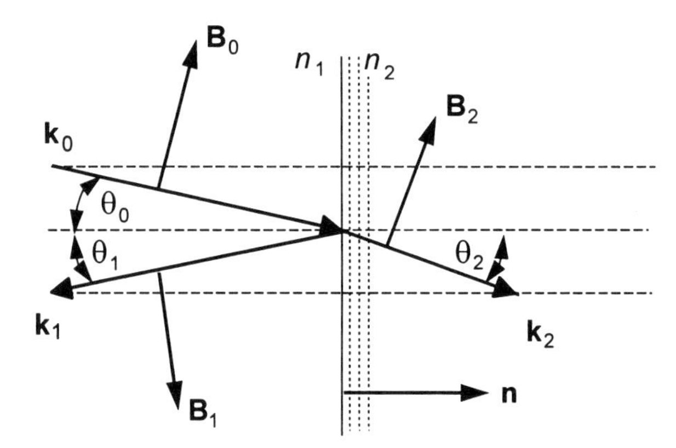

Fig. 9.4. Vectors for the case $\mathbf{E}\cdot\mathbf{n}=0$. Here, $\mathbf{E}$ points into the page.

Consider now the case in which $\mathbf{E}_0$ lies in the plane of the interface, but is perpendicular to the plane of incidence. Now

$$\begin{aligned} \mathbf{E}_j \cdot \mathbf{n} &= 0 \\ \mathbf{B}_j \cdot \mathbf{n} &= B_j \sin\theta_j \,, \end{aligned} \tag{9.35}$$

so that the first of the matching conditions (9.21) is trivially fulfilled, and the second condition can be replaced by the scalar equation

$$E_0 + E_1 = E_2 \tag{9.36}$$

since all the electric fields are parallel. The third condition combines with this to reproduce Snell's law (9.13), and noting that $\mathbf{E}$ is orthogonal to both $\mathbf{k}$ and $\mathbf{n}$, the fourth gives

$$Z_1^{-1}\left(\mathbf{n}\cdot\hat{\mathbf{k}}_0 E_0 + \mathbf{n}\cdot\hat{\mathbf{k}}_1 E_1\right) = Z_1^{-1}\mathbf{n}\cdot\hat{\mathbf{k}}_0\left(E_0 - E_1\right) = Z_2^{-1}\mathbf{n}\cdot\hat{\mathbf{k}}_2 E_2 \,, \tag{9.37}$$

which can be solved together with condition 2 to give

$$E_1 = E_0 \frac{Z_2\cos\theta_1 - Z_1\cos\theta_2}{Z_2\cos\theta_1 + Z_1\cos\theta_2} \tag{9.38}$$

and

$$E_2 = E_1 + E_0 = E_0 \frac{2Z_2\cos\theta_1}{Z_2\cos\theta_1 + Z_1\cos\theta_2} \,. \tag{9.39}$$

These equalities also reduce to the normal-incidence case when $\theta_1 = \theta_2 = 0$. Relations (9.31), (9.32), (9.38), and (9.39) are known as the *Fresnel equations*. They can be written in several different ways. For example, since the impedance factors always appear in ratios and

$$\frac{Z_1}{Z_2} = \frac{\mu_1 \sin\theta_1}{\mu_2 \sin\theta_2}, \tag{9.40}$$

one can replace Z_j everywhere by $\mu_j \sin\theta_j$ (see problem 4).

9.4 Conducting Surfaces

9.4.1 Conducting Screen

The linearity of Maxwell's equation permits a valuable simplification of many problems: if the electromagnetic fields $\mathbf{F}_1$ and $\mathbf{F}_2$ are known for two independent sources j_1 and j_2, then the field $\mathbf{F}$ for the combined source $j_1 + j_2$ is just the sum $\mathbf{F} = \mathbf{F}_1 + \mathbf{F}_2$. An important application of this *superposition principle* is to diffraction of radiation by objects and by holes in screens.

Consider here an ideal thin conducting screen that separates a linear isotropic medium into two halves and blocks radiation incident from one side, preventing its transmission. Let a wave be incident from the left. The action of the screen is mathematically equivalent to transmission of the incident radiation plus the superposition of plane-wave radiation from the screen toward the right of equal amplitude but opposite sign to the incident radiation. If we know the electromagnetic fields on the two sides of the screen, we can determine the current in the screen by Maxwell's equation.

Let the screen lie in the $\imath\mathbf{n}$ plane at $\langle x\mathbf{n}\rangle_S \equiv \mathbf{x}\cdot\mathbf{n} = 0$ so that $\mathbf{n}$ is the vector dual and hence normal to the screen. Take the field to be

$$\mathbf{F}(x) = \begin{cases} \mathbf{F}_-(x), & \langle x\mathbf{n}\rangle_S < 0 \\ \mathbf{F}_+(x), & \langle x\mathbf{n}\rangle_S > 0 \end{cases} \tag{9.41}$$

where both $\mathbf{F}_\pm(x)$ satisfy Maxwell's equation for source-free space:

$$\bar{\partial}\mathbf{F}_\pm(x) = 0\,. \tag{9.42}$$

Then $\bar{\partial}\mathbf{F}(x) = 0$ except in the plane of the screen at $\langle x\mathbf{n}\rangle_S = 0$. If we write

$$\mathbf{F}(x) = \mathbf{F}_-(x) + \theta\left(\langle x\mathbf{n}\rangle_S\right)\left[\mathbf{F}_+(x) - \mathbf{F}_-(x)\right], \tag{9.43}$$

then since the derivative of the *step function* θ is the Dirac delta function (see Section 5.5.1), Maxwell's equation is seen to give a current

$$\bar{j}(x) = \varepsilon c \bar{\partial} \mathbf{F}(x) = \varepsilon c \delta(\langle x\mathbf{n}\rangle_S)\, \mathbf{n}\,[\mathbf{F}_+(x) - \mathbf{F}_-(x)] \;. \tag{9.44}$$

Since j is real, the real components of $\mathbf{F}_+$ and $\mathbf{F}_-$ perpendicular to $\mathbf{n}$, as well as the imaginary components parallel to $\mathbf{n}$, must be equal at $\langle x\mathbf{n}\rangle_S = 0$:

$$(\mathbf{B}_+ - \mathbf{B}_-) \cdot \mathbf{n} = 0 = (\mathbf{E}_+ - \mathbf{E}_-) \times \mathbf{n} \,. \tag{9.45}$$

In particular, if $\mathbf{F}_+$ vanishes, then the components $\mathbf{B}_- \cdot \mathbf{n}$ and $\mathbf{E}_- \times \mathbf{n}$ of $\mathbf{F}_-$ must vanish in the screen.

The current j in the screen can be expressed as

$$j(x) \equiv K(x)\, \delta(\mathbf{x} \cdot \mathbf{n}) \;, \tag{9.46}$$

where $K(x) = \Sigma(x)\, c + \mathbf{K}(x)$ is the *surface current* in the screen

$$K(x) = \varepsilon c\, [\mathbf{F}_+(x) - \mathbf{F}_-(x)]\, \mathbf{n} \,, \; \langle x\mathbf{n}\rangle_S = 0 \,, \tag{9.47}$$

comprising both a vector current $\mathbf{K}(x)$ and c times the *surface charge* $\Sigma(x)$. Using the reality of $K(x)$, we can split this relation (9.47) into the homogeneous condition (9.45) plus scalar and vector parts:

$$\Sigma(x) = \varepsilon\, [\mathbf{E}_+(x) - \mathbf{E}_-(x)] \cdot \mathbf{n} \tag{9.48}$$

$$\mathbf{K}(x) = \mathbf{n} \times [\mathbf{B}_+(x) - \mathbf{B}_-(x)] \,/ \mu \,. \tag{9.49}$$

Note that the current in the screen depends only on the difference of the fields on the two sides: we could add the same external field to both $\mathbf{F}_+$ and $\mathbf{F}_-$ without changing the current. However, this does *not* mean that the current in the screen is independent of the external field since $\mathbf{F}_\pm$ themselves generally change when an external field is added.

Let $\mathbf{F}_0$ be the radiation field incident on the screen from the left, that is from the side with $\langle x\mathbf{n}\rangle_S < 0$, and assume the screen fully blocks the radiation: $\mathbf{F}_+ = 0$. The currents in the screen must radiate the field $-\mathbf{F}_0$ toward the right in order to cancel $\mathbf{F}_0$. By reflection symmetry, whatever field the current in the screen radiates to the right, it must also radiate to the left. That is, $\mathbf{F}_- - \mathbf{F}_0$ on the left of the screen must be the reflection in the screen of $-\mathbf{F}_0$ on the right. Remembering from chapter 2 that $\mathbf{n}\bar{x}^\dagger\mathbf{n}$ is the reflection of any element x in the $i\mathbf{n}$ plane, we find the reflected field

$$\mathbf{F}_1(x) = \mathbf{F}_-(x) - \mathbf{F}_0(x) \tag{9.50}$$

$$= -\mathbf{n}\bar{\mathbf{F}}_0^\dagger(\mathbf{n}\bar{x}\mathbf{n})\, \mathbf{n} = \mathbf{n}\mathbf{F}_0^\dagger(\mathbf{n}\bar{x}\mathbf{n})\, \mathbf{n} \,. \tag{9.51}$$

In the screen, $\langle x\mathbf{n}\rangle_S = \mathbf{x}\cdot\mathbf{n} = 0$ and $x = \mathbf{n}\bar{x}\mathbf{n}$. Therefore, the surface current in the screen (at $\langle x\mathbf{n}\rangle_S = 0$) is

$$\begin{aligned} K(x) &= -\varepsilon c\mathbf{F}_-(x)\,\mathbf{n} \\ &= -\varepsilon c\left[\mathbf{F}_0(x)\,\mathbf{n} + \mathbf{n}\mathbf{F}_0^\dagger(x)\right] \\ &= -2\varepsilon c\,\langle\mathbf{F}_0(x)\,\mathbf{n}\rangle_\Re\,. \end{aligned} \tag{9.52}$$

A plane wave $\mathbf{F}_0 = \left(1+\hat{\mathbf{k}}_0\right)\mathbf{E}_0(s)$, where $s = \langle k_0\bar{x}\rangle_S$, incident from the left will thus induce a current

$$\begin{aligned} K(x) &= -2\varepsilon c\left\langle\left(1+\hat{\mathbf{k}}_0\right)\mathbf{E}_0\mathbf{n}\right\rangle_\Re \\ &= 2\varepsilon c\left[\left(\hat{\mathbf{k}}_0\times\mathbf{E}_0\right)\times\mathbf{n} - \mathbf{E}_0\cdot\mathbf{n}\right]. \end{aligned} \tag{9.53}$$

The reflected wave is

$$\mathbf{F}_1(x) = -\mathbf{n}\bar{\mathbf{F}}_0^\dagger(\mathbf{n}\bar{x}\mathbf{n})\,\mathbf{n} = \left(1+\hat{\mathbf{k}}_1\right)\mathbf{E}_1(s_1)\,, \tag{9.54}$$

where $\mathbf{k}_1 = -\mathbf{n}\mathbf{k}_0\mathbf{n}$, $\mathbf{E}_1 = \mathbf{n}\mathbf{E}_0\mathbf{n}$, and $s_1 = \left\langle\bar{k}_0\mathbf{n}\bar{x}\mathbf{n}\right\rangle_S = \langle k_1\bar{x}\rangle_S$ with $k_1 = \omega/c + \mathbf{k}_1 = \mathbf{n}\bar{k}_0\mathbf{n}$. These results for a conducting screen can also be used as an approximation of a good planar conductor: practically all of the current induced resides in a thin layer at the surface, and the electromagnetic field is strongly attenuated in this layer.

Exercise 9.4. Verify the continuity equation for the current in the screen.

The simplest configuration to handle is normal incidence: let $\hat{\mathbf{k}}_0 = \mathbf{n}$. Then $\mathbf{k}_1 = -\mathbf{k}_0$, $\mathbf{E}_1 = -\mathbf{E}_0$, $s_1 = \omega t - \mathbf{k}_1\cdot\mathbf{x}$, and

$$\begin{aligned} \mathbf{F}_1 &= -\left(1-\hat{\mathbf{k}}_0\right)\mathbf{E}_0(s_1) \\ K(x) &= 2\varepsilon c\mathbf{E}_0(\omega t)\,. \end{aligned} \tag{9.55}$$

More interesting is the case of oblique incidence. Let $\mathbf{n} = \mathbf{e}_3$, $\hat{\mathbf{k}}_0 = \mathbf{e}_3\cos\theta - \mathbf{e}_2\sin\theta$, and $\mathbf{E}_0 = E_0(s)\,(\mathbf{e}_2\cos\theta + \mathbf{e}_3\sin\theta)$. Then

$$\begin{aligned} \mathbf{F}_0(x) &= \left(1+\hat{\mathbf{k}}_0\right)\mathbf{E}_0(s) \\ &= E_0(s)\,(1+\mathbf{e}_3\cos\theta - \mathbf{e}_2\sin\theta)\,(\mathbf{e}_2\cos\theta + \mathbf{e}_3\sin\theta) \\ &= E_0(s)\,(\mathbf{e}_2\cos\theta + \mathbf{e}_3\sin\theta - \mathbf{e}_2\mathbf{e}_3)\,, \end{aligned} \tag{9.56}$$

where

$$s = \langle k_0 \bar{x} \rangle_S = \omega t - \frac{\omega}{c} \left(x^3 \cos\theta - x^2 \sin\theta \right) . \tag{9.57}$$

The surface current in the screen is

$$\begin{aligned} K(x) &= -2\varepsilon c \left\langle \mathbf{F}_0(x)\, \mathbf{e}_3 \right\rangle_{\Re} \\ &= 2\varepsilon c E_0(s) \left(\mathbf{e}_2 - \sin\theta \right) , \end{aligned} \tag{9.58}$$

with $x^3 = 0$, and the reflected wave is

$$\begin{aligned} \mathbf{F}_1(x) &= -\mathbf{e}_3 \bar{\mathbf{F}}_0^{\dagger} \mathbf{e}_3 \\ &= -E_0(s_1) \left(\mathbf{e}_2 \cos\theta - \mathbf{e}_3 \sin\theta + \mathbf{e}_2\mathbf{e}_3 \right) \end{aligned} \tag{9.59}$$

where $s_1 = \omega t - \mathbf{k}_1 \cdot \mathbf{x} = \omega t + (x^3 \cos\theta + x^2 \sin\theta)\, \omega/c$. For a monochromatic wave with, for example, $E_0(s) = E_0(0) \cos s$, we thus find a ripple of current and a fluctuating charge density in the screen.

Exercise 9.5. Show that the same result can be obtained by boosting the current $j(x) = K(x)\, \delta(\langle x\mathbf{n} \rangle_S)$ and $\mathbf{F}$ for the case of normal incidence in the $-\mathbf{e}_2$ direction by an appropriate amount.

Exercise 9.6. Find the current and the field in the screen when the incident radiation has the orthogonal polarization, namely $\mathbf{E} = E_0(s)\, \mathbf{e}_1$, but the same propagation vector as in the oblique case above.
[Partial ans.: $K = 2\varepsilon_0 c E_0(s)\, \mathbf{e}_1 \cos\theta$.]

9.4.2 Real Conductors

Reflections from the interface of a good conductor occur over a finite skin depth close to the surface (see previous chapter). The reflections are nearly complete at "low" frequencies, which usually include the optical range, when the reduced wavelength $\beta = \lambda / (2\pi\delta)$ in units of the skin depth is large. In such cases, the approximation of a surface current is often useful: we neglect the skin depth δ and treat the reflection and induced current as though it occurred right at the surface of the conductor. In other words, we treat the conductor as though its surface was covered by the ideal conducting screen discussed in the previous subsection. We can then apply the matching condition (9.47), which includes effects of the surface current $K = \Sigma + \mathbf{K}$, with field $\mathbf{F}_- = 0$ and unit vector $\mathbf{n}$ specifying the outward normal of the surface.

While the surface-current approximation is excellent in a wide range of cases, its approximate nature and its limits of validity should be recognized. In real conductors, the current j and the conductivity σ are finite, and at sufficiently high frequencies, $\beta \lesssim 1$ and the skin depth should not be neglected. To determine the reflected field at a conductor surface, we apply matching conditions as before. Because the currents are finite, we can exclude effects of surface charge or current by shrinking the areas of the paths and the volumes of the pill boxes to zero. The matching conditions are then the same as in the dielectric case (9.19) and (9.20).

For simplicity, we consider an electromagnetic wave with normal incidence ($\hat{\mathbf{k}}_0 = \mathbf{n}$) on a conductor surface at $\mathbf{x} = \mathbf{n} \times \mathbf{r}$. (For convenience, we take the origin of coordinates to lie in the surface.) The conductor will be medium 2; its surroundings, medium 1.

From the last chapter, the field at the surface of the conductor is

$$\mathbf{F}_2 = \mathbf{E}_2 + ic_2\mathbf{B}_2 = \left[\cos\omega t + \sqrt{\alpha^2 + \beta^2}\hat{\mathbf{k}}_0 \cos(\omega t - \phi)\right] \mathbf{E}_2(0) \,, \tag{9.60}$$

where in terms of the dimensionless conductivity $\zeta = \sigma / (\varepsilon\omega)$,

$$\alpha = \left(\frac{\sqrt{1+\zeta^2}+1}{2}\right)^{1/2} = \sqrt{1+\beta^2} \,, \tag{9.61}$$

and

$$\tan\phi = \beta/\alpha \,. \tag{9.62}$$

The matching conditions specify that both $\mathbf{E}$ and $\mathbf{H}$ (which are tangential for normal incidence) must be continuous at the boundary. The sum of the incident and reflected waves at the surface must therefore be

$$\mathbf{F}_0 + \mathbf{F}_1 = \mathbf{E}_2 + i\,(Z_1/Z_2)\, c_2\mathbf{B}_2 \;\; \text{at } \mathbf{x} = \mathbf{n} \times \mathbf{r} \,. \tag{9.63}$$

Note the impedance ratio Z_1/Z_2 that multiplies $\mathbf{B}_2$; this is because it is not $c\mathbf{B}$ or $\mathbf{F}$ that is continuous at the interface, but rather $\mathbf{E}$ and $\mathbf{H} = \mathbf{B}/\mu = c\mathbf{B}/Z$. Now the fields $\mathbf{F}_0, \mathbf{F}_1$ are directed plane waves moving in the directions of $\mathbf{k}_0$ and $\mathbf{k}_1 = -\mathbf{k}_0$, respectively, and therefore, the only spacetime dependence of the waves is through the scalar parameters

$$s_j = \omega t - \mathbf{k}_j \cdot \mathbf{x} \,. \tag{9.64}$$

Thus,

$$\mathbf{F}_0 = \mathbf{F}_0(s_0)\,,\ \mathbf{F}_1 = \mathbf{F}_1(s_1)\ . \tag{9.65}$$

To separate the incident and reflected waves, we use the fact that directed plane waves are *right ideals,* that is they obey

$$\mathbf{F}_j = \left(P_j + \bar{P}_j\right)\mathbf{F}_j = P_j\mathbf{F}_j\,, \tag{9.66}$$

where $j = 1, 2$ and

$$P_j = \frac{1}{2}\left(1 + \hat{\mathbf{k}}_j\right) = P_j^2 = P_j^\dagger \tag{9.67}$$

is a real projector obeying

$$\begin{aligned} P_j + \bar{P}_j = 1,\ P_j\bar{P}_j = 0 \\ P_j\hat{\mathbf{k}}_j = P_j,\ \bar{P}_j\hat{\mathbf{k}}_j = -\bar{P}_j\,. \end{aligned} \tag{9.68}$$

Thus, the incident and reflected fields are linearly polarized in the direction of $\mathbf{E}_2(0)$:

$$\begin{aligned} \mathbf{F}_0 &= \left[\cos s_0 + \frac{Z_1}{Z_2}\sqrt{\alpha^2+\beta^2}\cos(s_0 - \phi)\right] P_0\mathbf{E}_2\,(0) \\ &= \sqrt{\left(\frac{Z_1}{Z_2}\alpha + 1\right)^2 + \left(\frac{Z_1}{Z_2}\beta\right)^2}\, P_0\mathbf{E}_2(0)\cos(s_0 - \psi_0) \end{aligned} \tag{9.69}$$

$$\begin{aligned} \mathbf{F}_1 &= \left[\cos s_1 - \frac{Z_1}{Z_2}\sqrt{\alpha^2+\beta^2}\cos(s_1 - \phi)\right] \bar{P}_0\mathbf{E}_2(0) \\ &= -\sqrt{\left(\frac{Z_1}{Z_2}\alpha - 1\right)^2 + \left(\frac{Z_1}{Z_2}\beta\right)^2}\, \bar{P}_0\mathbf{E}_2(0)\cos(s_1 - \psi_1)\,, \end{aligned} \tag{9.70}$$

where the phase angles ψ_0 and ψ_1 are given by

$$\tan\psi_0 = \frac{Z_1\beta}{Z_1\alpha + Z_2},\ \tan\psi_1 = \frac{Z_1\beta}{Z_1\alpha - Z_2}, \tag{9.71}$$

and we noted that $\sqrt{\alpha^2+\beta^2}\cos\phi = \alpha$. The reflection coefficient at the surface $\mathbf{k}_0\cdot\mathbf{x} = 0$ is

$$R = \left|\frac{\left\langle \mathbf{F}_1\mathbf{F}_1^\dagger\right\rangle_V}{\left\langle \mathbf{F}_0\mathbf{F}_0^\dagger\right\rangle_V}\right| = \frac{\left[\cos\omega t - \frac{Z_1}{Z_2}\sqrt{\alpha^2+\beta^2}\cos(\omega t - \phi)\right]^2}{\left[\cos\omega t + \frac{Z_1}{Z_2}\sqrt{\alpha^2+\beta^2}\cos(\omega t - \phi)\right]^2}, \tag{9.72}$$

which, when the incident and reflected intensities are independently averaged over rapid oscillations, gives

$$R_{av} = \frac{\left(1 - \frac{Z_1}{Z_2}\alpha\right)^2 + \left(\frac{Z_1}{Z_2}\beta\right)^2}{\left(1 + \frac{Z_1}{Z_2}\alpha\right)^2 + \left(\frac{Z_1}{Z_2}\beta\right)^2}, \tag{9.73}$$

Good conductors have $\alpha \approx \beta \gg 1,\ \cos\phi \approx 1/\sqrt{2}$, and the electric fields of the incident and reflected waves largely cancel one another while their magnetic fields add constructively and the reflection coefficient approaches unity:

$$R_{av} = 1 - 2\frac{Z_2}{Z_1}\alpha^{-1}. \tag{9.74}$$

The fields then approach those for the surface-current approximation discussed in the previous section.

The total wave in medium 1 is a standing wave with electric and magnetic components out of phase in both position and time by a quarter cycle:

$$\mathbf{F}_0 + \mathbf{F}_1 = \left[\cos\xi + \hat{\mathbf{k}}_0 \frac{Z_1}{Z_2}\sqrt{\alpha^2+\beta^2}\cos(\xi - \phi)\right] \mathbf{E}_2(0)\ , \tag{9.75}$$

where

$$\xi = \omega t - \hat{\mathbf{k}}_0 \mathbf{k}_0 \cdot \mathbf{x} \tag{9.76}$$

(see problem 6). Note that at the surface, $\xi = \omega t$ and the result (9.75) reduces to (9.69), (9.70), and (9.60). We have run into this dependence on $\omega t - \hat{\mathbf{k}}_0 \mathbf{k}_0 \cdot \mathbf{x}$ before, in section 6.6, when we looked at standing waves. In fact, $\omega t - \hat{\mathbf{k}}_0 \mathbf{k}_0 \cdot \mathbf{x}$ is invariant under the reversal of the direction of propagation $\hat{\mathbf{k}}_0$, as is required of standing waves.

Exercise 9.7. Expand $\cos\xi$ and $\cos(\xi - \phi)$ in (9.75) to obtain explicit expressions for the electric and magnetic fields in the standing wave. In particular, show that the electric field $\mathbf{E}_0 + \mathbf{E}_1$ is

$$\left[\cos\omega t \cos\mathbf{k}_0 \cdot \mathbf{x} + \frac{Z_1}{Z_2}\sqrt{\alpha^2+\beta^2}\sin(\omega t - \phi)\sin\mathbf{k}_0 \cdot \mathbf{x}\right] \mathbf{E}_2(0)\ ,$$

whereas the magnetic field $\mathbf{B}_0 + \mathbf{B}_1$ is given by

$$\left[\sin\omega t \sin\mathbf{k}_0 \cdot \mathbf{x} + \frac{Z_1}{Z_2}\sqrt{\alpha^2+\beta^2}\cos(\omega t - \phi)\cos\mathbf{k}_0 \cdot \mathbf{x}\right] \frac{\hat{\mathbf{k}}_0}{c_1} \times \mathbf{E}_2(0)\ .$$

As in the reflection from an ideal conducting screen, one might try to extend the reflection from real conductors to oblique incidence by boosting the normal-incidence case. This seems reasonable for a plane conducting surface, because Maxwell's equation $\bar{\partial}\mathbf{F} = Z\bar{j}$ is Lorentz covariant. However, the approximate relation $j = \sigma\mathbf{E}$ is valid only in the rest frame of the conductor. Instead, we need a more general solution than the damped plane wave found in the previous chapter. It has been shown that the phase of the reflected wave depends on its polarization, and as a result, the ellipticity of the wave can be changed by such oblique reflections. We will not pursue such solutions here.[1]

9.5 Wave Guides

Here we treat waves transmitted inside a hollow conductor, filled with an isotropic, homogeneous, linear dielectric of index of refraction $n = Z_0/Z = c_0/c = \sqrt{\varepsilon/\varepsilon_0}$, where $j = \sigma = 0$ and $\mu = \mu_0$. The conductor, on the other hand has very high conductivity so that all tangential electric and normal magnetic fields can be taken to vanish. With such boundary conditions, it is clear that directed plane waves cannot exist inside: recall that in directed plane waves, the field $\mathbf{F}$ is the same everywhere on spatial planes of infinite extent which move in the propagation direction at the speed of light; if $\mathbf{F}$ vanishes somewhere in such a wave (as it does on the conductor surface) it must vanish everywhere.

We must introduce a wave of more general form than the directed plane wave, one that is still directed but not everywhere the same in the perpendicular plane. Consider a straight wave guide of uniform cross section and take the z direction to lie along its axis. We posit an axially directed solution of the form[2]

$$\mathbf{F}(t + \mathbf{x}) = \mathbf{E} + ic\mathbf{B} = \mathbf{f}(x, y)\, g(s) \ , \tag{9.77}$$

where s is the scalar

$$s = \langle \bar{k}x \rangle_S = \omega t - \mathbf{k} \cdot \mathbf{x} = \omega t - k_z z \, . \tag{9.78}$$

For a monochromatic wave, $g(s)$ is a scalar harmonic function, obeying $g''(s) = -g(s)$, and the wave equation

$$\partial\bar{\partial}\mathbf{F} = 0 \tag{9.79}$$

[1] See, for example, M. Born and E. Wolfe (1975), chapter 13.

[2] For linearly polarized waves, the form (9.77) is too restrictive. We will analyze and cure this problem below.

gives the Helmholtz equation

$$\left(\nabla_{\perp}^{2}+\frac{\omega_{c}^{2}}{c^{2}}\right)\mathbf{f}(x,y)=0\,, \tag{9.80}$$

where

$$\nabla_{\perp}^{2}:=\partial_{x}^{2}+\partial_{y}^{2}=\nabla^{2}-\partial_{z}^{2} \tag{9.81}$$

is the transverse Laplacian operator and

$$\omega_{c}=\sqrt{\omega^{2}-\mathbf{k}^{2}c^{2}} \tag{9.82}$$

is the cut-off (angular) frequency. The scalar function $g(s)$ is also assumed to have the initial value $g(0)=1$. An arbitrary overall phase can be added by shifting the clock (i.e., redefining the point $t=0$). For circularly polarized waves of helicity κ, $g(s)=\exp(-i\kappa s)$, whereas for linearly polarized waves, $g(s)=\cos(s)$.

The free-space waves may be thought of as bouncing from side to side of the wave guide as they travel down its length, and the wave fronts make an oblique angle to the net direction $\hat{\mathbf{k}}$ of propagation. Furthermore, the (net) propagation vector $\mathbf{k}$ has a length that is less than ω/c. As a result, the *phase velocity* $\omega\mathbf{k}^{-1}$ has a magnitude greater than the speed of light c in the medium:

$$c_{\phi}=\omega/\left|\mathbf{k}\right|=c/\sqrt{1-\left(\omega_{c}/\omega\right)^{2}}>c. \tag{9.83}$$

The phase velocity gives the velocity of the wave fronts in the z direction as the obliquely moving plane wave moves down the wave guide. The wave fronts move at the speed of light *in the direction normal to the fronts*; in any other direction, the speed is greater, but then one sees different parts of successive fronts, and there is no possibility of relaying energy or information at such speeds. A phase velocity greater than the speed of light is no more physically significant than the superluminal velocities of a spot illuminated by a laser beam rapidly swept across the face of the moon by an earth-bound experimenter.

Information or energy in signals travel at the *group velocity* $\mathbf{c}_g=\nabla_{\mathbf{k}}\omega=(d\mathbf{k}/d\omega)^{-1}$ where

$$\nabla_{\mathbf{k}}=\mathbf{e}_{1}\frac{\partial}{\partial k_{x}}+\mathbf{e}_{2}\frac{\partial}{\partial k_{y}}+\mathbf{e}_{3}\frac{\partial}{\partial k_{z}}\,. \tag{9.84}$$

Since in wave guides, $\omega^2=c^2\mathbf{k}^2+\omega_c^2$,

$$2\omega\nabla_{\mathbf{k}}\omega=2c^{2}\mathbf{k}, \tag{9.85}$$

and the group velocity is just the inverse of the phase velocity and has a magnitude

$$\mathbf{c}_g = \nabla_{\mathbf{k}}\omega = c^2\mathbf{k}/\omega = c^2 c_\phi^{-1} < c\,. \tag{9.86}$$

Thus, signals cannot travel faster than the speed of light.

Maxwell's equation $\bar{\partial}\mathbf{F} = 0$ gives

$$\frac{g'}{g}\left(\frac{\omega}{c} - \mathbf{k}\right)\mathbf{f}(x,y) = -\nabla\mathbf{f}(x,y)\ , \tag{9.87}$$

where $\hat{\mathbf{k}} := \mathbf{e}_3$. The field at $t = z = 0$, namely $\mathbf{f}(x,y)$, is a complex vector and may in general have components along all three spatial axes. Consider just the transverse part

$$\mathbf{f}_\perp = \mathbf{f} - \mathbf{f}_\parallel := f_x\mathbf{e}_1 + f_y\mathbf{e}_2\,. \tag{9.88}$$

The gradient $\nabla\mathbf{f}_\perp(x,y)$ has only a scalar part and a vector part parallel to $\mathbf{e}_3$, and both are linear in the longitudinal part $\mathbf{f}_\parallel$:

$$\frac{g'}{g}\left(\frac{\omega}{c} - \mathbf{k}\right)\mathbf{f}_\parallel(x,y) = -\nabla\mathbf{f}_\perp(x,y)\ . \tag{9.89}$$

The difference of the equations (9.87) - (9.89) gives

$$\frac{g'}{g}\left(\frac{\omega}{c} - \mathbf{k}\right)\mathbf{f}_\perp(x,y) = -\nabla\mathbf{f}_\parallel(x,y)\ . \tag{9.90}$$

A wave is said to be *transverse electromagnetic* (TEM) if $\mathbf{f}_\parallel = 0$. However, then from (9.90) after multiplication by $(\omega/c + \mathbf{k})$, $g'\mathbf{f}_\perp = 0$, and thus either $\mathbf{f}_\perp = 0$ or $g' = 0$. If $\mathbf{f}_\perp = 0$,the total field vanishes. If $g' = 0$,the field is static and cannot form a wave for transmitting information. Thus, there are no TEM waves inside a hollow conductor. Either $\mathbf{E}_\parallel \neq 0$ or $\mathbf{B}_\parallel \neq 0$. If $\mathbf{E}_\parallel = 0$, but $\mathbf{B}_\parallel \neq 0$, the wave is said to be a *transverse electric* (TE) wave. If $\mathbf{B}_\parallel = 0$, but $\mathbf{E}_\parallel \neq 0$, it is said to be a *transverse magnetic* (TM) wave.

9.6 TE and TM Modes

To find the fields inside a wave guide, we must solve Maxwell's equation (9.87) subject to the boundary conditions

$$\mathbf{B}\cdot\mathbf{n} = 0 = \mathbf{E}\times\mathbf{n} \tag{9.91}$$

at the conductor surface. The first condition $\mathbf{B} \cdot \mathbf{n} = 0$ means that no magnetic-field lines can leave the wave guide through the walls, and the second condition, $\mathbf{E} \times \mathbf{n} = 0$, implies that the electric-field lines must be perpendicular to the walls when they leave. We will consider rectangular wave guides, for which the TE and TM solutions are linearly polarized and may be viewed as waves moving at a small angle to $\mathbf{e}_3$, bouncing back and forth between opposite walls as they move down the guide. In physical solutions representing linearly polarized fields, the complex phase of the oscillating part must be constant, since otherwise the fields $\mathbf{E} = \langle \mathbf{F} \rangle_{\Re}$ and $ic\mathbf{B} = \langle \mathbf{F} \rangle_{\Im}$ will change direction. Since a constant factor can be transferred from the oscillating term $g(s)$ to the coefficient field $\mathbf{f}(x, y)$, we can take the oscillating factor to be real and normalized to $g(0) = 1$. For a wave that is linearly polarized, monochromatic, and directed, we try a solution in the form (9.77) with $g(s) = \cos(s)$.

We run into an immediate problem with Maxwell's equation (9.87) since the left-hand side depends on s whereas the right-hand side depends only on x, y. Evidently, (9.87) can be true for all s only if both sides are independently zero. But it was the condition $\nabla_{\perp}\mathbf{f} = 0$ that led us to conclude that no TEM solutions exist inside a hollow conductor. Now we must conclude that no solutions of the form (9.77) exist inside a hollow conductor. The equations (9.89) and (9.90) hint at the problem: since the time dependence is given by the harmonic function $g(s)$ with $g''(s) = -g(s)$, the functions $g(s)$ and $g'(s)$,and hence the components $\mathbf{f}_{\parallel}$ and $\mathbf{f}_{\perp}$, are a quarter cycle out of phase. If in place of Eq. (9.77) we try a more general solution of the form[3]

$$\mathbf{F}(t + \mathbf{x}) = \mathbf{f}_{\parallel}(x, y) \cos s + \mathbf{f}_{\perp}(x, y) \sin s\,, \tag{9.92}$$

then equations (9.89) and (9.90) become

$$\left(\frac{\omega}{c} - \mathbf{k}\right) \mathbf{f}_{\parallel}(x, y) = \nabla \mathbf{f}_{\perp}(x, y) \tag{9.93}$$

and

$$-\left(\frac{\omega}{c} - \mathbf{k}\right) \mathbf{f}_{\perp}(x, y) = \nabla \mathbf{f}_{\parallel}(x, y)\ . \tag{9.94}$$

[3]Another approach is to take $g(s)$ to be the real part of a phasor $\exp(js)$ where $j^2 = -1$. Then different phases of oscillation enter simply as phase factors $\exp(j\phi)$. However, it is then important to observe that $j \neq i$. The j is introduced only to simplify the mathematics of the harmonic function $g(s)$ which is always meant to be the real part with respect to j. Its function is quite distinct from that of the geometric i, and the real part with respect to j is distinct from the real part with respect to i.

Since $\nabla \mathbf{f}_{\|} = \nabla \hat{\mathbf{k}} \left(\hat{\mathbf{k}} \cdot \mathbf{f} \right) = -\hat{\mathbf{k}} \nabla \left(\mathbf{f} \cdot \hat{\mathbf{k}} \right)$, (9.94) can also be written

$$\left(\frac{\omega}{c} - \mathbf{k} \right) \mathbf{f}_{\perp} = \hat{\mathbf{k}} \nabla \left(\mathbf{f} \cdot \hat{\mathbf{k}} \right) , \tag{9.95}$$

which is inverted with the help of Eq. (9.82) to give the transverse field amplitude $\mathbf{f}_{\perp}$ in terms of the longitudinal field $\mathbf{f} \cdot \hat{\mathbf{k}} = E_z^0 + icB_z^0$ at $s = 0$:

$$\mathbf{f}_{\perp} = c\omega_c^{-2} \left(\omega + \mathbf{k}c \right) \hat{\mathbf{k}} \nabla \left(\mathbf{f} \cdot \hat{\mathbf{k}} \right) . \tag{9.96}$$

In a TE mode, $E_z^0 = 0$, $\mathbf{f}_{\|} = ic\hat{\mathbf{k}} B_z^0$, and in fact the entire field (9.92) is determined by $B_z^0(x, y)$. From the imaginary part of Eq. (9.96), then

$$\mathbf{B}_{\perp} = \frac{|\mathbf{k}| \, c^2}{\omega_c^2} \nabla B_z^0 , \tag{9.97}$$

and the boundary condition on $\mathbf{B}$, namely $\mathbf{n} \cdot \mathbf{B} = 0$ is equivalent to one on the normal derivative:

$$\mathbf{n} \cdot \mathbf{B} = 0 = \mathbf{n} \cdot \nabla B_z^0 . \tag{9.98}$$

For a TE wave, (9.93) and (9.94) together require B_z^0 to satisfy the Helmholtz wave equation (9.80)

$$\left(\nabla_{\perp}^2 + \frac{\omega_c^2}{c^2} \right) B_z^0(x, y) = 0 \tag{9.99}$$

with the boundary conditions $\mathbf{n} \cdot \nabla B_z^0 = 0$.

Exercise 9.8. Prove (9.99) by multiplying (9.93) by $(\omega/c + \mathbf{k})$ and applying (9.94) to the right-hand side.

In a rectangular wave guide with conducting $i\mathbf{e}_1$ planes at $x = 0$ and $x = a$ and conducting $i\mathbf{e}_2$ planes at $y = 0$ and $y = b$, the separation of variables solution can be expressed

$$B_z^0 = B_0 \cos\left(\frac{m\pi x}{a} \right) \cos\left(\frac{n\pi y}{b} \right) , \tag{9.100}$$

where $m, n = 0, 1, 2, ...$ give the number of nodal lines cutting the horizontal (x) and vertical (y) axes, respectively, and from (9.99)

$$\frac{\omega_c^2}{c^2} := \frac{\omega^2}{c^2} - \mathbf{k}^2 = \pi^2 \left(\frac{m^2}{a^2} + \frac{n^2}{b^2} \right) . \tag{9.101}$$

The field for the TE$_{mn}$ mode is given by (9.92) with $\mathbf{f}_{\parallel} = icB_z^0\mathbf{e}_3$ and $\mathbf{f}_{\perp}$ is given by (9.96). The electric and magnetic fields are related by

$$\omega \mathbf{B}_{\perp} = -i\mathbf{k}\mathbf{E} = \mathbf{k} \times \mathbf{E}\,, \tag{9.102}$$

and from the real scalar part of (9.93), $\nabla \cdot \mathbf{E} = 0$. The TE$_{00}$ mode would have $\mathbf{f}_{\parallel} = icB_0\mathbf{e}_3$, which would mean that all the magnetic field lines cross a given $z =$ constant cross section in the same direction. However, since magnetic lines are continuous and, according to the boundary conditions (9.91) cannot leave through the walls of the wave guide, the same flux must exist at every point z. That would imply a wave of infinite length and hence $\mathbf{k} = 0 = \omega$. In other words, the TE$_{00}$ mode is possible only as a static solution. It does not exist as a wave in hollow conductors.

Similarly, in the TM mode, $\mathbf{f}_{\parallel} = \hat{\mathbf{k}}E_z^0$, $\nabla \cdot \mathbf{B} = 0$, and (9.96) gives the transverse field $\mathbf{f}_{\perp}$ in terms of E_z^0. In particular,

$$\mathbf{E}_{\perp} = \frac{|\mathbf{k}|\, c^2}{\omega_c^2} \nabla E_z^0\,, \tag{9.103}$$

and the transverse electric and magnetic fields are related by

$$\mathbf{E}_{\perp} = -\frac{c^2}{\omega}\mathbf{k} \times \mathbf{B}\,. \tag{9.104}$$

Evidently from (9.96), the tangential component $\left(\mathbf{n} \times \hat{\mathbf{k}}\right) \cdot \mathbf{E}$ of $\mathbf{E}_{\perp}$ vanishes only if $\left(\mathbf{n} \times \hat{\mathbf{k}}\right) \cdot \nabla E_z^0$ does. The orthogonality of $\mathbf{E}_{\perp}, \mathbf{B}_{\perp}$ for both TE and TM modes (9.104) and (9.102) ensures that the tangential component of $\mathbf{E}_{\perp}$ vanishes if and only if the normal component of $\mathbf{B}_{\perp}$ does.

The lowest-frequency (dominant) oscillating TE mode is therefore the TE$_{01}$ mode, where we assume $a < b$ (see Fig. 9.5). The longitudinal part of the field is

$$\mathbf{f}_{\parallel} = i\hat{\mathbf{k}}cB_0 \cos\frac{\pi y}{b}\,, \tag{9.105}$$

and the angular *cutoff frequency* [i.e., the frequency when the propagation vector is vanishingly small, see (9.101)] is

$$\omega_c = \sqrt{\omega^2 - c^2\mathbf{k}^2} = c\frac{\pi}{b}\,. \tag{9.106}$$

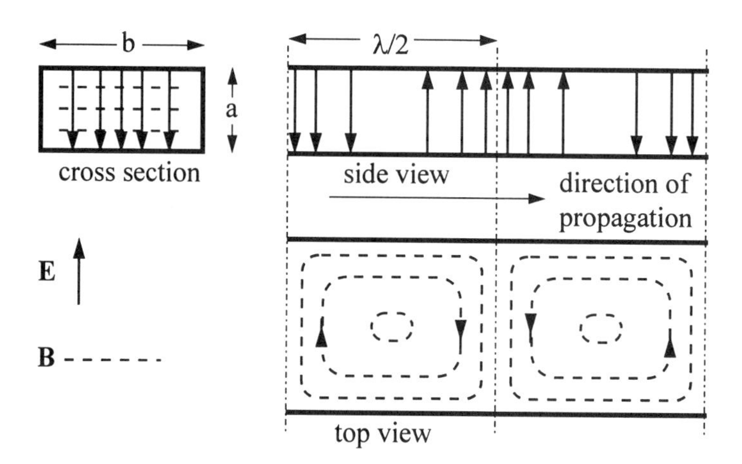

Fig. 9.5. The TE_{01} mode in a rectangular wave guide.

The transverse fields are found from (9.96):

$$\mathbf{f}_{\perp} = -\left(\frac{\omega}{c} + \mathbf{k}\right) \mathbf{e}_1 \frac{b}{\pi} c B_0 \sin \frac{\pi y}{b} . \tag{9.107}$$

The energy flow in the wave guide is given by the Poynting vector (see Chapt. 6)

$$\mathbf{S} = \frac{\varepsilon c}{2} \left\langle \mathbf{F}\mathbf{F}^{\dagger} \right\rangle_V , \tag{9.108}$$

where

$$\mathbf{F} = \mathbf{f}_{\parallel} \cos s + \mathbf{f}_{\perp} \sin s \tag{9.109}$$

is the total field. The result is a rapidly oscillating term in the y direction proportional to $\left\langle \mathbf{f}_{\parallel}\mathbf{f}_{\perp}^{\dagger} \right\rangle_{\Re V} \cos s \sin s$, which represents the wave bouncing from side to side in the wave guide, plus a relatively large term that varies as $\left\langle \mathbf{f}_{\perp}\mathbf{f}_{\perp}^{\dagger} \right\rangle_V \sin^2 s$ and represents the net motion of the wave travelling down the conducting enclosure. Only the second term survives a time average:

$$\mathbf{S}_{av} = \frac{c\varepsilon}{4} \left\langle \mathbf{f}_{\perp}\mathbf{f}_{\perp}^{\dagger} \right\rangle_V ,$$

which for the TE_{01} mode is

$$\mathbf{S}_{av} = \left(\frac{cB_0 b}{\pi}\right)^2 \frac{\varepsilon\omega\mathbf{k}}{2}\sin^2\frac{\pi y}{b}. \tag{9.110}$$

It is instructive to note that the waveguide modes can also be derived by boosting standing waves down the guide (see problems 8 and 9, below).

9.7 Problems

1. Find the reflection coefficient for normal incidence of light on an air/glass interface, where the indices of refraction are $n_{air} = 1,\ n_{glass} = 1.5$. Suppose a clear material is found with an index of refraction equal to 1.25, and that a sheet of this material is placed between the glass and the air. Find the total fraction of light reflected from both interfaces (you may ignore light that is reflected at more than once). This is the principle of optical impedance matching.

2. Fill in the steps in the derivation of the Fresnel equations (9.31), (9.32), (9.38), and (9.39) from the matching conditions (9.21).

3. Use *Fermat's principle*, which states that the light follows the path of least time, to derive Snell's law of refraction at a plane interface of two media.

4. Show that in the nonmagnetic case, where $\mu_j = \mu_0$ with $j = 1, 2$, the Fresnel equations can be written

$$E_1 = E_0\frac{\tan(\theta_1 - \theta_2)}{\tan(\theta_1 + \theta_2)} \tag{9.111}$$

$$E_2 = E_0\frac{2\cos\theta_1\sin\theta_2}{\sin(\theta_1 + \theta_2)\cos(\theta_1 - \theta_2)} \tag{9.112}$$

for the case $\mathbf{B}_j\!\cdot\!\mathbf{n} = 0$, and

$$E_1 = E_0\frac{\sin(\theta_2 - \theta_1)}{\sin(\theta_1 + \theta_2)} \tag{9.113}$$

$$E_2 = E_0\frac{2\cos\theta_1\sin\theta_2}{\sin(\theta_1 + \theta_2)} \tag{9.114}$$

for the case $\mathbf{E}_j\!\cdot\!\mathbf{n} = 0$.

5. Consider the explicit evanescent wave (9.16) with $\mathbf{n} = \mathbf{e}_3$ and $\mathbf{k}_{\Re} = |\mathbf{k}_{\Re}|\, \mathbf{e}_1$ for the two cases of real and imaginary $\mathbf{f}_0 = f_0\mathbf{e}_1$. Find the separate electric and magnetic fields. What polarizations of the incident field would excite these two waves under conditions of total internal reflection?

6. Derive equation (9.75) for the standing wave from the expressions (9.69) and (9.70) of the separate incident and reflected waves. [Hint: Use (9.69), (9.70), and the properties of projectors to prove that the incident wave is

$$\mathbf{F}_0 = P_0 \left[\cos\xi + \hat{\mathbf{k}}\frac{Z_1}{Z_2}\sqrt{\alpha^2 + \beta^2}\cos(\xi - \phi)\right] \mathbf{E}_2(0) \qquad (9.115)$$

and the reflected wave is

$$\mathbf{F}_1 = \bar{P}_0 \left[\cos\xi + \hat{\mathbf{k}}\frac{Z_1}{Z_2}\sqrt{\alpha^2 + \beta^2}\cos(\xi - \phi)\right] \mathbf{E}_2(0)\,. \qquad (9.116)$$

Then add to obtain Eq.(9.75).]

7. Refer to chapter 8 to find the total electromagnetic field in a good conductor with a normally incident wave, as a function of depth below the surface, and use this field to find the current $j = \sigma\mathbf{E}$. Integrate j over the depth and show that the result agrees with the surface current for the same wave normally incident on an ideal conducting screen.

8. Consider a standing-wave solution of lowest frequency between two parallel conductors located in the $i\mathbf{e}_2$ planes at $y = 0$ and $y = b$.

   (a) Show that a solution can be written

   $$\mathbf{F} = icB_0\mathbf{e}_3\cos(\omega_0 t_s - \mathbf{e}_2\omega_0 y_s/c)\ ,$$

   where t_s, x_s, y_s, z_s are the position variables in the frame of the standing wave.

   (b) Expand into electric plus magnetic parts.

   (c) Project out components directed along $\pm\mathbf{e}_2$.

   (d) Show that the standing wave also satisfies the appropriate boundary conditions for any conductor in an $i\mathbf{e}_1 = \mathbf{e}_2\mathbf{e}_3$ plane.

   (e) Boost the solution to a velocity $v\mathbf{e}_3$ and express the resulting field in terms of position variables in the lab frame. Show that the results give a TE_{01} wave in the guide.

(f) Verify that the longitudinal field is pure magnetic and is 90 degrees out of phase with respect to the transverse fields; that the group velocity $\mathbf{k}/\omega$ is $v\mathbf{e}_3$; that the cut-off frequency is $\omega_c = \omega_0$; and that the dispersion relation $\omega^2 = \omega_c^2 + c^2\mathbf{k}^2$ follows from the relation $\gamma^2 = 1 + \gamma^2 v^2$.

(g) Extend the above arguments to higher frequencies to produce TE_{0n} wave-guide modes.

9. What standing waves are needed to produce TE_{mn} modes with $m > 0$ by a boost transformation? TM_{mn} modes?

10. Show that the lowest-frequency TM wave in a rectangular wave guide is the TM_{11} mode. Find the cut-off wave number $\sqrt{\omega^2/c^2 - \mathbf{k}^2}$ for this mode.

11. For the TM_{11} wave in a rectangular wave guide, find

    (a) the total field, and

    (b) the average energy flow.

12. Consider a rectangular wave guide of dimensions $a = 6$ cm and $b = 3$ cm.

    (a) Find the lowest (linear) frequency (the linear cutoff frequency) of electromagnetic wave that can be propagated in the wave guide in the TE_{20} mode.

    (b) If a TE_{20} wave of twice the cutoff frequency is propagated along the guide, what is its phase velocity? Explain how this can be greater than the speed of light in a vacuum.

    (c) Find the total electromagnetic field $\mathbf{F}(x)$ in the guide for a TE_{20} wave. Sketch the fields.

    (d) Find the time-dependent intensity of the TE_{20} radiation in the guide. Interpret the rapidly fluctuating terms in terms of the wave motion.

Chapter 10

The Field of a Moving Charge

The Coulomb potential of a static charge is a well-known solution of Maxwell's equation for the paravector potential A,

$$\Box A = \mu_0 j \,, \tag{10.1}$$

where we assume the Lorenz gauge condition $\langle \bar{\partial} A \rangle_S = 0$. The static field can be Lorentz-transformed to provide the field of a charge moving at constant velocity. The spacetime planes of the electromagnetic field of a static charge are flat electric planes that contain the time axis. After a boost, the spacetime planes are tilted and pick up a spatial (magnetic) component, but they are still flat. The electric-field lines are given by a *time slice* of $\mathbf{F} = c\partial \bar{A}$, that is by the intersection of $\mathbf{F}$ with the spatial "hyperplane" of the observer at a given instant, and they all pass through the world line of the charge.

However, when the charge is accelerated, its Lorentz-transformed Coulomb field at any instant changes its tilt as the distance from the charge is increased. The reason for the change is *causality*: the field propagates at the speed of light and at greater distances from the world line, the tilt of the spacetime field planes arises from the charge at an earlier position and hence different velocity. As a consequence, the spacetime planes of the electromagnetic field of an accelerating charge are curved. It is the deviation of the field from flatness that constitutes the *radiation field*.

Thus, it is the fact that fields carry information about the motion of their sources "only" at a finite speed that is important in generating and determining the pattern of the radiation field. Any fields observed are the result of the sources at an earlier "retarded" time. The tools of geometric algebra are useful in understanding these geometrical spacetime relationships.

10.1 Liénard-Wiechert Potentials

The vector potential of a static point charge e, and thus of a static current density $j = ec\delta^{(3)}(\mathbf{x}-\mathbf{r})$ that is concentrated at the point $\mathbf{x}=\mathbf{r}$,[1] is

$$A_{rest}(x) = K\frac{e}{\langle R_{rest}\rangle_S}, \tag{10.2}$$

where it is convenient to use the constant K, which in vacuum has the value

$$K_0 = \frac{Z_0}{4\pi} = \frac{\mu_0 c}{4\pi} = \frac{1}{4\pi\varepsilon_0 c} = \dot{3}0 \text{ Ohm}, \tag{10.3}$$

with Z_0 the characteristic impedance of the vacuum, and where

$$R = R^0 + \mathbf{R} = x - r(\tau) \tag{10.4}$$

is the spacetime separation between the field point x and the charge at $r(\tau)$ at an earlier (*retarded*) proper time τ, and R_{rest} is the value of R in the rest frame of the charge. The scalar part, $\langle R_{rest}\rangle_S$, is just the distance between the charge and the field point x.

The retarded time τ is fixed by the *light-cone condition*

$$R\bar{R} = 0, \tag{10.5}$$

which means that R is proportional to a projector. In our case

$$R = R^0\left(1+\hat{\mathbf{R}}\right) = 2R^0 P_{\hat{\mathbf{R}}} \tag{10.6}$$

with the projector $P_{\hat{\mathbf{R}}} = \frac{1}{2}\left(1+\hat{\mathbf{R}}\right)$. Since $\langle R_{rest}\rangle_S$ is a scalar in a well-defined frame, it is a Lorentz scalar that can be written covariantly as $\langle R\bar{u}\rangle_S$. In terms of the proper velocity $u = \gamma(1+\mathbf{v}/c)$ of the charge, which is simply 1 when the charge is at rest,

$$\langle R_{rest}\rangle_S = \langle R\bar{u}\rangle_S = \gamma R^0\left(1-\hat{\mathbf{R}}\cdot\mathbf{v}/c\right). \tag{10.7}$$

By boosting (10.2), we obtain the paravector potential for a charge e moving at constant velocity:

$$A(x) = \Lambda A_{rest}\Lambda^\dagger = K\frac{eu}{\langle R\bar{u}\rangle_S}, \tag{10.8}$$

[1] The Dirac delta function in three dimensions, $\delta^{(3)}(\mathbf{x}-\mathbf{r})$, is discussed below in section 10.5.

where Λ is the eigenspinor of the charge and $u = \Lambda\Lambda^\dagger$. It is easy to see that this potential satisfies the Lorenz gauge condition, since in the rest frame the potential is a static scalar. Thus $\langle\partial\bar{A}\rangle_S = 0 = (\partial_t\phi_{rest})$, and because the Lorenz condition is Lorentz invariant, it holds in all inertial frames.

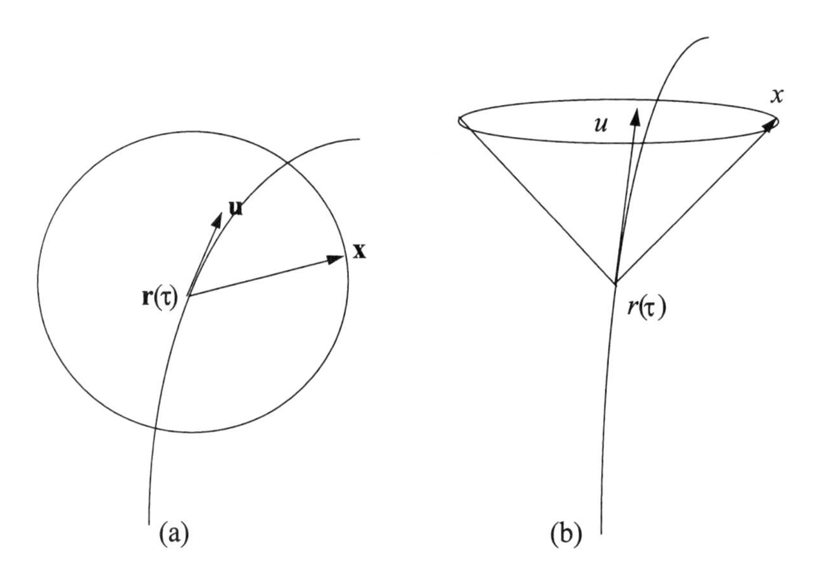

Fig. 10.1. The path $r(\tau)$ of a charge, with tangent spacetime vector *u* and observation point *x* as seen (a) in space and (b) in a spacetime diagram.

An accelerating charge moves on a world line $r(\tau)$ whose tangent spacetime vector is the time-like proper velocity $u(\tau)$ (see Fig. 10.1). Since the parapotential A propagates away from the particle at the speed of light, only the motion of the charge at the intersection of $r(\tau)$ with the backward light cone ($\langle R\rangle_S > 0$) from x can affect the field. For any field point x, there is only one point of intersection with the world line $r(\tau)$ of the point charge, and therefore the parapotential $A(x)$ depends on the position and velocity at only one proper time $\tau(x)$. The acceleration, on the other hand, can be observed only if the proper velocity $u(\tau)$ can be sampled at more than one point on the world line $r(\tau)$. One therefore expects $A(x)$ to be independent of the instantaneous acceleration of the charge; it should be the same for the accelerating charge as for a charge moving with constant velocity as long as the position and velocity are the same at the retarded time. The vector potential (10.8) therefore is that of an arbitrarily accelerating charge, as long as the right-hand side is evaluated at the retarded time τ, given by the light-cone condition (10.5). It is called the *Liénard-Wiechert potential* of the charge.

10.2 The Electromagnetic Field

The field $\mathbf{F}$ is found from the gradient of A:

$$\mathbf{F}(x) = \mathbf{E}+ic\mathbf{B} = c\partial\bar{A}\,, \tag{10.9}$$

since $\langle\partial\bar{A}\rangle_s = 0$ by the Lorenz gauge condition. However, there is now an important difference from cases we handled previously: when we vary the field position x, the retarded time τ also changes: $\tau = \tau(x)$ is a scalar field. It is this additional dependence on x that is ultimately responsible for radiation. We write explicitly

$$\partial = (\partial)_x + (\partial)_\tau\ , \tag{10.10}$$

where $(\cdots)_x$ indicates that part arising directly from changes in the field position x and $(\cdots)_\tau$ is that part arising indirectly through changes in the retarded time $\tau(x)$ which are correlated to changes in x by the light-cone condition (10.5). We can apply the chain rule for differentiation in the form

$$(\partial)_\tau = [\partial\tau(x)]\,\frac{d}{d\tau}\,, \tag{10.11}$$

and it is understood that $d/d\tau$ is applied to the source of the field at the retarded time.

In order to calculate $\mathbf{F}$ from the paravector potential of a moving charge (10.8), we need the gradients $\partial\tau(x)$ and $\partial\,\langle R\bar{u}\rangle_S$. These are given in the following two theorems.

Theorem 10.1 *The factor $\partial\tau(x)$ is*

$$\partial\tau(x) = \frac{R}{c\,\langle R\bar{u}\rangle_S}\,. \tag{10.12}$$

Proof: From the light-cone condition (10.5):

$$\begin{aligned} 0 &= \partial\left(R\bar{R}\right) = \partial\left(R^\mu R^\nu\right)\mathbf{e}_\mu\bar{\mathbf{e}}_\nu = 2\left(\partial R^\mu\right)R^\nu\eta_{\mu\nu} \\ &= 2\left[\mathbf{e}^\mu - c\left(\partial\tau\right)u^\mu(\tau)\right]R^\nu\eta_{\mu\nu} \qquad (10.13) \\ &= 2\left[R - c\left(\partial\tau\right)\langle R\bar{u}\rangle_S\right]\,, \qquad (10.14) \end{aligned}$$

where we used the metric tensor

$$\eta_{\mu\nu} = \frac{1}{2}\left(\mathbf{e}_\mu \bar{\mathbf{e}}_\nu + \mathbf{e}_\nu \bar{\mathbf{e}}_\mu\right) = \begin{cases} 1, & \mu = \nu = 0 \\ -1, & \mu = \nu = 1,2,3 \\ 0, & \text{otherwise} \end{cases} \tag{10.15}$$

and noted $u^\mu R^\nu \eta_{\mu\nu} = \left\langle u\bar{R}\right\rangle_S = \left\langle R\bar{u}\right\rangle_S$. ■

It follows that (10.10) can be written

$$\partial = (\partial)_x + \frac{R}{\left\langle R\bar{u}\right\rangle_S} \frac{d}{cd\tau} . \tag{10.16}$$

For example,

$$\partial R^\mu = \mathbf{e}^\mu - \frac{R}{\left\langle R\bar{u}\right\rangle_S} u^\mu(\tau) , \tag{10.17}$$

where the position $r(\tau)$ and proper velocity $u(\tau) = dr/cd\tau$ of the charge are evaluated at the retarded proper time τ.

Theorem 10.2 *The gradient of the Lorentz scalar* $\left\langle R\bar{u}\right\rangle_S$ *is*

$$\partial \left\langle R\bar{u}\right\rangle_S = u + \frac{R}{\left\langle R\bar{u}\right\rangle_S} \left\langle \frac{R\bar{\dot{u}}}{c} - 1 \right\rangle_S . \tag{10.18}$$

Proof: From (10.10) and (10.12), we find

$$\begin{aligned} \partial \left\langle R\bar{u}\right\rangle_S &= \partial \left\langle (x - r)\,\bar{u}\right\rangle_S \\ &= (\partial)_x \left\langle x\bar{u}\right\rangle_S + (\partial\tau) \left\langle \frac{d}{d\tau}\left(R\bar{u}\right)\right\rangle_S \\ &= u + \frac{R}{\left\langle R\bar{u}\right\rangle_S} \left\langle -u\bar{u} + \frac{R\bar{\dot{u}}}{c} \right\rangle_S \end{aligned} \tag{10.19}$$

where we used

$$(\partial)_x \left\langle x\bar{u}\right\rangle_S = (\partial)_x \left\langle u\bar{x}\right\rangle_S = u . \tag{10.20}$$

Since $u\bar{u} = 1$, the theorem follows. ■

The field of an accelerating charge is thus [see equations (10.8), (10.9)]

$$\begin{aligned} \mathbf{F}(x) &= Ke\left[R\bar{\dot{u}} - c\partial\langle R\bar{u}\rangle_S \bar{u}\right] / \langle R\bar{u}\rangle_S^2 \\ &= Ke\left[\left(R\bar{\dot{u}} - c\right)\langle R\bar{u}\rangle_S - \left\langle R\bar{\dot{u}} - c\right\rangle_S R\bar{u}\right] / \langle R\bar{u}\rangle_S^3 . \end{aligned} \tag{10.21}$$

By expanding the scalar parts in the numerator, for example

$$\langle R\bar{u}\rangle_S = \frac{1}{2}\left(R\bar{u} + u\bar{R}\right), \tag{10.22}$$

and noting the light-cone condition $R\bar{R} = 0$, the field can be put in the form

$$\mathbf{F}(x) = \frac{Ke}{\langle R\bar{u}\rangle_S^3}\left(c\langle R\bar{u}\rangle_V + \frac{1}{2}R\bar{\dot{u}}u\bar{R}\right). \tag{10.23}$$

Exercise 10.1. Verify the derivation of (10.23) from (10.21).

It is worthwhile to analyze equation (10.23) in some detail. Note that $\mathbf{F}(x)$ depends on r, u, and $\dot{u}$ of the charge at a single retarded spacetime event. We will see that the two terms are spacetime bivectors that represent the transformed Coulomb field $\mathbf{F}_c$ and the radiation field $\mathbf{F}_r$:

$$\mathbf{F} = \mathbf{F}_c + \mathbf{F}_r \tag{10.24}$$

with

$$\mathbf{F}_c = \frac{cKe}{\langle R\bar{u}\rangle_S^3}\langle R\bar{u}\rangle_V, \ \mathbf{F}_r = \frac{Ke}{2\langle R\bar{u}\rangle_S^3}R\bar{\dot{u}}u\bar{R}. \tag{10.25}$$

We note immediately that $\mathbf{F}_c$ is independent of the acceleration and falls off at large distances like the square of the distance, whereas $\mathbf{F}_c$ is linear in the proper acceleration and falls off as the inverse distance.

Exercise 10.2. Verify that $\mathbf{F}_r^2 = 0$ and $\mathbf{F}_r \cdot \mathbf{F}_c = 0$, and consequently that the Lorentz-invariant scalar $\mathbf{F}^2$ is given entirely by the Coulomb contribution: $\mathbf{F}^2 = \mathbf{F}_c^2$.

10.2.1 Covariant Coulomb Field

In the rest frame of the charge at the retarded proper time τ, $u = 1$, so that the Coulomb part there becomes

$$(\mathbf{F}_c)_{rest} = \frac{cKe\mathbf{R}_{rest}}{\left(R_{rest}^0\right)^3}, \tag{10.26}$$

where $R_{rest} = R^0_{rest} + \mathbf{R}_{rest} = \bar{\Lambda} R \bar{\Lambda}^\dagger$. Since

$$\Lambda \mathbf{R}_{rest} \bar{\Lambda} = \frac{1}{2} \Lambda \left(R_{rest} - \bar{R}_{rest} \right) \bar{\Lambda} = \frac{1}{2} \left(R\bar{u} - u\bar{R} \right) = \langle R\bar{u} \rangle_V \tag{10.27}$$

and remembering that $R^0_{rest} = \langle R\bar{u} \rangle_S$ is a Lorentz invariant, we see that $(\mathbf{F}_c)_{rest}$, when transformed to the lab frame, does indeed give $\mathbf{F}_c$:

$$\Lambda \left(\mathbf{F}_c \right)_{rest} \bar{\Lambda} = \mathbf{F}_c = \frac{cKe}{\langle R\bar{u} \rangle_S^3} \langle R\bar{u} \rangle_V \,. \tag{10.28}$$

The Lorentz-invariant square of $\mathbf{F}_c$ is the square of the radial Coulomb field in the rest frame:

$$\mathbf{F}_c^2 = \left[\frac{cKe}{\langle R\bar{u} \rangle_S^2} \right]^2 \tag{10.29}$$

as may be verified by noting that because of the light-cone condition $R\bar{R} = 0$,

$$\langle R\bar{u} \rangle_V^2 = \frac{1}{4} \left(R\bar{u}R\bar{u} + u\bar{R}u\bar{R} \right) = \langle R\bar{u} \rangle_S^2 \,, \tag{10.30}$$

so that in the expression (10.28) for $\mathbf{F}_c$, the ratio $\langle R\bar{u} \rangle_V / \langle R\bar{u} \rangle_S$ is a timelike unit biparavector. It represents a spacetime plane spanned by the lightlike R and the timelike u, both evaluated at the retarded position of the charge.

10.2.2 Lab-Frame Expression of $\mathbf{F}_c$

Any observer will naturally split the field $\mathbf{F}_c$ into electric and magnetic parts. To obtain explicit expressions of the electric and magnetic fields in the lab frame, we need

$$R\bar{u} = R^0 \left(1 + \hat{\mathbf{R}} \right) \gamma \left(1 - \mathbf{v}/c \right) \tag{10.31}$$

with a complex vector part

$$\langle R\bar{u} \rangle_V = \gamma R^0 \left(\hat{\mathbf{R}} - \mathbf{v}/c - i\hat{\mathbf{R}} \times \mathbf{v}/c \right) \tag{10.32}$$

and a real scalar part

$$\langle R\bar{u} \rangle_S = \gamma R^0 \left(1 - \hat{\mathbf{R}} \cdot \mathbf{v}/c \right) = \hat{\mathbf{R}} \cdot \langle R\bar{u} \rangle_V \,. \tag{10.33}$$

The transformed Coulomb field can thus be written

$$\begin{aligned} \mathbf{F}_c &= \mathbf{E}_c + ic\mathbf{B}_c \\ &= \left\langle \left(1+\hat{\mathbf{R}}\right)\mathbf{E}_c \right\rangle_V , \end{aligned} \tag{10.34}$$

where

$$\mathbf{E}_c = \frac{Ke\left(\hat{\mathbf{R}}c - \mathbf{v}\right)}{(\gamma R^0)^2\left(1 - \hat{\mathbf{R}}\cdot\mathbf{v}/c\right)^3},\ c\mathbf{B}_c = \hat{\mathbf{R}}\times\mathbf{E}_c\,. \tag{10.35}$$

Exercise 10.3. As a generalization of the above split of $\langle R\bar{u}\rangle_V$ into real and imaginary parts, prove by direct expansion of the RHS that

$$\left\langle p^0 p\bar{q}\right\rangle_V = \langle p\,\langle \bar{q}p\rangle_{\Re V}\rangle_V\ , \tag{10.36}$$

where p and q are any two real paravectors.

Example 10.4. Calculate the electric field lines of a charge moving with constant velocity and show that they are straight and pass through the instantaneous position (*not* some retarded position) of the charge. Find the $\mathbf{E}_c^2 =$ constant surface and sketch the field lines within this surface for the case $v = 0.9\,c$.

Solution: Consider the field lines at the instant $x^0 = 0$, and let the charge e pass through the origin at that instant. We can take the proper time of the charge at this point on its path to be $\tau = 0$. The spacetime position of the charge is then $r = cu\tau$, where $u = \gamma + \mathbf{u}$ is constant, and at the instant $x^0 = 0$,

$$R = x - r = -\gamma c\tau + \mathbf{x} - \mathbf{u}c\tau\,. \tag{10.37}$$

The retarded proper time τ is the $c\tau < x^0$ solution of the light-cone condition

$$R\bar{R} = 0 = (\gamma c\tau)^2 - (\mathbf{x} - \mathbf{u}c\tau)^2 = c^2\tau^2 - \mathbf{x}^2 + 2\mathbf{x}\cdot\mathbf{u}c\tau \tag{10.38}$$

since $\gamma^2 - \mathbf{u}^2 = 1$. Thus

$$c\tau = -\mathbf{x}\cdot\mathbf{u} - \sqrt{(\mathbf{x}\cdot\mathbf{u})^2 + \mathbf{x}^2}\,. \tag{10.39}$$

Noting that

$$\hat{\mathbf{R}} = \frac{\mathbf{x} - \mathbf{u}c\tau}{-\gamma c\tau} = \frac{\mathbf{v}}{c} - \frac{\mathbf{x}}{\gamma c\tau} \tag{10.40}$$

and

$$\begin{aligned} 1-\hat{\mathbf{R}}\cdot\frac{\mathbf{v}}{c} &= 1-\frac{\mathbf{v}^2}{c^2}+\frac{\mathbf{x}\cdot\mathbf{v}}{\gamma c^2\tau}=\tau^{-1}\gamma^{-2}\left(\tau+\frac{\mathbf{x}\cdot\mathbf{u}}{c}\right) \\ &= -(c\tau)^{-1}\gamma^{-2}\sqrt{(\mathbf{x}\cdot\mathbf{u})^2+\mathbf{x}^2}\,, \end{aligned} \tag{10.41}$$

we find from (10.35) the electric field

$$\mathbf{E}_c=\frac{cKe\gamma\mathbf{x}}{\left[(\mathbf{x}\cdot\mathbf{u})^2+\mathbf{x}^2\right]^{3/2}}=\frac{cKe\hat{\mathbf{x}}}{\mathbf{x}^2}\gamma\left[(\hat{\mathbf{x}}\cdot\mathbf{u})^2+1\right]^{-3/2}. \tag{10.42}$$

The constant $\mathbf{E}_c^2=\left(cKe/x_0^2\right)^2$ surface is given by

$$|\mathbf{x}|=x_0\gamma^{1/2}\left[(\hat{\mathbf{x}}\cdot\mathbf{u})^2+1\right]^{-3/4} \tag{10.43}$$

and is invariant under reflection in or perpendicular to $\mathbf{u}$. The field lines are plotted in Fig. 10.2 for $v=0$ and $v=0.9\,c$.

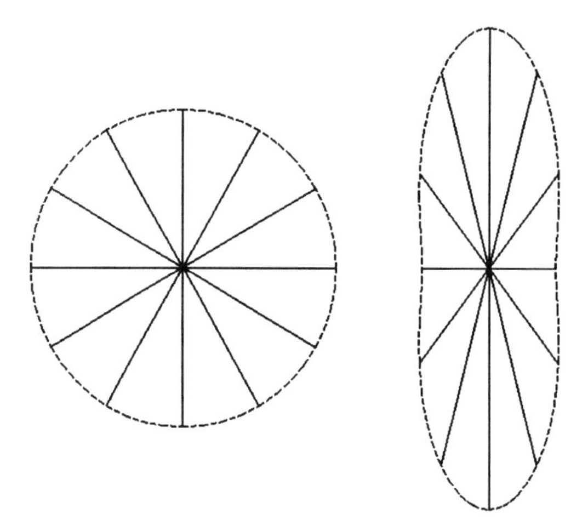

Fig. 10.2. A comparison of the field lines of a charge at rest (on the left) with one moving at speed $v=0.9\,c$ along the horizontal axis.

Note that the field lines pass through the instantaneous position of the charge, namely $\mathbf{x}=0$, and are radial. That is, at the instant the charge crosses the origin, the field line at $\mathbf{x}$ points in the direction of $\mathbf{x}$. The fact that the field lines are straight and pass through the charge is a necessary consequence of these properties for the static charge together with the effect of a *linear* transformation. The field lines are spacelike slices of spacetime bivectors (planes) which contain the path of the charge, and a linear transformation transforms flat planes into flat planes. Spacelike slices of the transformed planes are again straight lines that pass through the charge. (See also the section on Virtual-Photon Streams later in this chapter.)

The electric-field line is generally seen to lie along the direction of the difference vector $\mathbf{R} - \mathbf{v}R^0/c = \mathbf{x} - (\mathbf{r} + \mathbf{v}R^0/c)$, where the position $\mathbf{r}$ and velocity $\mathbf{v}$ of the charge are evaluated at the retarded time $t_{ret} = r^0/c = t - R^0/c$. The direction of $\mathbf{E}_c$ at the observation point x is thus always directly away from the position the charge *would* occupy if it were to continue traveling at constant velocity from its retarded position r. Thus, for a charge with constant velocity, $\mathbf{r} = \mathbf{r}(0) + \mathbf{v}(t - R^0/c)$, the field direction

$$\begin{aligned} \mathbf{R} - \mathbf{v}R^0/c &= \mathbf{x} - \mathbf{r} - \mathbf{v}R^0/c \\ &= \mathbf{x} - \mathbf{r}(0) - \mathbf{v}t \end{aligned} \tag{10.44}$$

is radially away from $\mathbf{r}(0) + \mathbf{v}t$ at all field positions $\mathbf{x}$.

For an accelerating charge, the Coulomb field lines emanate from a position $\mathbf{r} + \mathbf{v}R^0/c$ that depends on the retarded time and hence on the field position $\mathbf{x}$. There are two important consequences: the field lines of the transformed Coulomb field are curved and they are generally *not continuous.* We must add the radiation field to make the lines continuous.

10.2.3 Covariant Radiation Field

The radiation field,

$$\mathbf{F}_r = \frac{Ke}{2\langle R\bar{u}\rangle_S^3} R\bar{\dot{u}}u\bar{R}\,, \tag{10.45}$$

on the other hand, may be seen to be a null element known as a flag (see Chapt. 6) that falls off at large distances as $(R^0)^{-1}$ and is proportional to the acceleration of the charge at the retarded time. From the lightlike property of the flagpole $R = R^0\left(1 + \hat{\mathbf{R}}\right)$, the field must be null

$$\mathbf{F}_r^2 = 0 \tag{10.46}$$

and satisfy the relations

$$\left(1-\hat{\mathbf{R}}\right)\mathbf{F}_r = \mathbf{F}_r\left(1+\hat{\mathbf{R}}\right) = 0\,. \tag{10.47}$$

It follows that $\mathbf{F}_r$ represents a directed wave in the direction $\hat{\mathbf{R}}$, with transverse fields:

$$\left\langle \hat{\mathbf{R}}\mathbf{F}_r \right\rangle_S = \hat{\mathbf{R}}\cdot\mathbf{F}_r = 0\,. \tag{10.48}$$

The unimodularity of the proper velocity u guarantees that the scalar part of $R\bar{\dot{u}}u\bar{R}$ vanishes:

$$\overline{R\bar{\dot{u}}u\bar{R}} = R\bar{u}\dot{u}\bar{R} = -R\bar{\dot{u}}u\bar{R} \tag{10.49}$$

since $\bar{u}u = 1$ implies $\bar{\dot{u}}u + \bar{u}\dot{u} = 0$. (This result is independent of the nature of R.) If the biparavector $\bar{\dot{u}}u$ is expanded in parts parallel and perpendicular to $\hat{\mathbf{R}}$,

$$\bar{\dot{u}}u = \left(\bar{\dot{u}}u\right)_{\parallel} + \left(\bar{\dot{u}}u\right)_{\perp}\,, \tag{10.50}$$

only the perpendicular part survives the R, $\bar{R}$ sandwich:

$$R\bar{\dot{u}}u\bar{R} = R^2\left(\bar{\dot{u}}u\right)_{\perp}\,. \tag{10.51}$$

10.2.4 Lab-Frame Expression of $\mathbf{F}_r$

From the relation (10.47), the radiation field can be written

$$\begin{aligned}
\mathbf{F}_r &= \frac{1}{2}\left(1+\hat{\mathbf{R}}\right)\mathbf{F}_r \\
&= \frac{1}{2}\left(1+\hat{\mathbf{R}}\right)\mathbf{F}_r + \frac{1}{2}\left[\mathbf{F}_r\left(1+\hat{\mathbf{R}}\right)\right]^{\dagger} \\
&= \left(1+\hat{\mathbf{R}}\right)\mathbf{E}_r\,,
\end{aligned} \tag{10.52}$$

where $\mathbf{E}_r = \langle \mathbf{F}_r \rangle_{\Re}$ is the electric field. The relation

$$\mathbf{F}_r = \hat{\mathbf{R}}\mathbf{F}_r\,. \tag{10.53}$$

relates electric and magnetic fields in the familiar right-handed way to the propagation direction:

$$c\mathbf{B}_r = \hat{\mathbf{R}}\times\mathbf{E}_r,\ \mathbf{E}_r = -\hat{\mathbf{R}}\times c\mathbf{B}_r\,. \tag{10.54}$$

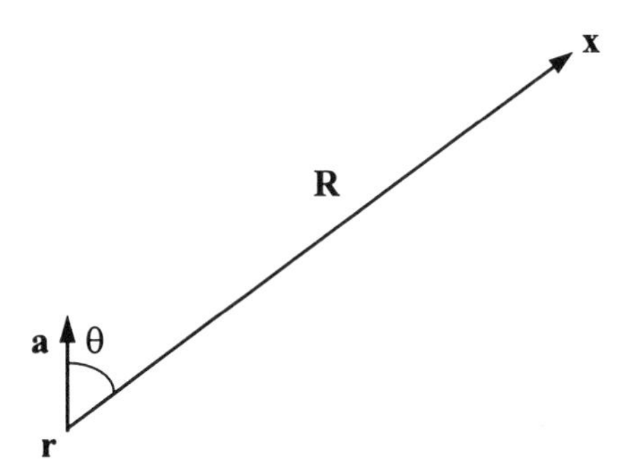

Fig. 10.3. The field radiated from an accelerating charge is $\mathbf{F}_r = \left(1+\hat{\mathbf{R}}\right)\mathbf{E}_r$, where $\mathbf{E}_r = (\gamma/c)^2\,\mathbf{R}\times(\mathbf{E}_c\times\mathbf{a})$, with $\mathbf{E}_c$ the Coulomb field and $\mathbf{a}$ the lab-frame acceleration of the charge at the retarded time.

In the rest frame, $R^2\left(\bar{\dot{u}}u\right)_\perp$ reduces to $-R^2_{rest}\mathbf{a}_\perp/c$. A more general calculation (see problem 1) for a charge with retarded velocity $\mathbf{v}$ gives

$$\begin{aligned} R^2\left(\bar{\dot{u}}u\right)_\perp &= -R^2\gamma^3c^{-1}\left(\mathbf{a}+i\mathbf{a}\times\mathbf{v}/c\right)_\perp \\ &= 2\left(1+\hat{\mathbf{R}}\right)\gamma^3c^{-1}\left(R^0\right)^2\hat{\mathbf{R}}\times\left[\mathbf{a}\times\left(\mathbf{v}/c-\hat{\mathbf{R}}\right)\right], \end{aligned} \tag{10.55}$$

where $\mathbf{a} = d\mathbf{v}/dt = d^2\mathbf{r}/dt^2$ is the (noncovariant) acceleration (see Fig. 10.3). Thus,

$$\mathbf{E}_r = \frac{Ke}{R^0\left(1-\hat{\mathbf{R}}\cdot\mathbf{v}/c\right)^3}\hat{\mathbf{R}}\times\left[\frac{\mathbf{a}}{c}\times\left(\frac{\mathbf{v}}{c}-\hat{\mathbf{R}}\right)\right] = \frac{\gamma^2}{c^2}\mathbf{R}\times\left(\mathbf{E}_c\times\mathbf{a}\right). \tag{10.56}$$

In the low velocity limit, this reduces to

$$\mathbf{E}_r = -\frac{Ke}{R^0}\frac{\mathbf{a}_\perp}{c}, \quad \mathbf{a}_\perp = \hat{\mathbf{R}}\times\left(\mathbf{a}\times\hat{\mathbf{R}}\right), \quad \frac{\mathbf{v}^2}{c^2} \ll 1. \tag{10.57}$$

10.3 Example: Uniformly Accelerated Charge

Let's calculate the field $\mathbf{F}$ of a uniformly accelerated charge. Consider a charge e in "hyperbolic" motion, whose world line is

$$r(\tau) = \mathbf{e}_2\exp(a\tau\mathbf{e}_2/c) = \mathbf{e}_2 e^{c\tau\mathbf{e}_2}, \tag{10.58}$$

where the unit of length is $l_0 = c^2/a = 1$. Differentiating, we have

$$\begin{aligned} r &= \mathbf{e}_2 u \\ u &= \exp(c\tau\mathbf{e}_2) \\ \dot{u} &= cr \\ \bar{\dot{u}}u/c &= -\mathbf{e}_2 \, . \end{aligned}$$

The proper time τ will generally be taken at its retarded value. We note $u\bar{u} = 1 = -r\bar{r}$. The field position is $x = x^\mu \mathbf{e}_\mu = x_0 + \mathbf{x}$ and the relative position $R = x - r$ is lightlike:

$$R\bar{R} = R_0^2 - \mathbf{R}^2 = x\bar{x} - 1 - 2\langle x\bar{r}\rangle_S = 0 \, .$$

Some other useful values, all in units with $c^2/a = 1$:

$$\begin{aligned} R\bar{u} &= x\bar{u} - \mathbf{e}_2 = -(x\bar{r} + 1)\,\mathbf{e}_2 \\ \langle R\bar{u}\rangle_S &= \langle x\bar{u}\rangle_S = \gamma R^0 \left(1 - \mathbf{v}\cdot\hat{\mathbf{R}}/c\right) \\ \langle R\bar{u}\rangle_V &= \langle x\bar{u}\rangle_V - \mathbf{e}_2 \\ c^{-1}R\bar{\dot{u}} &= -R\bar{u}\mathbf{e}_2 = R\bar{r} = x\bar{r} + 1 \\ \left\langle c^{-1}R\bar{\dot{u}} - 1\right\rangle_S &= \langle x\bar{r}\rangle_S \\ \partial\langle R\bar{u}\rangle_S &= \partial\langle x\bar{u}\rangle_S = u + \frac{R\langle x\bar{r}\rangle_S}{\langle x\bar{u}\rangle_S} \, . \end{aligned}$$

The Liénard-Wiechert field is

$$\begin{aligned} \mathbf{F} &= \frac{Kce}{\langle x\bar{u}\rangle_S^3}\left(\langle R\bar{u}\rangle_V + \frac{1}{2c}R\bar{\dot{u}}u\bar{R}\right) \\ &= \frac{Kce}{2\langle x\bar{u}\rangle_S^3}\left[R\bar{u} + \left(c^{-1}R\bar{\dot{u}} - 1\right)u\bar{R}\right] \\ &= \frac{Kce}{2\langle x\bar{u}\rangle_S^3}\left[-(x\bar{r}+1)\,\mathbf{e}_2 + x\bar{r}\mathbf{e}_2\,(r\bar{x}+1)\right] \\ &= -\frac{Kce}{2\langle x\bar{u}\rangle_S^3}\left(\mathbf{e}_2 + x\mathbf{e}_2\bar{x}\right) \, . \end{aligned} \tag{10.59}$$

Exercise 10.5. Evaluate $-(\mathbf{e}_2 + x\mathbf{e}_2\bar{x})$ to determine the direction of the field. In particular, show that at $t = 0$, the total magnetic field vanishes. (The magnetic components of the Coulomb and radiation fields are nonzero but cancel at $t = 0$.)

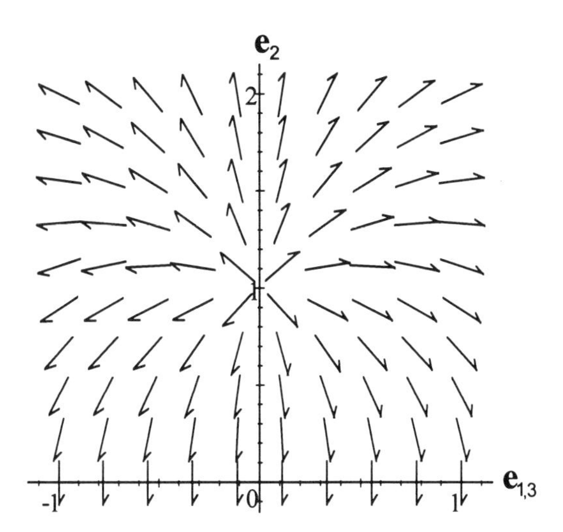

Fig. 10.4. A field plot of the electric field of a uniformly accelerating charge. At the instant of the plot, the charge is reversing its velocity and has come momentarily to rest at $r = \mathbf{e}_2$.

It may be shown that although $\mathbf{F}$ satisfies Maxwell's equation, the Coulomb and radiation fields separately do not. Thus, while field lines can be drawn for the total field (see Fig. 10.4), connected and continuous lines do not exist for the Coulomb and radiation fields separately.

The field lines of $\mathbf{F}$ are seen to droop down toward the $\mathbf{x} \cdot \mathbf{e}_2 = 0$ spatial plane. At $t = 0$, the lines do not cross the $\mathbf{x} \cdot \mathbf{e}_2 = 0$ plane because the region $\mathbf{x} \cdot \mathbf{e}_2 < 0$ lies outside the light cones of the charge formed at all earlier times from points on its world line. *Einstein's principle of equivalence* equates local observations made by a stationary observer in a uniform gravitational field to those by a uniformly accelerating observer in the absence of gravitation. Since accelerating charges radiate, the principle naturally raises the question of whether stationary charges in a gravitational field also radiate. If they do, where does the energy for the radiation come from? If they don't, how can the principle of equivalence be satisfied?

From our calculation, it appears that the fields are the same for both cases, but that the two observers would describe the result differently. For an observer in a uniform gravitational field, the drooping of the lines would be ascribed to the gravitational field, whereas for an inertial observer, the curvature of the lines would evidently be evidence of radiation.

Exercise 10.6. The dimensions in expression (10.59) may seem strange because of the unit of length c^2/a adopted there. Reinsert this unit in order to express $\mathbf{F}$ in SI units.
Solution: If we continue to express x in units of c^2/a, the SI value of $\mathbf{F}$ is

$$\mathbf{F} = -\frac{Kea^2}{2c^3 \langle x\bar{u}\rangle_S^3} \left(\mathbf{e}_2 + x\mathbf{e}_2\bar{x}\right) . \tag{10.60}$$

On the other hand, if x is expressed in SI units, then

$$\mathbf{F} = -\frac{Kce}{2 \langle x\bar{u}\rangle_S^3} \left(\frac{\mathbf{e}_2 c^2}{a} + \frac{x\mathbf{e}_2\bar{x}a}{c^2}\right) . \tag{10.61}$$

Exercise 10.7. Find the limit when $R^0 \ll c^2/a$ and show that as $a \to 0$, $\mathbf{F}$ approaches the Coulomb field.
Solution: We can continue to work in dimensionless units, in which the unit of length $c^2/a = 1$, but the limit now becomes $R^0 \to 0$. Since from (10.58) $r\mathbf{e}_2\bar{r} = -\mathbf{e}_2$,

$$\begin{aligned}
\mathbf{e}_2 + x\mathbf{e}_2\bar{x} &= \mathbf{e}_2 + (R+r)\,\mathbf{e}_2\overline{(R+r)} \\
&= 2\left(R^0\right)^2\left(1+\hat{\mathbf{R}}\right)\left(\mathbf{e}_2 - \mathbf{e}_2\cdot\hat{\mathbf{R}}\,\hat{\mathbf{R}}\right) + u\bar{R} - R\bar{u} \\
&\to -2\left\langle R\bar{u}\right\rangle_V .
\end{aligned} \tag{10.62}$$

Consequently, at low velocities and accelerations

$$\mathbf{F} \to \frac{Kce}{\langle R\bar{u}\rangle_S^3}\left\langle R\bar{u}\right\rangle_V \to \frac{Kce}{\mathbf{R}^2}\hat{\mathbf{R}} . \tag{10.63}$$

10.4 Radiated Power

Since the Coulomb field $\mathbf{F}_c$ falls off as $\langle R\rangle_S^{-2}$ whereas the radiation field decreases as $\langle R\rangle_S^{-1}$, the radiation field dominates at large distances from the charge. The energy-momentum density of the radiation field is

$$\mathcal{E}_r + \mathbf{S}_r/c = \frac{1}{2}\varepsilon_0\mathbf{F}_r\mathbf{F}_r^\dagger = \varepsilon_0\left(1+\hat{\mathbf{R}}\right)\mathbf{E}_r^2 . \tag{10.64}$$

In the low-velocity limit, the radiated intensity reduces to

$$\mathbf{S}_r = \frac{Ke^2}{4\pi\left(R^0\right)^2}\frac{\hat{\mathbf{R}}\mathbf{a}_\perp^2}{c^2} = \frac{Ke^2\mathbf{a}^2}{4\pi\left(R^0\right)^2c^2}\hat{\mathbf{R}}\sin^2\theta , \tag{10.65}$$

where θ is the angle between the radiation direction $\hat{\mathbf{R}}$ and the acceleration $\mathbf{a}$ (see Fig. 10.3). Note the characteristic $\sin^2\theta$ dependence and the proportionality to the square of the acceleration. The radiated power per unit solid angle in the low-velocity limit is

$$\frac{d\mathcal{P}}{d\Omega} = \left(R^0\right)^2 \mathbf{S}_r \cdot \hat{\mathbf{R}} = \frac{Ke^2\mathbf{a}^2}{4\pi c^2} \sin^2\theta\,, \tag{10.66}$$

and the total radiated power is

$$\mathcal{P} = \int_{4\pi} d\Omega \frac{d\mathcal{P}}{d\Omega} = \frac{2}{3} K \frac{e^2\mathbf{a}^2}{c^2}\,, \tag{10.67}$$

which is known as the *Larmor power formula*.

The radiated intensity for relativistic motion can be expressed in terms of the coordinate velocity $\mathbf{v}$ and acceleration $\mathbf{a}$ with the help of (10.56) and (10.64):

$$\mathbf{S}_r = \frac{Ke^2\hat{\mathbf{R}}}{4\pi c^2 \left(R^0\right)^2} \frac{\left\{\hat{\mathbf{R}}\times\left[\mathbf{a}\times\left(\mathbf{v}/c-\hat{\mathbf{R}}\right)\right]\right\}^2}{\left(1-\hat{\mathbf{R}}\cdot\mathbf{v}/c\right)^6}\,. \tag{10.68}$$

The radiation from accelerating charges is known as *Bremsstrahlung* (see exercise 3). Note how the radiation is thrown forward at high velocities: because of the denominator, the intensity is strongly enhanced in the directions around $\hat{\mathbf{R}} = \hat{\mathbf{v}}$. Part of the enhancement is a result of the compression of the radiation from an extended period of motion of the charge into a smaller time interval at the observer when the charge is moving at high velocity toward the observer. For example, if the charge is moving directly toward the observer at a speed of 0.9 times the speed of light, then after ten seconds, the radiation will have traveled a distance of 10 light-s (of $r^0 =: ct$) and the charge will have moved 9 light-s: all the radiation emitted over the ten-second period will be compressed in the forward direction into a one second interval (of x^0/c).

In calculating the total power radiated by a charge, we are interested not in the intensity observed but in how much energy is radiated per second of travel of the charge. Thus we want

$$\frac{d\mathcal{P}}{d\Omega} = \left(R^0\right)^2 \hat{\mathbf{R}} \cdot \mathbf{S}_r \frac{dx^0}{dr^0}\,. \tag{10.69}$$

Differentiating the light-cone condition

$$\left(x^0 - r^0\right)^2 = \mathbf{R}^2 \tag{10.70}$$

with respect to r^0 gives

$$2\left(\frac{dx^0}{dr^0}-1\right)R^0=-2\frac{\mathbf{R}\cdot\mathbf{v}}{c}\,, \tag{10.71}$$

and thus

$$\frac{dx^0}{dr^0}=1-\frac{\hat{\mathbf{R}}\cdot\mathbf{v}}{c}=\frac{\langle R\bar{u}\rangle_S}{\gamma R^0}\,. \tag{10.72}$$

Exercise 10.8. Consider the case discussed above: a charge moves at 0.9 c directly toward the observer. Show that

$$\frac{dx^0}{dr^0}=\frac{1}{10}\,. \tag{10.73}$$

Therefore, the power radiated (10.69) is

$$\frac{d\mathcal{P}}{d\Omega}=\frac{Ke^2}{4\pi c^2}\frac{\left\{\hat{\mathbf{R}}\times\left[\mathbf{a}\times\left(\mathbf{v}/c-\hat{\mathbf{R}}\right)\right]\right\}^2}{\left(1-\hat{\mathbf{R}}\cdot\mathbf{v}/c\right)^5}\,. \tag{10.74}$$

Integration over angles gives (with some work!) the total power radiated:

$$\mathcal{P}=\frac{2}{3}K\frac{e^2\gamma^6}{c^2}\left[\mathbf{a}^2-\left(\frac{\mathbf{v}}{c}\times\mathbf{a}\right)^2\right]\,. \tag{10.75}$$

This is a relativistic generalization of the Larmor power formula, first obtained by Liénard in 1898 (before Einstein's theory of relativity!). In its simple covariant form

$$\mathcal{P}=-\frac{2}{3}Ke^2\dot{u}\bar{\dot{u}}\,, \tag{10.76}$$

it is seen to be a Lorentz scalar.

Example 10.9. Integrate $d\mathcal{P}/d\Omega$ over all directions $\hat{\mathbf{R}}$ to obtain the Larmor result (10.75) for the total power $\mathcal{P}$.
Solution:

$$\mathcal{P}=\frac{1}{4\pi}\frac{Ke^2}{c^2}\int_{4\pi}d\hat{\mathbf{R}}\,\eta^{-5}\left\{\cdots\right\}^2\,, \tag{10.77}$$

where in spherical coordinates (see Appendix) $d\hat{\mathbf{R}} = d\phi d\cos\theta$,

$$\eta := 1 - \hat{\mathbf{R}} \cdot \mathbf{v}/c \tag{10.78}$$

and

$$\begin{aligned} \{\cdots\} &= \hat{\mathbf{R}} \times \left[\mathbf{a} \times \left(\frac{\mathbf{v}}{c} - \hat{\mathbf{R}}\right)\right] \\ &= \hat{\mathbf{R}} \cdot \frac{\mathbf{v}}{c} \mathbf{a} - \hat{\mathbf{R}} \cdot \mathbf{a} \frac{\mathbf{v}}{c} - \mathbf{a} + \mathbf{a} \cdot \hat{\mathbf{R}} \hat{\mathbf{R}} \\ &= -\mathbf{a}\eta + \mathbf{a} \cdot \hat{\mathbf{R}} \left(\hat{\mathbf{R}} - \mathbf{v}/c\right) . \end{aligned} \tag{10.79}$$

Squaring, one finds

$$\begin{aligned} \{\cdots\}^2 &= \mathbf{a}^2\eta^2 - 2\eta\, \mathbf{a} \cdot \hat{\mathbf{R}} \left(\mathbf{a} \cdot \hat{\mathbf{R}} - \mathbf{a} \cdot \frac{\mathbf{v}}{c}\right) + \left(\mathbf{a} \cdot \hat{\mathbf{R}}\right)^2 \left(1 - 2\hat{\mathbf{R}} \cdot \frac{\mathbf{v}}{c} + \frac{\mathbf{v}^2}{c^2}\right) \\ &= \mathbf{a}^2\eta^2 + 2\eta\, \mathbf{a} \cdot \hat{\mathbf{R}}\, \mathbf{a} \cdot \frac{\mathbf{v}}{c} - \gamma^{-2} \left(\mathbf{a} \cdot \hat{\mathbf{R}}\right)^2 . \end{aligned} \tag{10.80}$$

To proceed further, we need to specify a coordinate system. In spherical coordinates,

$$\hat{\mathbf{R}} = (\mathbf{e}_1 \cos\phi + \mathbf{e}_2 \sin\phi) \sin\theta + \mathbf{e}_3 \cos\theta , \tag{10.81}$$

where $\mathbf{e}_3$ is the unit vector in the direction of $\mathbf{v}$ and $\mathbf{a}$ is taken to lie in the $\mathbf{e}_3\mathbf{e}_1$ plane:

$$\mathbf{v} = v\mathbf{e}_3 \tag{10.82}$$

$$\mathbf{a} = a\,(\mathbf{e}_1 \sin\alpha + \mathbf{e}_3 \cos\alpha) . \tag{10.83}$$

After averaging of the azimuthal angle ϕ, we find

$$\frac{1}{2\pi} \int_{2\pi} d\phi \{\cdots\}^2 = a^2 \left[\eta^2 + 2\eta \frac{v}{c} \cos\theta \cos^2\alpha - (\cdots)\right] , \tag{10.84}$$

where here

$$(\cdots) = \gamma^{-2} \left(\cos^2\theta \cos^2\alpha + \frac{1}{2} \sin^2\theta \sin^2\alpha\right) \tag{10.85}$$

$$= \beta \left(\frac{v}{c} \cos\theta\right)^2 + \frac{1}{2}\gamma^{-2} \sin^2\alpha , \tag{10.86}$$

with

$$\beta = \left(\gamma \frac{v}{c}\right)^{-2} \left(1 - \frac{3}{2} \sin^2\alpha\right) . \tag{10.87}$$

We can eliminate θ by using the relation $(v/c)\cos\theta = 1 - \eta$ to obtain

$$\frac{1}{2\pi}\int_{2\pi} d\phi\,\{\cdots\}^2 = a^2\left[A\eta^2 + B\eta + C\right], \tag{10.88}$$

with

$$\begin{aligned} A &= 1 - 2\cos^2\alpha - \beta = 2\sin^2\alpha - 1 - \beta \\ B &= 2\cos^2\alpha + 2\beta = -2\sin^2\alpha + 2 + 2\beta \\ C &= -\frac{1}{2}\gamma^{-2}\sin^2\alpha - \beta. \end{aligned} \tag{10.89}$$

The power $\mathcal{P}$ radiated is thus

$$\mathcal{P} = \frac{Ke^2a^2}{c^2}\left[AI_2 + BI_3 + CI_4\right], \tag{10.90}$$

where I_n is the integral

$$\begin{aligned} I_n &= \frac{1}{2}\int_{-1}^{+1} d\cos\theta\,\eta^{-n-1} = \frac{c}{2v}\int_{1-v/c}^{1+v/c} d\eta\,\eta^{-n-1} \\ &= \frac{c\gamma^{2n}}{2nv}\left[(1+v/c)^n - (1-v/c)^n\right] \\ &= \gamma^{2n}\left[1 + \frac{(n-1)(n-2)}{3!}\frac{v^2}{c^2}\right], \quad n = 2,3,4. \end{aligned} \tag{10.91}$$

Substitution gives the final result:

$$\begin{aligned} \mathcal{P} &= \frac{Ke^2a^2\gamma^6}{c^2}\left[A\gamma^{-2} + B\left(1 + \frac{v^2}{3c^2}\right) + \gamma^2 C\left(1 + \frac{v^2}{c^2}\right)\right] \\ &= \frac{2Ke^2a^2\gamma^6}{3c^2}\left[1 - \frac{v^2}{c^2}\sin^2\alpha\right] \\ &= \frac{2Ke^2\gamma^6}{3c^2}\left[\mathbf{a}^2 - (\mathbf{a}\times\mathbf{v}/c)^2\right]. \end{aligned} \tag{10.92}$$

Exercise 10.10. Show that the two expressions (10.75) and (10.76) for the power P are equal.

10.4.1 Thomson Scattering

The agent that accelerates a charge is usually the electromagnetic field, and for nonrelativistic motion, the acceleration is

$$\mathbf{a} = \frac{e}{m}\mathbf{E}. \tag{10.93}$$

Consider an electron in the field of an incident directed plane wave (see Chapt. 6). The ratio of the radiated power (10.67) of an accelerated charge to the incident energy flux $\mathbf{S}_{\text{in}} = \mathbf{E} \times \mathbf{B}/\mu_0$ (power per unit area) of the field accelerating the charge has the dimensions of area and is known as the *Thomson scattering cross section*:

$$\sigma_{Th} = \frac{\mathcal{P}}{|\mathbf{S}_{\text{in}}|} = \frac{\frac{2}{3}Ke^2\,(e/mc)^2\,\mathbf{E}^2}{(4\pi K)^{-1}\,\mathbf{E}^2} = \frac{8\pi}{3}\left(\frac{Ke^2}{mc}\right)^2 . \tag{10.94}$$

The quantity[2] $(Ke^2/mc) = 2.81\,794$ fm is known as the classical electron radius and the cross section σ_{Th} has the value $0.66\,524\,6$ b, where the unit $1\text{b} = 10^{-28}\text{m}^2$ is called a barn.[3]

For linearly polarized waves, one can use (10.66) to define a *differential cross section*

$$\frac{d\sigma_{Th}}{d\Omega} = \frac{1}{|\mathbf{S}|}\frac{d\mathcal{P}}{d\Omega} = \left(\frac{Ke^2}{mc}\right)^2 \sin^2\theta\,, \tag{10.95}$$

where θ is the angle between the field $\mathbf{E}$ and $\hat{\mathbf{R}}$. To find the differential cross section for *unpolarized* light, we average the cross section for two orthogonal linear polarizations:

$$\begin{aligned}\frac{d\sigma_{Th}}{d\Omega} &= \frac{1}{2}\left(\frac{Ke^2}{mc}\right)^2\left(\sin^2\theta_1 + \sin^2\theta_2\right)\\ &= \frac{1}{2}\left(\frac{Ke^2}{mc}\right)^2\left(2 - \cos^2\theta_1 - \cos^2\theta_2\right)\\ &= \frac{1}{2}\left(\frac{Ke^2}{mc}\right)^2\left(1 + \cos^2\theta_3\right)\,,\end{aligned} \tag{10.96}$$

where θ_3 is the scattering angle relative to the incident direction, and we used, for an orthogonal system,

$$\cos^2\theta_1 + \cos^2\theta_2 + \cos^2\theta_3 = 1\,. \tag{10.97}$$

[2]The unit here is fm = femtometer $= 10^{-15}$m.

[3]The name "barn" is said to have been taken by accelerator physicists from the expression, "he couldn't hit the side of a barn."

10.5 Green Functions

10.5.1 Potentials of Static Charge Distributions

Most radiation we observe comes not from the motion of a single charge, but from that of many, described in the form of a current density $j(r)$. Since the field equation (10.1) is linear, the total field is simply a sum of the fields (10.8) from all the contributing charges:

$$A(x) = K \sum_n \frac{e_n u_n}{\langle R_n \bar{u}_n \rangle_S}, \tag{10.98}$$

where $r_n = R_n - x$ is the retarded position of the charge. When there are many charges, it is often more convenient to consider them as part of a continuous distribution. The sum is then replaced by an integral. This is easiest to carry through in the static case where $u_n = 1$. The potential is then

$$A(x) = K \sum_n \frac{e_n}{R_n^0} = K \sum_n \frac{e_n}{|\mathbf{x} - \mathbf{r}_n|}. \tag{10.99}$$

The basic rule for going to the continuum limit is

$$\sum_n [...] \to \int d^3\mathbf{y} \frac{dn}{d^3\mathbf{y}} [...], \tag{10.100}$$

where $dn/d^3\mathbf{y}$ is the number density of charges. We can incorporate the charges e_n into the charge density $\rho(y_{ret})$ at $y_{ret} = ct_{ret} + \mathbf{y}$:

$$\sum_n e_n [...] \to \int d^3\mathbf{y} \rho(y_{ret}) [...], \tag{10.101}$$

which gives, as the vector potential for a static charge distribution $\rho(\mathbf{y})$:

$$A(x) = K \int d^3\mathbf{y} \frac{\rho(\mathbf{y})}{|\mathbf{x} - \mathbf{y}|}. \tag{10.102}$$

The result (10.102) may be considered a generalization of the potential (10.2) of a single static charge. To close the loop, we should be able to regain the potential (10.2) of a single static charge by an appropriate choice of the charge density

ρ. The distribution for a single point charge e at $\mathbf{r}$ vanishes everywhere except at $\mathbf{y} = \mathbf{r}$, but its value at $\mathbf{r}$ is so large that the total charge is e:

$$e = \int d^3\mathbf{y}\, \rho(\mathbf{y}) \, . \tag{10.103}$$

We put

$$\rho(\mathbf{y}) = e\delta^{(3)}(\mathbf{y} - \mathbf{r}) \, , \tag{10.104}$$

where the three-dimensional Dirac delta function $\delta^{(3)}(\mathbf{y} - \mathbf{r})$ is zero if $\mathbf{y} \neq \mathbf{r}$ and is infinite (undefined, unbounded) at $\mathbf{y} = \mathbf{r}$. The delta function can be considered the limit of any number of sharply peaked functions, all of which have the same integral as $\delta^{(3)}$ itself

$$\int_{\mathcal{V}} d^3\mathbf{y}\, \delta^{(3)}(\mathbf{y} - \mathbf{r}) = 1 \tag{10.105}$$

when the domain of integration $\mathcal{V}$ is all space. However, since $\delta^{(3)}$ vanishes if its argument is nonzero, we can be more precise: if $f(\mathbf{y})$ is any bounded function of the position,

$$\int_{\mathcal{V}} d^3\mathbf{y}\, f(\mathbf{y})\, \delta^{(3)}(\mathbf{y} - \mathbf{r}) = \begin{cases} f(\mathbf{r})\, , & \mathbf{r} \in \mathcal{V} \\ 0, & \mathbf{r} \notin \mathcal{V} \end{cases} . \tag{10.106}$$

With the delta-function distribution (10.104), the potential (10.102) becomes the static potential of a point charge:

$$A(x) = K \int d^3\mathbf{y} \frac{e\delta^{(3)}(\mathbf{y} - \mathbf{r})}{|\mathbf{x} - \mathbf{y}|} = K \frac{e}{|\mathbf{x} - \mathbf{r}|} \, . \tag{10.107}$$

The procedure of finding the solution for a charge distribution from that for a point charge is formalized in the Green-function method. The Green function $G(\mathbf{x}, \mathbf{y})$ is the solution at $\mathbf{x}$ when the source is a unit point charge at $\mathbf{y}$. In the static case, the wave-equation form (10.1) of Maxwell's equation becomes *Poisson's equation* with a δ-function density:

$$\nabla^2 G(\mathbf{x}, \mathbf{y}) = -4\pi K \delta^{(3)}(\mathbf{x} - \mathbf{y}) \, , \tag{10.108}$$

and its solution is the scalar Green function

$$G(\mathbf{x}, \mathbf{y}) = \frac{K}{|\mathbf{x} - \mathbf{y}|} \, . \tag{10.109}$$

The paravector potential of any superposition of sources

$$\rho(\mathbf{x}) = \int d^3\mathbf{y}\, \rho(\mathbf{y})\, \delta^{(3)}(\mathbf{x} - \mathbf{y}) \tag{10.110}$$

is just the same superposition of the solutions $G(\mathbf{x}, \mathbf{y})$:

$$A(x) = \int d^3\mathbf{y}\, \rho(\mathbf{y})\, G(\mathbf{x}, \mathbf{y}) = K \int d^3\mathbf{y}\, \frac{\rho(\mathbf{y})}{|\mathbf{x} - \mathbf{y}|}\,, \tag{10.111}$$

and this reproduces equation (10.102).

10.5.2 Potentials from Currents

We want now to extend the above treatment to moving charges. The retarded-time condition, required by the usual *causality* argument that the fields are caused by the motion of the charges, is obviously important: the charges at different points in space will contribute to the potential $A(x)$ at the single spacetime point x at different retarded times. We now look for a scalar Green function $G(x, y)$ of spacetime variables x, y which represents a static unit point charge that exists for only an instant at y. It obeys

$$\partial\bar{\partial} G(x, y) = \Box G(x, y) = 4\pi K \delta^{(4)}(x - y)\ , \tag{10.112}$$

where $\delta^{(4)}(x - y)$ is a four-dimensional Dirac delta function and ∂ and $\Box$ differentiate with respect to the observer point x. The physical solution to (10.112) is a potential that exists only on the forward light cone of y and has the form

$$G(x, y) = K \frac{\delta(x^0 - y^0 - |\mathbf{x} - \mathbf{y}|)}{x^0 - y^0}\,, \tag{10.113}$$

which is known as the *retarded Green function.* Linear superpositions of this solution (10.113)

$$A(x) = \int d^4y\, G(x, y)\ j(y)\, /c \tag{10.114}$$

give the potential for an arbitrary current distribution satisfying Maxwell's equation (10.1)

$$\Box A(x) = \mu_0 j(x) = 4\pi K j(x)\, /c\,. \tag{10.115}$$

Thus,

$$A(x) = K \int d^4y \frac{j(y)/c}{x^0 - y^0} \delta(x^0 - y^0 - |\mathbf{x} - \mathbf{y}|) = K \int d^3\mathbf{y} \left[\frac{j(y)/c}{x^0 - y^0} \right]_{ret}, \tag{10.116}$$

where we have eliminated the one-dimensional delta function by integrating over y^0, and $[...]_{ret}$ means that the contents of the brackets are to be evaluated at the retarded time. This relation for the spacetime vector potential is the starting point for discussions of Liénard-Wiechert potentials in many texts.

10.5.3 Coulomb Gauge

The simple form (10.1) of Maxwell's equation and the covariant solution (10.116) are valid in any Lorenz gauge. However, other gauges are occasionally useful. If no gauge condition is imposed, the equations

$$\mathbf{F} = c \langle \partial \bar{A} \rangle_V = c\partial \bar{A} - c \langle \partial \bar{A} \rangle_S \tag{10.117}$$

$$\bar{\partial} \mathbf{F} = (\varepsilon_0 c)^{-1} \bar{j} \tag{10.118}$$

combine to yield

$$\Box A - \partial \langle \partial \bar{A} \rangle_S = \mu_0 j\,. \tag{10.119}$$

By choosing the Lorenz-gauge condition, the second term of the LHS vanishes, leaving the result (10.115). More generally, however, equation (10.119) can be expanded to yield the scalar and vector parts

$$\nabla^2 \phi + \frac{\partial}{\partial t} \nabla \cdot \mathbf{A} = -\varepsilon_0^{-1} \rho \tag{10.120}$$

$$\Box \mathbf{A} + \nabla \left(\frac{\partial}{c^2 \partial t} \phi + \nabla \cdot \mathbf{A} \right) = \mu_0 \mathbf{j}\,. \tag{10.121}$$

The gauge choice $\nabla \cdot \mathbf{A} = 0$ is popular for low-energy interactions, for example, in the interaction of atoms with external fields. It is known as the *Coulomb gauge.* In the Coulomb gauge, the scalar equation above reduces to the Poisson equation:

$$\nabla^2 \phi = -\varepsilon_0^{-1} \rho\,, \tag{10.122}$$

whose solution is the *instantaneous Coulomb potential*

$$\phi(x) = \frac{1}{4\pi\varepsilon_0} \int d^3\mathbf{r}\, \frac{\rho(r)}{|\mathbf{x} - \mathbf{r}|}, \tag{10.123}$$

with $\langle r \rangle_S = \langle x \rangle_S$. In other words, $\rho(r)$ is evaluated not at the retarded time but at the instant of observation! What happened to the finite propagation time required by causality and the special theory of relativity?

Expressions for the field $\mathbf{F}$ are functions of both ϕ and $\mathbf{A}$, and $\mathbf{F}$ does not depend on the gauge choice. The field is needed to convey signals or energy, and it still depends on the sources at the retarded times. Causality is still intact.

In the Coulomb gauge, the differential equation for the vector potential is modified by ϕ :

$$\Box \mathbf{A} + \nabla \left(\frac{\partial}{c^2 \partial t} \phi \right) = \mu_0 \mathbf{j}\,. \tag{10.124}$$

The divergence of (10.124), together with the gauge condition $\nabla \cdot \mathbf{A} = 0$ and the Poisson equation (10.122), gives the continuity equation $\langle \partial \bar{j} \rangle_S = 0$, whereas the curl of (10.124) eliminates the ϕ term:

$$\Box \mathbf{B} = \mu_0 \nabla \times \mathbf{j}\,. \tag{10.125}$$

Any vector field $\mathbf{A}$ can be split $\mathbf{A} = \mathbf{A}_l + \mathbf{A}_t$ into *longitudinal* and *transverse* components, where the longitudinal part is divergence-free ($\nabla \cdot \mathbf{A}_t = 0$) and the transverse part is irrotational ($\nabla \times \mathbf{A}_l = 0$).[4] The Coulomb-gauge condition makes $\mathbf{A}$ transverse and is sometimes called the *transverse gauge.* We can, for this gauge, take the longitudinal and transverse parts of Maxwell's equation (10.124) for the potential to obtain

$$\Box \mathbf{A} = \mu_0 \mathbf{j}_t \tag{10.126}$$

$$\nabla \left(\frac{\partial}{c^2 \partial t} \phi \right) = \mu_0 \mathbf{j}_l\,. \tag{10.127}$$

10.5.4 Dipole Radiation

As a simple application of (10.116), consider the radiation from a nonrelativistic oscillating dipole

$$\mathbf{d}(t) = \mathbf{d}_0 \cos(\omega t) \tag{10.128}$$

[4]This split may be established convincingly in the wave-vector space of Fourier transforms of $\mathbf{A}$ (an extension to three dimensions of the Fourier integrals in section 5.5).

at large distances (far-field region, radiation zone) $|\mathbf{x}| \gg |\mathbf{d}_0|$. We assume that $\mathbf{y}$ =0 at the center of the dipole. Then $R^0 = x^0 - y^0 = |\mathbf{x} - \mathbf{y}| \approx |\mathbf{x}|$ and $\hat{\mathbf{R}} \approx \hat{\mathbf{x}}$ so that $A(x)$ (10.116) becomes

$$A(x) = \frac{K}{R^0 c} \int d^3\mathbf{y}\, j(y_{ret}) \,. \tag{10.129}$$

Consider first the vector part and assume that the distribution falls to zero on or outside some finite surface $\partial\mathcal{V}$ surrounding the origin. Then apply the divergence theorem: for any constant vector $\mathbf{a}$,

$$0 = \oint_{\partial\mathcal{V}} d\mathbf{s} \cdot \mathbf{j}\,\mathbf{y} \cdot \mathbf{a} = \int d^3\mathbf{y}\, \nabla \cdot (\mathbf{j}\,\mathbf{y} \cdot \mathbf{a}) = \int d^3\mathbf{y}\, (\mathbf{j} \cdot \mathbf{a} + \mathbf{y} \cdot \mathbf{a} \nabla \cdot \mathbf{j}) \,. \tag{10.130}$$

The continuity equation $\nabla \cdot \mathbf{j} + \partial_t \rho = 0$ then gives

$$\int d^3\mathbf{y}\, (\mathbf{j} \cdot \mathbf{a} - \mathbf{y} \cdot \mathbf{a} \partial_{t_{ret}} \rho) = 0 \,, \tag{10.131}$$

and therefore,

$$\int d^3\mathbf{y}\, \mathbf{j}\,(y_{ret}) = \int d^3\mathbf{y}\, \mathbf{y} \partial_{t_{ret}} \rho =: \dot{\mathbf{d}} \tag{10.132}$$

where

$$\mathbf{d} := \int d^3\mathbf{y}\, \mathbf{y} \rho \tag{10.133}$$

is the dipole moment of the distribution and for nonrelativistic motion, we can ignore the difference between the coordinate time t_{ret} at the source and the (retarded) proper time τ. The dominant effect of evaluating $\mathbf{j}$ at the retarded time is to have $\dot{\mathbf{d}}$ also evaluated then. Next consider the scalar part of (10.129). If the variation of the retarded time across the distribution is ignored, the integral

$$\int d^3\mathbf{y}\, \rho(\mathbf{y}, t_0) = Q \tag{10.134}$$

gives the total charge of the distribution, which we assume vanishes. Evidently the smallest nonvanishing term arises from the first-order correction for retarded-time effects. Let's expand the charge density about the retarded time t_0 at the origin:

$$\rho(\mathbf{y}, t_{ret}) \approx \rho(\mathbf{y}, t_0) + [\partial_t \rho(\mathbf{y}, t)]_{t=t_0} \Delta t \,, \tag{10.135}$$

where

$$c\Delta t = ct_{ret} - ct_0 = |\mathbf{x}| - |\mathbf{x} - \mathbf{y}| \approx \mathbf{y} \cdot \hat{\mathbf{R}} . \tag{10.136}$$

Thus, with $Q = 0$ we obtain

$$\begin{aligned} \int d^3\mathbf{y}\, \rho(\mathbf{y}, t_{ret}) &\approx \int d^3\mathbf{y}\, \left[\partial_t \rho(\mathbf{y}, t)\ \mathbf{y}\right]_{t=t_0} \cdot \hat{\mathbf{R}}/c \\ &= \left[\frac{d}{dt} \int d^3\mathbf{y}\, \rho(\mathbf{y}, t)\ \mathbf{y}\right]_{t=t_0} \cdot \hat{\mathbf{R}}/c \\ &= \dot{\mathbf{d}}_{ret} \cdot \hat{\mathbf{R}}/c . \end{aligned} \tag{10.137}$$

The full vector potential (10.129),

$$A(x) = \frac{K}{R_0} \int d^3\mathbf{y}\ \left[\rho(\mathbf{y}, t_{ret}) + \mathbf{j}(\mathbf{y}, t_{ret})/c\right] , \tag{10.138}$$

to lowest order in retarded-time effects, is therefore

$$A(x) = \frac{K}{R^0 c} \left(\dot{\mathbf{d}} \cdot \hat{\mathbf{R}} + \dot{\mathbf{d}}\right)_{ret} . \tag{10.139}$$

To lowest order in the velocity v the gradient (10.16) is

$$\partial = (\partial)_x + (\partial\tau) \frac{d}{d\tau} = (\partial)_x + \frac{R}{\langle R\bar{u}\rangle_S} \frac{d}{cd\tau} \tag{10.140}$$

$$\approx (\partial)_x + \frac{R}{R^0} \frac{d}{cd\tau} = (\partial)_x + \left(1 + \hat{\mathbf{R}}\right) \frac{d}{cd\tau} , \tag{10.141}$$

and the radiation field $\mathbf{F}_r$ is the part of the field linear in the acceleration:

$$\begin{aligned} \mathbf{F}_r(x) &= \langle c\left[\partial - (\partial)_x\right] \bar{A}\rangle_V \\ &\approx \frac{K}{R^0 c} \left\langle \left(1 + \hat{\mathbf{R}}\right) \left(\ddot{\mathbf{d}} \cdot \hat{\mathbf{R}} - \ddot{\mathbf{d}}\right)_{ret} \right\rangle_V \\ &= -\frac{K}{R^0 c} \left(1 + \hat{\mathbf{R}}\right) \left(\ddot{\mathbf{d}}_\perp\right)_{ret} , \end{aligned} \tag{10.142}$$

where the perpendicular component is

$$\ddot{\mathbf{d}}_\perp = \ddot{\mathbf{d}} - \ddot{\mathbf{d}} \cdot \hat{\mathbf{R}}\hat{\mathbf{R}} . \tag{10.143}$$

Substituting dipole oscillations (10.128) into (10.142), we obtain a plane-polarized radiation field

$$\mathbf{F}_r(x) = \frac{K}{R^0 c} \left(1 + \hat{\mathbf{R}}\right) \mathbf{d}_{0\perp} \omega^2 \cos(\omega t - \mathbf{k}' \cdot \mathbf{R}) , \tag{10.144}$$

where t is the time variable at the field point and we have defined $\mathbf{k}' := (\omega/c)\, \hat{\mathbf{R}}$.

10.6 Fields from Currents

A general expression for the electromagnetic field can be obtained by differentiating the parapotential (10.116):

$$\mathbf{F}(x) = c\left\langle \partial \bar{A} \right\rangle_V = K \int d^3\mathbf{y} \left\langle \partial_x \left[\bar{j}(y) / R^0 \right]_{ret} \right\rangle_V , \tag{10.145}$$

where ∂_x is the gradient operator with respect to the 4-dimensional field position $x = R + y$. As above in equations (10.10) and (10.11), the factor $\left[\bar{j}(y) / R^0 \right]_{ret}$ depends on the field position x through the retarded time. Now, however, the source is the current density $j(y)$ rather than a single particle following a world line. We therefore modify the previous treatment by using the retarded coordinate time $y^0/c = t_{ret}$ in place of the retarded proper time τ :

$$\partial_x = \left(\partial_x \right)_x + \left(\partial_x t_{ret} \right) \frac{\partial}{\partial t_{ret}} . \tag{10.146}$$

The gradient $\left(\partial_x t_{ret} \right)$ of the scalar field $t_{ret}\left(x \right)$ is determined by the retardation condition

$$\left(R^0 \right)^2 = \left(x^0 - c t_{ret} \right)^2 = \left(\mathbf{x} - \mathbf{y} \right)^2 = \mathbf{R}^2 . \tag{10.147}$$

Applying ∂_x (10.146) we find

$$\begin{aligned} \partial_x \left(x^0 - c t_{ret} \right)^2 &= \partial_x \left(R^0 \right)^2 = 2 R^0 \partial_x R^0 \\ &= 2 R^0 \left[1 - c \left(\partial_x t_{ret} \right) \right] , \end{aligned} \tag{10.148}$$

whose equality with

$$\partial_x \left(\mathbf{x} - \mathbf{y} \right)^2 = -2 \left(\mathbf{x} - \mathbf{y} \right) = -2 \mathbf{R} \tag{10.149}$$

gives

$$c \left(\partial_x t_{ret} \right) = 1 + \hat{\mathbf{R}} = R / R^0 \tag{10.150}$$

$$\partial_x R^0 = -\hat{\mathbf{R}} . \tag{10.151}$$

Note also that

$$\partial_x \left(R^0 \right)^{-1} = - \left(R^0 \right)^{-2} \partial_x R^0 = \hat{\mathbf{R}} \left(R^0 \right)^{-2} . \tag{10.152}$$

Thus, (10.145) can be expanded to give

$$\mathbf{F}(x) = K \int d^3\mathbf{y} \left\langle \frac{\hat{\mathbf{R}}\bar{j}(y) + R\partial_0\bar{j}(y)}{\mathbf{R}^2} \right\rangle_{ret,V} , \tag{10.153}$$

where

$$\partial_0\bar{j}(y) = \frac{\partial}{\partial y^0}\bar{j}(y) = \frac{\partial}{c\partial t_{ret}}\bar{j}(y) . \tag{10.154}$$

The real and imaginary parts of (10.153) are known as the generalized Coulomb-Faraday and generalized Biot-Savart laws, respectively.[5] For another derivation, see problem 10 at the end of this chapter.

10.7 Virtual-Photon Streams

The electromagnetic field $\mathbf{F}$ of an accelerating charge in a vacuum can be found from virtual "photon streams" that are emitted continuously and isotropically from the charge in its rest frame (*i.e.*, in the inertial frame instantaneously commoving with the charge). The virtual photons move along straight rays on the light cones of the charge. Although the virtual photons are radiated isotropically, it is useful to think of them, in the nature of field lines, as being emitted along fixed rays in the instantaneous rest frame of the charge. We thus imagine the virtual photons as concentrated in a number $e/\varepsilon_0 = 4\pi Kce$ of rays, evenly distributed around the charge in its rest frame. Although each virtual photon moves in a fixed spatial direction at the speed of light, the *stream* comprises photons emitted at different times from a moving source. If the source oscillates, the stream develops kinks, much as would a stream of water from an oscillating hose, even though each water molecule follows a ballistic trajectory.

In spacetime terms, as the charge progresses along its world line, the rays sweep out continuous spacetime *sheets* that cut the light cones of the charge. As we will see below, the Liénard-Wiechert field of the charge is the tangent plane to the sheet at the field position and it is the slice of such sheets at an instant in time that gives the electric field lines. (See figure 10.5.)

Consider first the simple case that the charge is moving at constant velocity on its world line $r(\tau)$. The virtual ray which in the rest frame is directed along the

[5] See, for example, D. J. Griffiths and Mark A. Heald, "Time-dependent generalizations of the Biot-Savart and Coulomb laws", *Am. J. Phys.* **59**, 111-117 (1991); **60**, 393-394 (1992).

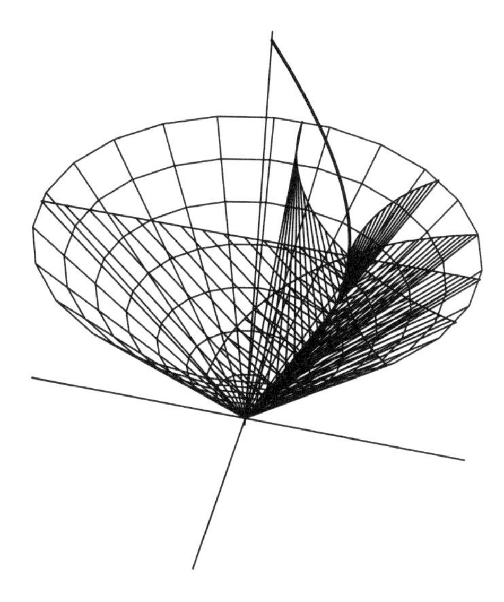

Fig. 10.5. The virtual-photon sheets of a charge undergoing uniform acceleration (in hyperbolic motion) are seen here in a spacetime diagram. The world line of the charge is the heavy bow-shaped line. The sheets emanate from the charge. In the view here, they are cut at the time when the charge is instantaneously at rest.

lightlike paravector R_{rest} will generate a sheet

$$r(\tau) + \alpha R = r(\tau) + \alpha \Lambda R_{\text{rest}} \Lambda^{\dagger}, \tag{10.155}$$

where the $\alpha \geq 0$ and the proper time τ are the real scalar parameters of the sheet and Λ is the eigenspinor of the charge. Since both the spacetime direction R and the proper velocity u are constant, the sheet is coincident with the biparavector $\langle R\bar{u}\rangle_V$ and hence with the electromagnetic field $\mathbf{F} = \mathbf{F}_c$. In any inertial frame, a spacelike slice of the sheets (at a given time in that frame) gives the electric-field lines of the charge along $\langle\langle R\bar{u}\rangle_V \mathbf{e}_0\rangle_{\Re} = \langle R\bar{u}\rangle_{\Re V}$.

More generally, the sheets swept out by the rays of an accelerating charge are not flat but curved, and the plane tangent to the sheet at $x = r + R$ on the ray along R emitted from the charge at $r(\tau)$ is given by the biparavector

$$\left\langle R\frac{d}{d\tau}\left(\bar{r} + \bar{R}\right)\right\rangle_V = \left\langle R\left(\bar{u}c + \frac{d}{d\tau}\bar{R}\right)\right\rangle_V. \tag{10.156}$$

Now in addition to the change in the position r of the base of the ray, the spatial direction of the lightlike ray R may also change. Since the direction of the ray is fixed in the rest frame,

$$\begin{aligned} \frac{d}{d\tau}\bar{R} &= \frac{d}{d\tau}\overline{\Lambda R_{\text{rest}}\Lambda^\dagger} \\ &= \overline{\langle \Omega R\rangle}_{\Re}\,, \end{aligned} \tag{10.157}$$

where $\Omega = 2\dot{\Lambda}\bar{\Lambda}$ is the spacetime rotation rate of the charge. Since $R\bar{R} = 0$, and the scalar part of Ω vanishes, the plane tangent to the electromagnetic field at x is

$$\langle R\bar{u}\rangle_V\, c - \frac{1}{2}R\Omega^\dagger\bar{R}\,. \tag{10.158}$$

Consider the case that the acceleration is spatially collinear with the velocity. Then Ω is real and commutes with u, so that

$$\begin{aligned} \dot{u} &= \dot{\Lambda}\Lambda^\dagger + \Lambda\dot{\Lambda}^\dagger = \langle \Omega u\rangle_{\Re} \\ &= \Omega u\,. \end{aligned} \tag{10.159}$$

In this case, $\Omega = \dot{u}\bar{u}$ and the tangent plane becomes

$$\langle R\bar{u}\rangle_V\, c + \frac{1}{2}R\bar{\dot{u}}u\bar{R}\,, \tag{10.160}$$

which is just the spacetime plane of the electromagnetic field $\mathbf{F}$. The electric field lines lie in the virtual photon sheets. They are generally no longer in the propagation direction $\hat{\mathbf{R}}$, but are formed by the slice of the sheet at a given instant in time. The field line connects virtual-photon rays emitted in a given rest-frame direction from different positions on the world line of the charge.

When the charge undergoes a brief period of acceleration, both $\hat{\mathbf{R}}$ and u change and the virtual photon sheets become kinked. The Coulomb field lines are "broken" by the acceleration; they do not meet and cannot be joined by the Coulomb field during the acceleration period. However, the virtual photon sheet is continuous. A slice of the sheet at a constant time gives continuous field lines, with the Coulomb segments joined by the radiation field.

Relatively simple programs can be written to display and animate the field in time for cases of collinear motion. The examples here are frames generated by a relatively simple MapleV[6] worksheet for arbitrary accelerated motion of a charge

[6] Maple is a trademark of Waterloo Maple Software, Waterloo, Ontario, Canada.

along a straight line. The worksheet, `rad3.mws`, is given in Appendix B and is also one of several relevant to this text that are available through the Birkhäuser web site.[7] It contains explicit equations for three types of motion: an oscillator, an initially static charge hit by a short impulse, and a charge in hyperbolic (uniformly accelerated) motion.

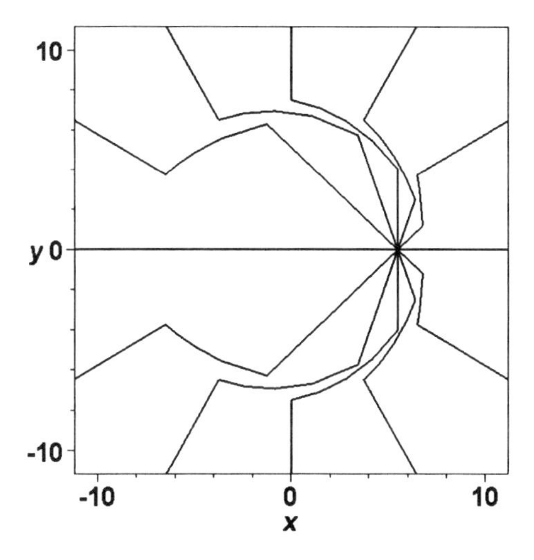

Fig. 10.6. The electric field lines of a charge that remained at rest at the origin for many time units before being accelerated to $\pi/4\,c$ during a single time unit, and thereafter continuing at constant speed along the x-axis. This figure is one frame calculated by a MapleV worksheet (see text).

Figure 10.6 shows the Maple calculation of the electric field of a charge that was initially at rest at the origin and then accelerated suddenly to $v = \pi c/4$. The lines inside the circular region are the Coulomb field lines of the charge moving at constant velocity after acceleration. The straight lines outside the circular region are the Coulomb lines from the static charge. The circular region represents the shell of radiation emitted while the charge was being accelerated.

Figure 10.7 shows virtual photons from a uniformly accelerated charge; the photon streams form the electric field lines. The photons are shown both when the charge is instantaneously at rest at the origin and later, when it has accelerated some distance toward the right. The field lines may be compared to those shown

[7]Search the site http://www.birkhauser.com for the title of this text to locate worksheets written for recent releases of Maple.

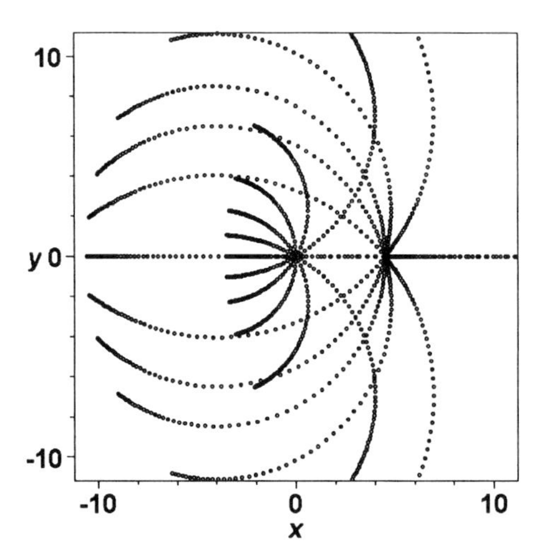

Fig. 10.7. The virtual photons that form the electric field lines of a uniformly accelerated charge while it is accelerating along the $\mathbf{e}_1$ axis, as calculated in a Maple V worksheet (see text). The photons are shown both at the instant the charge is turning around and later, as it has accelerated some distance along the x axis.

above in Fig. 10.4 (but note that the direction of acceleration and the turning point are different).

10.8 Review Questions

This chapter presented fundamental ideas about the nature and sources of electromagnetic radiation, as well as many practical details about their calculation. To be sure that the basic concepts aren't lost in the details, the reader may wish to consider the following review questions:

1. Where does radiation come from?

2. Define the retarded proper time of a moving charge. What mathematical relation can be used to specify the retarded proper time?

3. The retarded proper time may be thought of as a scalar field $\tau(x)$. Derive the spacetime gradient $\partial\tau(x)$.

4. How do radiation fields arise that are linear in the acceleration of the charge when the Liénard-Wiechert potential depends only on the retarded position and velocity?

10.9 Problems

1. Use the relations $R\bar{R} = 0,\ u = \gamma(1 + \mathbf{v}/c),\ u\bar{u} = 1$, and $dt = \gamma d\tau$ to prove equation (10.55).

2. Boost a null propagation vector $\mathbf{k}$ from the rest frame of the charge to the lab frame in which the charge has proper velocity $u = \gamma(1 + \mathbf{v}/c)$. Show that the angle between $\mathbf{k}$ and the boost direction $\hat{\mathbf{u}}$ decreases under the boost, and that if $\mathbf{k}'$ is the propagation vector after the boost,

$$\hat{\mathbf{k}}' \cdot \hat{\mathbf{u}} = \frac{v + c\hat{\mathbf{k}} \cdot \hat{\mathbf{u}}}{c + v\hat{\mathbf{k}} \cdot \hat{\mathbf{u}}}$$

Verify that for speeds v approaching the speed of light, an approximately isotropic distribution of photons radiated by a charge in its rest frame get thrown forward into a cone of angular opening $\theta \approx \gamma^{-1}$.

3. **Bremsstrahlung** is radiation from accelerating charges.[8] It literally means "braking radiation" and arises whenever a rapidly moving charge plows into a stationary target. The acceleration $\mathbf{a}$ is typically in the opposite direction of the velocity $\mathbf{v}$. Show that even though the radiation for a rapidly moving charge is thrown forward in the direction of $\mathbf{v}$, when $\mathbf{a}$ and $\mathbf{v}$ are aligned, no radiation is directed exactly in the direction of $\mathbf{v}$. Give the intensity distribution for this case in terms of the angle θ between $\mathbf{R}$ and $\mathbf{v}$.

4. **Synchrotron radiation** is an important source of energy loss for particle accelerators and storage rings. Show that the fraction of the particle energy γmc^2 required every cycle to keep an electron in a circular orbit is of radius r_c is

$$\frac{2\pi}{\omega}\frac{P}{\gamma mc^2} = \frac{4\pi}{3}\frac{r_e}{r_c}|\mathbf{u}|^3, \tag{10.161}$$

[8]In German, *bremsen* means to brake and *Strahlung* is radiation.

where $\omega = v/r_c$ is the angular velocity, $r_e = Ke^2/(mc)$ is the classical electron radius, and $|\mathbf{u}| = \gamma v$ is the magnitude of the proper velocity vector. Show that for a 1 TeV electron in a storage ring of 80 km circumference, this ratio is about 7.

5. Consider the radiation from a charge moving in uniform circular motion. Assume that the observer is located a large distance from the charge relative to the orbital radius and that the motion is nonrelativistic. Show that the radiation is elliptically polarized with an ellipticity that depends on the angle of viewing with respect to the plane of the orbit. What type of polarization does the radiation have that is emitted along the axis of the orbit? What type in the plane of the orbit?

6. Find the average distribution of intensity at large distances emitted by nonrelativistic charges (a) in uniform circular motion and (b) in linear harmonic motion. Show that the average distribution from uniform circular motion is the same as from two orthogonal linear oscillators, and specify how the linear oscillators should be oriented with respect to the orbit of circular motion.

7. Consider an oscillating dipole $\mathbf{d} = e\mathbf{r}$ formed by an oscillating charge e at $\mathbf{r}$ and a stationary charge $-e$ at the average position of the oscillating charge. Show that the radiation field (10.52) in the nonrelativistic limit (10.57) agrees with equation (10.142). Note that the potentials used in the two derivations have different Lorenz gauges.

8. Check the "virtual-photon stream" method for a charge moving with constant velocity $\mathbf{v} = 0.8c\mathbf{e}_1$ by calculating the positions of virtual photons emitted every nanosecond at rest-frame angles $\theta = n\pi/6,\ n = 0, 1, \dots, 6$ over a period of five nanoseconds. Show that the virtual photons emitted at the rest-frame angle $\theta = \pi/2$ form a field line in the plane perpendicular to $\mathbf{v}$ in the lab frame, and that the $\mathbf{e}_1$ component of the lab velocity of such photons is equal to the speed of the charge.

9. Consider a charge e that is at rest at the origin for times $t < 0$ but moves at constant velocity $0.8\,c\mathbf{e}_1$ at all times $t > 0$. Consider the field existing over several units of distance around the origin at $t = 1$. The radiation field exists only on the sphere of radius 1 and connects the Coulomb fields inside and outside the sphere. Sketch the radiation field in red and the Coulomb

fields in blue.[9]

10. Derive the field $\mathbf{F}$ (10.153) directly from the Green-function solution

$$\mathbf{F}(x) = K \int d^4y \frac{\partial \bar{j}(y)}{R^0} \delta\left(R^0 - |\mathbf{R}|\right)$$

of the second-order equation obtained by differentiating Maxwell's equation:

$$\Box \mathbf{F} = 4\pi K \, \partial \bar{j}/c$$

where $R = x - y$. [Hint: The simplest way to keep track of the influence of the retarded-time constraint is to leave the integral as a four-dimensional integral with a delta function until the very last step. Note that $\langle \partial \bar{j} \rangle_S = 0$ and consequently $\partial \bar{j} = \langle \partial \bar{j} \rangle_V$. Here, $\partial = \partial/\partial y^0 - \nabla_{\mathbf{y}}$ and the gradient of $\bar{j}$ can be eliminated by integration by parts. This in turn gives a factor

$$\begin{aligned} \nabla_{\mathbf{y}} \delta\left(R^0 - |\mathbf{R}|\right) &= -\left(\nabla_{\mathbf{y}} |\mathbf{R}|\right) \delta'\left(R^0 - |\mathbf{R}|\right) \\ &= \hat{\mathbf{R}} \frac{\partial}{\partial R^0} \delta\left(R^0 - |\mathbf{R}|\right) . \end{aligned} \tag{10.162}$$

One more integration by parts, plus the recognition that since x^0 is fixed, $\partial/\partial R^0 = -\partial/\partial y^0$, gives the final result (10.153), provided one respects the order of noncommuting terms.]

11. Derive the Liénard-Wiechert field (10.23) of an accelerating charge e on the world line $r(\tau)$, where τ is the proper time of the charge, from the expression (10.153) by identifying

$$\bar{j}(y) = ec\delta^{(3)}[\mathbf{y} - \mathbf{r}(t_{ret})] \, \bar{u}/\gamma$$

where $u(\tau) = c^{-1} dr(\tau)/d\tau$ is the proper velocity of the charge and $\gamma = \langle u \rangle_S$. [Hint: Consider the Jacobian

$$\begin{aligned} |\partial(\mathbf{s})/\partial(\mathbf{y})| &= \left|\partial\left(s^1, s^2, s^3\right)/\partial\left(y^1, y^2, y^3\right)\right| \\ &= \begin{vmatrix} \partial s^1/\partial y^1 & \partial s^1/\partial y^2 & \partial s^1/\partial y^3 \\ \partial s^2/\partial y^1 & \partial s^2/\partial y^2 & \partial s^2/\partial y^3 \\ \partial s^3/\partial y^1 & \partial s^3/\partial y^2 & \partial s^3/\partial y^3 \end{vmatrix} \end{aligned} \tag{10.163}$$

[9] You may wish to try other velocities. Note that the speeds do not need to be relativistic to generate radiation. Try sketching the fields, for example, when the velocity of the charge changes suddenly from $0.01c\mathbf{e}_1$ to $-0.01c\mathbf{e}_1$. Note that the electric fields are symmetric under rotations about $\mathbf{e}_1$ and that the spacing between the radiation field lines falls off as R_0^{-1},as required. This construction (the nonrelativistic version) to understand the origin of radiation goes back to J. J. Thomson.

where $\mathbf{s} = \mathbf{y} - \mathbf{r}\left(y^0\right)$ and use it to show

$$\int d^3\mathbf{y}\,\delta^{(3)}[\mathbf{y} - \mathbf{r}\left(t_{ret}\right)] = \frac{\gamma R^0}{\langle R\bar{u}\rangle_S}\,. \tag{10.164}$$

It is probably simplest to recombine (10.153) into (10.145), to integrate over the Dirac delta function and relate the result to the derivation at the beginning of this chapter.]

12. As discussed in the text, the shape of the electric field lines of a stationary charge in a uniform gravitational field is related by the equivalence principle to the field lines of a uniformly accelerated charge. Consider the case considered in section 10.3, namely a charge e in "hyperbolic" motion, whose world line is

$$r(\tau) = \mathbf{e}_3 \exp(a\tau\mathbf{e}_3/c)\ , \tag{10.165}$$

where the unit of length is c^2/a.

  (a) Show that if a has the magnitude of 10 m/s^2,that the unit length is roughly one light year.

  (b) Sketch the spacetime diagram of the world line and indicate that part of spacetime that can receive no signals (or field lines) from the charge.

  (c) Prove that the Schott term in the Lorentz-Dirac equation exactly compensates for the Larmor power lost by the charge, so that the Lorentz force law holds precisely.

  (d) Demonstrate that even though $\mathbf{F}_r{\cdot}\mathbf{F}_c = 0$, there is an interference term between the radiated field and the Coulomb field that does contribute to the energy density and Poynting vectors of the total field.

13. Calculate the direction of the electric field at several positions in the $\mathbf{e}_1\mathbf{e}_3$ plane for the accelerating charge of the previous problem to estimate the shape of the electric field at the instant $t = 0 = \tau$, when the charge is instantaneously at rest.

  (a) Find the directions of the electric field at $\mathbf{x} = 0.5\,\mathbf{e}_3,\ \mathbf{e}_3,\ 2\,\mathbf{e}_3,\ \mathbf{e}_3{\pm}\mathbf{e}_1$, and at $\mathbf{x} = z\mathbf{e}_3 + x\mathbf{e}_1$ in the limit of $z \to 0$.

  (b) Sketch the field lines of the total electric field.

(c) Also sketch the direction of the transformed Coulomb field, $\mathbf{F}_c$ in the $\mathbf{e}_1\mathbf{e}_3$ plane.

(d) Would you expect the Coulomb field to be represented by field lines? Defend your answer.

(e) Comment on the shape of the total electric field and on the identification of the Coulomb and radiation fields for a charge at rest in a uniform gravitational field.

Chapter 11

The Model Atom and Other Applications

Many of the radiative properties of atoms can be modeled classically by a three-dimensional harmonic oscillator comprising a charge e isotropically bound to the center of a region of equal and opposite charge by a linear restoring force

$$\mathbf{F} = -m\omega_0^2\mathbf{r}\,. \tag{11.1}$$

Here m is the mass of the negative charge and ω_0 is the natural frequency of the oscillator. In this chapter, we investigate the behavior of such a model atom in external fields and interacting with each other.

11.1 Static Polarizability

In a slowly varying uniform electric field $\mathbf{E}$ the net force on the negative charge is

$$\mathbf{F} = e\mathbf{E} - m\omega_0^2\mathbf{r}\,. \tag{11.2}$$

The equilibrium displacement of the charge is given by the condition $\mathbf{F} = 0$:

$$\mathbf{r}_{\text{eq}} = \frac{e\mathbf{E}}{m\omega_0^2} \tag{11.3}$$

and it produces an electric dipole moment

$$\mathbf{d} = e\mathbf{r}_{\text{eq}} = \alpha\mathbf{E}\,, \tag{11.4}$$

where

$$\alpha = \frac{e^2}{m\omega_0^2} \tag{11.5}$$

is the *static dipole polarizability* of our model atom.

The potential energy of the polarized system is the negative of the work done by the force $\mathbf{F}$

$$\begin{aligned} W &= -\int_0^{\mathbf{r}_0} \mathbf{F} \cdot d\mathbf{r} && (11.6) \\ &= -e\mathbf{E} \cdot \mathbf{r}_0 + \frac{1}{2} m\omega_0^2 r_0^2 && (11.7) \\ &= -\frac{1}{2}\alpha \mathbf{E}^2 . && (11.8) \end{aligned}$$

A number density N of such atoms in a dilute gas will have a polarization $\mathbf{P} = N\mathbf{d} = N\alpha\mathbf{E}$ and therefore a displacement field of

$$\mathbf{D} = \varepsilon\mathbf{E} = \varepsilon_0\mathbf{E} + \mathbf{P} \tag{11.9}$$

and therefore a permittivity of[1]

$$\varepsilon = \varepsilon_0 + N\alpha \, . \tag{11.10}$$

Assuming negligible magnetization, the polarizability α will give rise to an index of refraction

$$n = \sqrt{\varepsilon/\varepsilon_0} = \sqrt{1 + N\alpha/\varepsilon_0} \tag{11.11}$$

for electromagnetic radiation at low frequencies $(\omega \ll \omega_0)$. We will investigate the interaction of electromagnetic waves in such media in more detail below.

11.2 Atoms in Anisotropic Media

The electrons on atoms in crystals of low symmetry, for example trigonal or rhombic crystals, usually experience anisotropic restoring forces. In a rhombohedral crystal like quartz, for example, the restoring force along one axis is different

[1] This expression is valid for dilute gases. At higher density, corrections are needed for the local field. See problem 1.

from that in the perpendicular plane: in place of equation (11.1) the restoring force of the model atom has the form

$$\mathbf{F} = -m\omega_{\|}^2 \mathbf{r}_{\|} - m\omega_{\perp}^2 \mathbf{r}_{\perp}. \tag{11.12}$$

In these circumstances, the equilibrium displacement in an electric field $\mathbf{E}$ is not always in the direction of $\mathbf{E}$:

$$e\mathbf{E} = m\omega_{\|}^2 \mathbf{r}_{\|} + m\omega_{\perp}^2 \mathbf{r}_{\perp} \tag{11.13}$$

$$\mathbf{r}_0 = \frac{e\mathbf{E}_{\|}}{m\omega_{\|}^2} + \frac{e\mathbf{E}_{\perp}}{m\omega_{\perp}^2}. \tag{11.14}$$

The polarizability is no longer a scalar, but must be a *dyadic*:

$$\overleftrightarrow{\boldsymbol{\alpha}} = \frac{e^2}{m\omega_{\|}^2} \overleftarrow{\mathbf{e}_3} \overrightarrow{\mathbf{e}_3} + \frac{e^2}{m\omega_{\perp}^2} \left(\overleftrightarrow{\mathbf{1}} - \overleftarrow{\mathbf{e}_3} \overrightarrow{\mathbf{e}_3} \right). \tag{11.15}$$

The meaning of a dyadic polarizability is that the electric dipole moment $\mathbf{d}$ induced by an external field $\mathbf{E}$ is generally not parallel to $\mathbf{E}$. Instead,

$$\mathbf{d} = \overleftrightarrow{\boldsymbol{\alpha}} \cdot \mathbf{E} \tag{11.16}$$

$$\begin{aligned} &= \alpha^{11}\mathbf{e}_1\,\mathbf{e}_1 \cdot \mathbf{E} + \alpha^{12}\mathbf{e}_1\,\mathbf{e}_2 \cdot \mathbf{E} + \alpha^{13}\mathbf{e}_1\,\mathbf{e}_3 \cdot \mathbf{E} \\ &+ \alpha^{21}\mathbf{e}_2\,\mathbf{e}_1 \cdot \mathbf{E} + \alpha^{22}\mathbf{e}_2\,\mathbf{e}_2 \cdot \mathbf{E} + \alpha^{23}\mathbf{e}_2\,\mathbf{e}_3 \cdot \mathbf{E} \\ &+ \alpha^{31}\mathbf{e}_3\,\mathbf{e}_1 \cdot \mathbf{E} + \alpha^{32}\mathbf{e}_3\,\mathbf{e}_2 \cdot \mathbf{E} + \alpha^{33}\mathbf{e}_3\,\mathbf{e}_3 \cdot \mathbf{E}, \end{aligned} \tag{11.17}$$

which, for the cylindrically symmetric case described above, reduces to

$$\mathbf{d} = \frac{e^2}{m\omega_{\|}^2} \mathbf{E}_{\|} + \frac{e^2}{m\omega_{\perp}^2} \mathbf{E}_{\perp}. \tag{11.18}$$

Similarly, the dielectric constant is a dyadic:

$$\overleftrightarrow{\boldsymbol{\varepsilon}} = \varepsilon_0 \overleftrightarrow{\mathbf{1}} + n \overleftrightarrow{\boldsymbol{\alpha}}. \tag{11.19}$$

In anisotropic media, the displacement field is not necessarily in the same direction as the electric field $\mathbf{E}$:

$$\mathbf{D} = \overleftrightarrow{\boldsymbol{\varepsilon}} \cdot \mathbf{E}, \tag{11.20}$$

and the medium can be *birefringent,* that is the index of refraction can depend on the linear polarization of the incident electromagnetic wave.

11.3 Scattering of Radiation by Bound Charge

Let a charge e be isotropically bound with a linear restoring force $-m\omega_0^2\mathbf{r}$. The equation of motion of the charge in an external electric field $\mathbf{E}$ in the nonrelativistic limit is given by Newton's second law to be

$$\ddot{\mathbf{r}} + \omega_0^2\mathbf{r} = \frac{e}{m}\mathbf{E}. \tag{11.21}$$

We look for steady-state solutions of the form $\mathbf{r}(t) = \mathbf{r}(0)\exp(-i\omega t)$ when $\mathbf{E}(t) = \mathbf{E}(0)\exp(-i\omega t)$:

$$\left(\omega_0^2 - \omega^2\right)\mathbf{r}(0)\,e^{-i\omega t} = \frac{e}{m}\mathbf{E}(0)\,e^{-i\omega t}. \tag{11.22}$$

Now the complex solution is not physical, but since the equation of motion (11.21) is real and linear, any linear combination of solutions and their complex conjugates is also a solution. In particular, for an exciting field $\mathbf{E}(t) = \mathbf{E}(0)\cos\omega t$, the steady-state solution is $\mathbf{r}(t) = \mathbf{r}(0)\cos\omega t$ where the amplitude of oscillation is

$$\mathbf{r}(0) = \frac{e\mathbf{E}(0)/m}{\omega_0^2 - \omega^2}. \tag{11.23}$$

The result can be described as a *dynamic polarizability*

$$\begin{aligned}\alpha(\omega) &= e\mathbf{r}(0)\,\mathbf{E}^{-1}(0)\\ &= \frac{e^2/m}{\omega_0^2 - \omega^2},\end{aligned} \tag{11.24}$$

which leads to a frequency-dependent permittivity

$$\varepsilon(\omega) = \varepsilon_0 + N\alpha(\omega) \tag{11.25}$$

and index of refraction

$$\begin{aligned}n(\omega) &= \sqrt{\varepsilon(\omega)/\varepsilon_0}\\ &= \sqrt{1 + N\alpha(\omega)/\varepsilon_0}.\end{aligned} \tag{11.26}$$

All of the dynamic values reduce to the static limits as $\omega \to 0$.

The amplitude of the oscillation has a *resonance* at the natural frequency $\omega = \omega_0$ of the oscillator. Indeed our result predicts an infinite amplitude of oscillation at $\omega = \omega_0$. In fact , the amplitude will be limited by the damping effects of the radiation it emits. We will estimate the effects of such damping below. Note that the displacement of the charge is in the direction of the electric field at frequencies $\omega < \omega_0$ below the natural resonance frequency but is in the opposite direction, in other words 180 degrees out of phase with the field, at frequencies above resonance.

From the previous chapter, we know that the oscillating charge will radiate an electromagnetic field

$$\mathbf{F}_r = \left(1 + \hat{\mathbf{R}}\right) \mathbf{E}_r \,, \tag{11.27}$$

where in the nonrelativistic limit,

$$\mathbf{E}_r = \frac{Ke}{cR^0}\hat{\mathbf{R}}\times\left(\hat{\mathbf{R}} \times \mathbf{a}\right) = -\frac{Ke}{cR^0}\mathbf{a}_\perp . \tag{11.28}$$

The power of this radiation radiated into the solid angle $d\Omega$ is given by

$$\mathbf{S}_r \cdot \hat{\mathbf{R}}\mathbf{R}^2 d\Omega = c\varepsilon_0 \mathbf{R}^2 \mathbf{E}_r^2 d\Omega = \frac{Ke^2\mathbf{a}^2}{4\pi c^2} \sin^2\theta \, d\Omega \,, \tag{11.29}$$

where θ is the angle between $\mathbf{R}$ and $\mathbf{a}$, as in Fig. 10.3. Since the acceleration $\mathbf{a}$ is given by $(e/m)\,\mathbf{E}{-}\omega_0^2\mathbf{r}$, the radiated power per steradian is

$$\frac{d\mathcal{P}}{d\Omega} = \frac{Ke^2\mathbf{a}^2}{4\pi c^2} \sin^2\theta \tag{11.30}$$

$$= \frac{Ke^4\mathbf{E}^2\omega^4 \sin^2\theta}{4\pi m^2 c^2 \left(\omega^2 - \omega_0^2\right)^2}, \tag{11.31}$$

which, with $K = (4\pi\varepsilon_0 c)^{-1}$ and the incident intensity $\mathbf{S}_{\text{in}} = \varepsilon_0 c \hat{\mathbf{R}}\mathbf{E}^2$, gives a cross section

$$\frac{d\sigma}{d\Omega} = \frac{1}{|\mathbf{S}_{\text{in}}|} \frac{d\mathcal{P}}{d\Omega} \tag{11.32}$$

$$= \frac{Ke^4\omega^4 \sin^2\theta}{4\pi m^2 c^3 \varepsilon_0 \left(\omega^2 - \omega_0^2\right)^2} \tag{11.33}$$

$$= \left(\frac{Ke^2\omega^2}{mc\left(\omega^2 - \omega_0^2\right)}\right)^2 \sin^2\theta \tag{11.34}$$

$$\to \left(\frac{Ke^2}{mc}\right)^2 \sin^2\theta \,, \tag{11.35}$$

where the last line is the Thomson-scattering limit, derived in the last chapter for free electrons, and valid for the bound charge when $\omega \gg \omega_0$.

The opposite limit is the *Rayleigh-scattering* limit, valid for low frequencies $\omega \ll \omega_0$. In Rayleigh scattering the cross section increases as the fourth power of the frequency:

$$\frac{d\sigma}{d\Omega} \to \left(\frac{Ke^2\omega^2}{mc\omega_0^2}\right)^2 \sin^2\theta\,. \tag{11.36}$$

More generally, the scattered intensity from the bound electron has a strong resonance at $\omega = \omega_0$. Real atoms and molecules have many resonances. One can model them with a classical oscillator with many modes and with an oscillator strength that is distributed among them.

Note from (11.28) that if the incident radiation is linearly polarized in a fixed direction $\mathbf{E}$, the scattered radiation will be linearly polarized along $\hat{\mathbf{R}}\times\left(\hat{\mathbf{R}}\times\mathbf{E}\right)$, that is along the projection of the incident polarization vector onto the plane perpendicular to the propagation direction. The scattered radiation is generally polarized even when the incident radiation is unpolarized. The scattering cross section for unpolarized exciting radiation should be averaged over the initial polarizations. As shown in Chapt. 10, the result is

$$\frac{d\sigma}{d\Omega} = \left(\frac{Ke^2\omega^2}{mc\left(\omega^2-\omega_0^2\right)}\right)^2 \left(\frac{1+\left(\hat{\mathbf{k}}\cdot\hat{\mathbf{R}}\right)^2}{2}\right), \tag{11.37}$$

where $\hat{\mathbf{k}}$ is the incident propagation direction. When $\hat{\mathbf{k}}$ and $\hat{\mathbf{R}}$ are perpendicular, the scattered radiation must be polarized normal to the $\hat{\mathbf{k}}\hat{\mathbf{R}}$ plane. Because of this effect, light scattered in the sky from a direction that makes a right angle to the line from the sun is strongly polarized. It will not be completely polarized because of effects we have ignored, such as multiple scattering and scattering from anisotropic molecules.

11.4 Blue Sky

The Rayleigh-scattering cross section varies as the fourth power of the frequency. One can therefore expect media to scatter blue light more strongly than red. The increased scattering cross section at high frequencies is a major factor explaining the blue color of the sky and, for light passing through the medium, a progressive reddening of the transmitted color.

However, the Rayleigh-scattering cross section is not the whole story behind blue sky and red sunsets. If the medium is homogeneous, most scattering not in the forward direction is wiped out by interference. Consider, for example, a thin layer of medium in a plane perpendicular to the propagation vector $\mathbf{k}$ of the incident directed plane wave. At any given instant t, the field $\mathbf{F}$ is the same everywhere in the layer and the driven oscillations of all the electrons in the layer will be in phase. By symmetry, the net propagation direction of the total field scattered by the electrons in the layer must be normal to the layer: components of $\mathbf{k}$ in the plane arising from electrons in one half of the plane will be canceled by symmetrically placed electrons in the other half. The medium can produce a net scattering of radiation only if it is not homogeneous.

To account for coherence effects in the scattering of radiation from a distribution of identical scattering centers, the scattering cross section for a single center must be multiplied by a *structure factor* that adds the relative phases from the distribution of scatterers.[2] The phase factor contributed by the path from the source at $\mathbf{r}_s$ to the scattering center at $\mathbf{r}_j$ along incident propagation direction $\mathbf{k}$ and then onto the observer at $\mathbf{x}$ is

$$\exp\left[i\mathbf{k}\cdot(\mathbf{r}_j - \mathbf{r}_s) + i\,|\mathbf{k}|\,\hat{\mathbf{R}}\cdot(\mathbf{x} - \mathbf{r}_j)\right] . \tag{11.38}$$

The structure factor $\mathcal{F}(\mathbf{q})$ is the absolute magnitude of the sum

$$\mathcal{F}(\mathbf{q}) = \left|\sum_j e^{-i\mathbf{q}\cdot\mathbf{r}_j}\right|^2 , \tag{11.39}$$

where $\mathbf{q} = |\mathbf{k}|\,\hat{\mathbf{R}} - \mathbf{k}$. If the positions $\mathbf{r}_j$ of the molecules are random, the interference in the structure factor will cancel for $\mathbf{q} \neq \mathbf{0}$, leaving $\mathcal{F}(\mathbf{q}) = NV$, the number of molecules. The molecules then scatter independently. On the other hand, if the molecules occupy a regular array whose spacing is much less than the wavelength of the radiation, the structure factor can disappear except for forward scattering, for which $\mathbf{q} = 0$. Inhomogeneities arise from density fluctuations in the air. Let N be the average density of air molecules in a given region of the atmosphere. From basic statistics, we expect a volume V of air with an *average*

[2]The treatment here ignores intensity differences arising from slightly different total path lengths. Such differences are usually small and their inclusion, while straightforward, significantly complicates the expressions.

number of NV molecules to have fluctuations in the actual number of molecules in V on the order of $\sqrt{NV}$. In other words, if many measurements were made of the number of molecules in the volume V, a distribution of values would be found with an average value of NV and a standard deviation on the order of $\sqrt{NV}$. Consequently, typical *density* fluctuations in the volume V will be $\sqrt{N/V}$, that is, inversely proportional to the square root of V. However, as Einstein discovered in one of his early studies, thermodynamics also plays an important role, and the fluctuations become large near the triple point, where critical opalescence can set in. The total result of density fluctuations and the Rayleigh-scattering cross section is that radiation at the extreme blue end of the visual spectrum is roughly $(1.6)^4 = 6.5$ times more likely to be scattered than radiation at lower frequency near the extreme red end. It is the combined effect that is responsible for our blue skies and red sunsets.

11.5 Scattering in a Magnetic Field

One can extend the scattering problem above by adding (i) the effect of a constant magnetic field $\mathbf{B}$ on the motion of the charge and (ii) a small damping force of the form $-m\Gamma\dot{\mathbf{r}}$.

The damping force arises from the radiation reaction averaged over an oscillator cycle. The total power radiated by a nonrelativistic charge of acceleration $\mathbf{a}$ was found in the last chapter to be given by the Larmor result

$$\mathcal{P} = \frac{2Ke^2\mathbf{a}^2}{3c^2}. \tag{11.40}$$

For the harmonic oscillator, the average value $\langle \mathbf{a}^2 \rangle_{av} = \omega^2 \langle \dot{\mathbf{r}}^2 \rangle_{av}$ and the average power radiated is the same as that from a radiation force

$$-\mathbf{f}_{rad} = \frac{2Ke^2\omega^2\dot{\mathbf{r}}}{3c^2} = m\Gamma\dot{\mathbf{r}}. \tag{11.41}$$

The force $\mathbf{f}_{rad} = -m\Gamma\dot{\mathbf{r}}$, with

$$\Gamma = \frac{2Ke^2\omega^2}{3mc^2} \tag{11.42}$$

is the net *radiation reaction force* acting on the radiating charge.[3]

[3] The more general problem of radiation reaction is addressed by the Lorentz-Dirac equation in the next chapter.

The magnetic field $\mathbf{B}$ adds a force $e\mathbf{v} \times \mathbf{B}$ to the moving charge. When this and the radiation reaction force are added to Newton's second law, the equation of motion takes the form

$$\ddot{\mathbf{r}} + \Gamma\dot{\mathbf{r}} - \boldsymbol{\omega}_c \times \dot{\mathbf{r}} + \omega_0^2\mathbf{r} = \frac{e}{m}\mathbf{E}, \tag{11.43}$$

where $\boldsymbol{\omega}_c = -(e/m)\,\mathbf{B}$ is the cyclotron frequency. To separate the equation into component form, apply the projectors $P = \frac{1}{2}(1 + \mathbf{e}_3)$ and $\bar{P}$ to both sides of the equation, where $\mathbf{e}_3$ is the direction of $\mathbf{B}$, and then taking scalar components with $\mathbf{e}_1$ and $\mathbf{e}_3$. Since

$$\begin{aligned} P\mathbf{E}P &= \left(P\mathbf{E} + \mathbf{E}\bar{P}\right)P = 2\left\langle P\mathbf{E}\right\rangle_S P \\ &= \mathbf{E}\cdot\mathbf{e}_3\,P, \end{aligned} \tag{11.44}$$

then

$$\begin{aligned} \frac{e}{m}\mathbf{E}\cdot\mathbf{e}_3 &= \frac{e}{m}\left(P\mathbf{E}P + \bar{P}\mathbf{E}\bar{P}\right) &(11.45)\\ &= \left(\ddot{\mathbf{r}} + \Gamma\dot{\mathbf{r}} + \omega_0^2\mathbf{r}\right)\cdot\mathbf{e}_3 . &(11.46)\end{aligned}$$

Similarly,

$$\begin{aligned} P\mathbf{E}\bar{P} &= \mathbf{E}_\perp\bar{P} = \left(E^1\mathbf{e}_1 + E^2\mathbf{e}_2\right)\bar{P} &(11.47)\\ &= \left(E^1 - iE^2\right)\mathbf{e}_1\bar{P} &(11.48)\\ &= \mathbf{E}\cdot\mathbf{n}^1\,\mathbf{n}_1, &(11.49)\end{aligned}$$

where $\mathbf{n}_1 = \mathbf{e}_1 + i\mathbf{e}_2$ and $\mathbf{n}^1 = \frac{1}{2}\mathbf{n}_1^\dagger = \frac{1}{2}\mathbf{n}_{-1}$ is the corresponding reciprocal basis vector (see Appendix). The $\mathbf{n}^1$ component of $\boldsymbol{\omega}_c \times \mathbf{r}$ is

$$\begin{aligned} \left(\boldsymbol{\omega}_c \times \mathbf{r}\right)\cdot\mathbf{n}^1\,\mathbf{n}_1 &= -\frac{i}{2}P\left[\boldsymbol{\omega}_c, \mathbf{r}\right]\bar{P} \\ &= -i\omega_c P\mathbf{r}\bar{P} \\ &= -i\omega_c\mathbf{r}\cdot\mathbf{n}^1\,\mathbf{n}_1, \end{aligned} \tag{11.50}$$

where by definition, ω_c is positive for (negatively charged) electrons with $e < 0$. Solving for steady-state oscillating solutions as for the equation of motion (11.43), we obtain the dependence

$$\mathbf{r}(t)\cdot\mathbf{n}^M = \frac{\frac{e}{m}\mathbf{E}(t)\cdot\mathbf{n}^M}{\omega_0^2 - \omega^2 - M\omega\omega_c - i\Gamma\omega}, \tag{11.51}$$

where $M = \pm 1, 0$ and

$$\mathbf{n}_M = \begin{cases} \mathbf{e}_1 \pm i\mathbf{e}_2 \,, & M = \pm 1 \\ \mathbf{e}_3 \,, & M = 0 \end{cases} \,. \tag{11.52}$$

The physical significance of the complex phase that multiplies $\mathbf{E}(t)$ is that it shows that the oscillations of $\mathbf{r}(t)$ are generally not in phase with those of $\mathbf{E}(t)$. Observation of this dependence can be used to measure atomic lifetimes (see problem 4).

11.6 Drude Model of Van der Waals Forces

The Drude model of interacting atoms considers two identical isotropic oscillators, separated by $\mathbf{R}_{12}$ and coupled by the dipole-dipole interaction between their oscillating dipole moments:

$$V = -\frac{e^2}{4\pi\varepsilon_0 \left|\mathbf{R}_{12}\right|^3}\left(\mathbf{r}_1 \cdot \mathbf{r}_2 - 3\mathbf{r}_1 \cdot \hat{\mathbf{R}}_{12}\,\mathbf{r}_2 \cdot \hat{\mathbf{R}}_{12}\right) \,. \tag{11.53}$$

The Hamiltonian is

$$H = \frac{\mathbf{p}_1^2}{2m} + \frac{\mathbf{p}_2^2}{2m} + \frac{1}{2}m\omega^2\left(\mathbf{r}_1^2 + \mathbf{r}_2^2\right) + V. \tag{11.54}$$

The oscillators can be uncoupled by a change in variables. First, choose the z-axis to be parallel to $\mathbf{R}_{12}$ and split the Hamiltonian into

$$H = H_x + H_y + H_z \,, \tag{11.55}$$

where

$$H_x = \frac{1}{2}m\left(\dot{x}_1^2 + \dot{x}_2^2\right) + \frac{1}{2}m\omega^2\left(x_1^2 + x_2^2 + 2\beta_x x_1 x_2\right) \tag{11.56}$$

$$\beta_x = -\frac{e^2}{4\pi\varepsilon_0 m\omega^2}\left|\mathbf{R}_{12}\right|^{-3}, \tag{11.57}$$

and similar expressions for H_y and H_z except with

$$\beta_z = -2\beta_x = -2\beta_y \,. \tag{11.58}$$

Now put

$$\begin{aligned} x_\pm &= x_1 \pm x_2 & (11.59)\\ x_1 &= \frac{1}{2}(x_+ + x_-) & (11.60)\\ x_2 &= \frac{1}{2}(x_+ - x_-) & (11.61)\end{aligned}$$

so that

$$\begin{aligned} x_1^2 + x_2^2 &= \frac{1}{2}\left(x_+^2 + x_-^2\right) & (11.62)\\ \dot{x}_1^2 + \dot{x}_2^2 &= \frac{1}{2}\left(\dot{x}_+^2 + \dot{x}_-^2\right) & (11.63)\\ x_1 x_2 &= \frac{1}{4}\left(x_+^2 - x_-^2\right) & (11.64)\end{aligned}$$

and

$$H_x = \frac{1}{4}m\left(\dot{x}_+^2 + \dot{x}_-^2\right) + \frac{1}{4}m\omega^2\left[x_+^2(1+\beta_x) + x_-^2(1-\beta_x)\right]. \qquad (11.65)$$

Elementary quantum theory predicts zero-point fluctuations of energy

$$\begin{aligned} E_x &= \frac{1}{2}\hbar\omega\left(\sqrt{1+\beta_x} + \sqrt{1-\beta_x}\right)\\ &\approx \frac{1}{2}\hbar\omega\left(2 - \frac{1}{4}\beta_x^2\right), & (11.66)\end{aligned}$$

where the second line is valid to order β_x^4 for small couplings: $\beta_x^2 \ll 1$. Summing the energies of the x, y, and z oscillators, we find

$$E = 3\hbar\omega\left(1 - \frac{1}{4}\beta_x^2\right). \qquad (11.67)$$

To lowest nonvanishing order, the dipole-dipole coupling of the oscillators has thus lowered the total ground-state energy by

$$\frac{3}{4}\hbar\omega\beta_x^2 = \frac{3\hbar e^4}{4\left(4\pi\varepsilon_0 m\right)^2\omega^3}\left|\mathbf{R}_{12}\right|^{-6}. \qquad (11.68)$$

This is the van der Waals interaction.

11.7 Diffraction

Light passing a sharp edge or small opaque object displays a series of light and dark fringes at the edge of the shadow. The fringes are an aspect of *diffraction* and are characteristic of waves of all sorts. We can understand the phenomenon by considering the radiation from the current in part of a conducting screen, as treated in Chapter 9. Consider a plane wave

$$\mathbf{F}_0 = \left(1 + \hat{\mathbf{k}}_0\right) \mathbf{E}_0 \cos s\,, \quad s = \langle k\bar{y}\rangle_S = \omega t - \mathbf{k}_0 \cdot \mathbf{y} \tag{11.69}$$

at spacetime position y normally incident on a screen located at $\mathbf{y} \cdot \mathbf{k}_0 = 0$. The current generated in the screen was found to be

$$j(y) = 2\varepsilon c \mathbf{E}_0(\omega t)\, \delta\left(\hat{\mathbf{k}}_0 \cdot \mathbf{y}\right)\,, \tag{11.70}$$

and from Chapter 10 we know that this current will radiate a field with vector potential

$$\begin{aligned} A(x) &= \frac{\mu}{4\pi} \int d^3\mathbf{y} \left[\frac{j(y)}{|\mathbf{x} - \mathbf{y}|}\right]_{ret} &(11.71)\\ &= \frac{\mathbf{E}_0}{2\pi c} \int d^2\mathbf{y}\, \frac{\cos(\omega t - \omega\,|\mathbf{x} - \mathbf{y}|\,/c)}{|\mathbf{x} - \mathbf{y}|}\,, &(11.72) \end{aligned}$$

where the integration in the last expression is over the screen. At this point, we must distinguish between *Fresnel diffraction* and *Fraunhofer diffraction.* In the Fresnel case, the source and projection (observation) points are relatively close to the screen, whereas in the Fraunhofer case, either both are at a distance large compared to the size of the screen or a collimating lens is used so that only radiation on parallel paths can interfere.

To calculate Fraunhofer diffraction from a single rectangular section of screen of length L, assumed large compared to a wavelength $2\pi c/\omega$ so that interference from the $y^{(2)}$ integration is negligible, and of width $2a$, we note that for parallel radiation at an angle θ to the normal, the distance to the screen is

$$R^0 = |\mathbf{x} - \mathbf{y}| \simeq |\mathbf{x}| - y^1 \sin\theta\,, \tag{11.73}$$

where $|\mathbf{x}| \gg a$ is the distance from the center of the section of screen (see

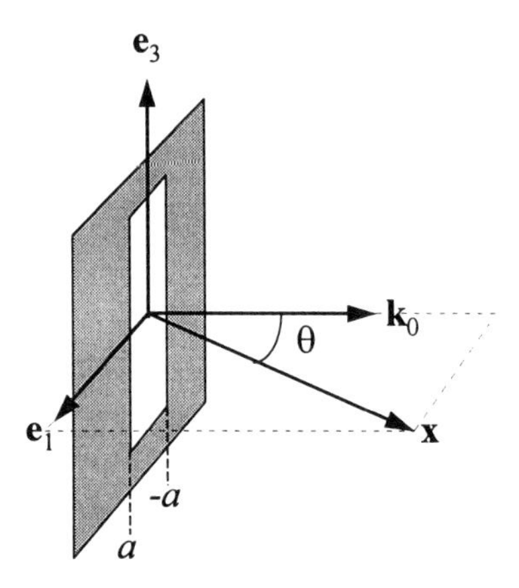

Fig. 11.1. Diffraction from a rectangular conducting screen or, equivalently, from a rectangular aperture in a conducting screen.

Fig. 11.1). We find

$$\begin{aligned} A(x) &= \frac{\mathbf{E}_0 L}{2\pi c}\int_{-a}^{a} dy^1\, \frac{\cos[\omega t - \omega\left(|\mathbf{x}| - y^1 \sin\theta\right)/c]}{|\mathbf{x}-\mathbf{y}|} \\ &\simeq \frac{\mathbf{E}_0 L}{2\pi c\,|\mathbf{x}|}\int_{-a}^{a} dy^1\, \cos\big[\omega t - \omega\left(|\mathbf{x}| - y^1 \sin\theta\right)/c\big] \\ &\simeq -\frac{\mathbf{E}_0 L}{\pi\omega\,|\mathbf{x}|\sin\theta}\cos(\omega t - \omega\,|\mathbf{x}|/c)\sin\Big(\frac{\omega a}{c}\sin\theta\Big)\,, \end{aligned} \tag{11.74}$$

where $s = \omega t - \mathbf{k}\cdot\mathbf{x}$ and $\mathbf{k} = \omega\hat{\mathbf{x}}/c$. The intensity is proportional to the square of the amplitude, and hence to

$$I \sim \left(\frac{\sin\xi}{\xi}\right)^2, \tag{11.75}$$

where $\xi = (\omega a/c)\sin\theta$ gives the position from the center of the image of the rectangular screen. as we will see in the following subsection, this diffraction pattern is also the Fraunhofer pattern for a slit.

11.7.1 Babinet's and Huygen's Principles

To represent the effect of a hole in a conducting screen, we can simply omit the current from that region, or equivalently, we can add compensating current of the opposite sign. This equivalence is the basis of *Babinet's principle,* which states that the diffraction pattern of a conduction screen with a hole in it is the same as without the screen but with the "plug" that would just fill the hole. However, it is significant that the phase of the driven oscillators in the plug produces an electromagnetic field with a 90° phase shift from that field that passes through the hole. Babinet's principle is a refinement of *Huygen's principle*, which simply states that wave propagation is equivalent to each point on any wave front acting as a new point source. Huygen's principle does not address the important question of phase.

11.8 Linear Antenna

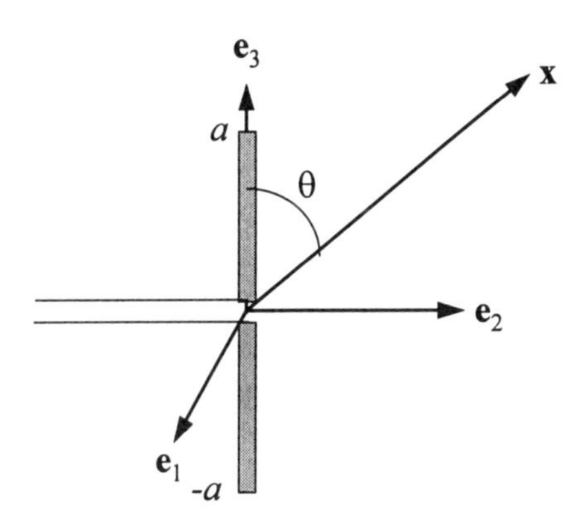

Fig. 11.2. A linear antenna of total length $2a$ aligned along the $\mathbf{e}_3$ axis is driven by a harmonic current.

Consider a linear antenna of length $2a$ centered at the origin, aligned with $\mathbf{e}_3$, and carrying a vector current in the form of a standing wave

$$\mathbf{j}(y) = I\mathbf{e}_3 \cos(\omega t) \sin\left[\frac{\omega}{c}\left(a - \left|y^{(3)}\right|\right)\right] \delta^{(2)}(\mathbf{y}_\perp) \ , \ -a < y^{(3)} < a \, , \qquad (11.76)$$

where $\mathbf{y}_\perp \equiv \mathbf{y} - y^{(3)}\mathbf{e}_3$. The current is appropriate for a center-fed antenna when radiation reaction is ignored.[4] The corresponding vector potential (11.71) is a vector aligned with $\mathbf{e}_3$:

$$\mathbf{A}(x) = \frac{\mu I \mathbf{e}_3}{4\pi}\left\langle \int_{-a}^{a} dy^{(3)}\, e^{-i\omega t'} \frac{\sin\left[\frac{\omega}{c}\left(a - \left|y^{(3)}\right|\right)\right]}{\left|\mathbf{x}-\mathbf{y}\right|_{ret}} \right\rangle_{\Re}, \tag{11.77}$$

where $t' = t - R^0/c$ is the retarded time. At large distances corresponding to the Fraunhofer limit, $R^0 \simeq |\mathbf{x}| - y^{(3)}\cos\theta \gg a$, where $\cos\theta = \hat{\mathbf{x}}\cdot\mathbf{e}_3$, and the integral simplifies to

$$\begin{aligned} \mathbf{A}(x) &= \frac{\mu I \mathbf{e}_3}{4\pi\,|\mathbf{x}|}\left\langle e^{is}\int_{-a}^{a} dy^{(3)}\, e^{i\mathbf{k}\cdot\mathbf{e}_3 y^{(3)}} \sin\left[\frac{\omega}{c}\left(a - \left|y^{(3)}\right|\right)\right]\right\rangle_{\Re} && (11.78) \\ &= \frac{\mu c I \mathbf{e}_3 \cos s}{2\pi\omega\,|\mathbf{x}|}\,\frac{\cos\left(\frac{\omega a}{c}\cos\theta\right) - \cos\left(\frac{\omega a}{c}\right)}{\sin^2\theta} && (11.79) \end{aligned}$$

where $s = \omega t - \mathbf{k}\cdot\mathbf{x}$ and $\mathbf{k} = \hat{\mathbf{x}}\omega/c$. The time dependence is contained in the factor $\cos s$, which therefore governs the propagation of the radiation. It is seen that $\mathbf{A}$ propagates in the $\hat{\mathbf{x}} = \hat{\mathbf{k}}$ direction. Although we have not calculated the scalar part ϕ/c of the paravector potential, we know from Chapter 6 that with the Lorenz gauge condition, only the vector component $\mathbf{A}_\perp$ perpendicular to the propagation direction contributes to the radiation field. Thus

$$\begin{aligned} \mathbf{F} &= c\left\langle \partial \bar{A}\right\rangle_V = -c\left\langle \partial \mathbf{A}_\perp\right\rangle_V && (11.80) \\ &= \left(1+\hat{\mathbf{k}}\right)\mathbf{E} && (11.81) \end{aligned}$$

with an electric field in the direction of $\hat{\boldsymbol{\varepsilon}}\sin\theta = \mathbf{e}_3 - \mathbf{e}_3\cdot\hat{\mathbf{k}}\,\hat{\mathbf{k}}$, given at large $|\mathbf{x}|$ by

$$\begin{aligned} \mathbf{E} &= -\omega \mathbf{A}'_\perp(s) && (11.82) \\ &= \frac{\mu c I \hat{\boldsymbol{\varepsilon}}\sin s}{2\pi\,|\mathbf{x}|}\left[\frac{\cos\left(\frac{\omega a}{c}\cos\theta\right) - \cos\left(\frac{\omega a}{c}\right)}{\sin\theta}\right]. && (11.83) \end{aligned}$$

The energy density and Poynting vector are

$$\begin{aligned} \mathcal{E} + \mathbf{S}/c &= \frac{1}{2}\varepsilon \mathbf{F}\mathbf{F}^\dagger = \varepsilon\left(1+\hat{\mathbf{k}}\right)\mathbf{E}^2 && (11.84) \\ &= \mu\left(1+\hat{\mathbf{k}}\right)\left(\frac{I\sin s}{2\pi\,|\mathbf{x}|}\right)^2\left[\frac{\cos\left(\frac{\omega a}{c}\cos\theta\right) - \cos\left(\frac{\omega a}{c}\right)}{\sin\theta}\right]^2 && (11.85) \end{aligned}$$

[4]In practical antennas, the radiated energy is maximized by choosing a length $2a \simeq \lambda/2$. The fraction of stored energy that is radiated per cycle is then large and the approximation to neglect radiation reaction is poor.

The average power per unit solid angle is thus

$$\frac{dP}{d\Omega} = \langle \mathbf{S} \cdot \hat{\mathbf{x}} \rangle |\mathbf{x}|^2 \tag{11.86}$$

$$= \frac{I^2 Z}{8\pi^2} \left[\frac{\cos\left(\frac{\omega a}{c} \cos\theta\right) - \cos\left(\frac{\omega a}{c}\right)}{\sin\theta} \right]^2, \tag{11.87}$$

where $Z = \mu c = (\varepsilon c)^{-1}$ is the impedance of the medium.[5]

11.9 Problems

1. The relation (11.10) is valid for dilute gases, but at higher density corrections are needed to the local field. Show that the *Clausius-Mossotti* equation,

$$\frac{\varepsilon - \varepsilon_0}{\varepsilon + 2\varepsilon_0} = \frac{1}{3} N\alpha \tag{11.88}$$

   gives a correction

$$\varepsilon = \varepsilon_0 + N\alpha + \frac{1}{3}(N\alpha)^2 + \cdots . \tag{11.89}$$

2. **Scattering amplitude.** Consider the scattering of radiation by a harmonically bound charge e (see section 11.3). Let the charge motion be nonrelativistic, driven by an incident wave of complex (unphysical) electric field $\mathbf{E}(t) = \mathbf{E}(0)\exp(-i\omega t)$. Show that the radiated electric field at spacetime position $R = x - r(t_{\text{ret}}) = R^0 + \mathbf{R}$ relative to the retarded position $r(t_{\text{ret}})$ of the charge can be expressed by the complex spherical wave

$$\mathbf{E}_r(t) = A\frac{e^{-i\omega t + i\mathbf{k}'\cdot\mathbf{R}}}{R^0}\hat{\boldsymbol{\varepsilon}}' |\mathbf{E}(0)|, \tag{11.90}$$

   where $\mathbf{k}' = \hat{\mathbf{R}}\omega/c$ is the propagation vector of the scattered field and $\hat{\boldsymbol{\varepsilon}}'$ is the real unit polarization vector in the direction of $\mathbf{E}_\perp = \hat{\mathbf{R}}\times\left(\mathbf{E}\times\hat{\mathbf{R}}\right)$. The scalar coefficient A with dimensions of length is called the *scattering amplitude.* Determine A as a function of the angle θ between the incident field direction ($\hat{\mathbf{E}}$) and scattered ($\hat{\mathbf{k}}'$) propagation direction.

[5]The topics of diffraction theory and antennas are vast, with many possible configurations, analyses, and practical applications. See, for example, J. D. Jackson (1998) and references therein.

3. **Optical Theorem**. Add the radiation damping force $\mathbf{f}_{rad} = -m\Gamma\dot{\mathbf{r}}$ to the oscillator in the previous question in order to balance the average radiated energy with the energy absorbed from the incident field. Show that the scattering amplitude now becomes complex:

$$A = -\left(\frac{Ke^2}{mc}\right)\frac{\omega^2 \sin\theta}{\omega^2 - \omega_0^2 + i\omega\Gamma}. \tag{11.91}$$

Show further that the imaginary part of A for forward scattered radiation ($\hat{\mathbf{k}}' = \hat{\mathbf{k}}$) is related to the total scattering cross section σ by

$$\sigma = \frac{4\pi}{k}\Im\left\{A\left(\hat{\mathbf{k}}' = \hat{\mathbf{k}}\right)\right\}. \tag{11.92}$$

This is a special case of the *optical theorem,* which can be shown by conservation of energy to be true more generally for complex scattered waves with a harmonic dependence of the form $e^{-i\omega t}$.

4. Consider an isotropic oscillator in a constant magnetic field $\mathbf{B}$, driven by an electric field $\mathbf{E}(t)$.

   (a) Show that close to resonance, $\omega \approx \omega_0$, the frequency dependence of the oscillator displacement can be approximated by

$$\mathbf{r}(t)\cdot\mathbf{n}^M \approx \frac{e\mathbf{E}(t)\cdot\mathbf{n}^M}{2m\omega_0}\left(\omega_0 - M\omega_c/2 - \omega - i\Gamma/2\right)^{-1}. \tag{11.93}$$

   (b) Determine the scattered intensity in the direction of $\mathbf{E}(0)$ as a function of the magnetic-field strength when $\mathbf{B}\cdot\mathbf{E}(0) = 0$ for the case of "white-light" excitation, where the incident intensity is nearly constant at all frequencies in the neighborhood of the relatively sharp resonance line. Show that the scattered intensity varies as

$$\frac{\omega_c^2}{\omega_c^2 + \Gamma^2} \tag{11.94}$$

   and sketch the dependence as a function of ω_c. [Hint: this part requires contour integration in the complex-ω plane. You may find the relation

$$\left(\frac{1}{\omega - a}\right)\left(\frac{1}{\omega - b}\right) = \left(\frac{1}{a - b}\right)\left(\frac{1}{\omega - a} - \frac{1}{\omega - b}\right) \tag{11.95}$$

   useful. You may assume that the incident intensity falls off rapidly at large frequency separations from the resonance line.]

(c) Could you use a measurement of such intensity to determine the damping constant Γ? [Note: this problem is a classical model of the *Hanle* effect, which has been used to determine the lifetimes of atomic levels.]

5. Expand the square-root terms to determine the next higher-order corrections to the van der Waals interaction.

6. Van der Waals interactions also occur between atoms of different types and hence of different natural oscillator frequencies. Extend the Drude model to the lowest-order van der Waals interaction between isotropic oscillators of natural frequencies ω_1 and ω_2. Show that the interaction (11.68) is regained in the limit $\omega_2 \to \omega_1 = \omega$.

7. Consider the radiation pattern for the linear antenna in three cases.

(a) Show that the pattern has the dipole dependence $\sin^2\theta$ in the limit that $a \ll \lambda = 2\pi c/\omega$.

(b) Show that in the usual case of a half-wave antenna, in which $2a = \lambda/2$, the power distribution is

$$\frac{dP}{d\Omega} = \frac{I^2 Z}{8\pi^2}\left[\frac{\cos\left(\frac{\pi}{2}\cos\theta\right)}{\sin\theta}\right]^2. \tag{11.96}$$

(c) Show that if the antenna length is a full wavelength, $2a = \lambda$, the power per steradian is

$$\frac{dP}{d\Omega} = \frac{I^2 Z}{2\pi^2}\left[\frac{\cos^2\left(\frac{\pi}{2}\cos\theta\right)}{\sin\theta}\right]^2, \tag{11.97}$$

and demonstrate that in the equatorial plane ($\theta = \pi/2$), the power is equal to that from the coherent superposition of fields from two have-wave antennas, oscillating in phase.

(d) Compare the angular dependencies for the three cases (a), (b), and (c), above.

Chapter 12

The Lorentz-Dirac Equation

12.1 Introduction

Maxwell's equation together with the Lorentz-force equation form a consistent set of relations that conserve energy and momentum, as seen for example in Poynting's theorem. From the previous chapter, we know that any accelerating charge radiates electromagnetic fields and, hence, energy and momentum.[1] There must be compensating changes in the motion of the charge.

In the instantaneous *rest frame* of a point charge e in an external field $\mathbf{F} = \mathbf{E} + ic\mathbf{B}$, the time-rate of change of the spacetime momentum of the charge is

$$\dot{p} = e\left\langle \mathbf{F}u\right\rangle_{\Re} = e\mathbf{E}\,. \tag{12.1}$$

This accelerates the charge at the rate $\mathbf{a} = (e/m)\,\mathbf{E}$ but provides no rate of change in the particle energy:

$$\dot{E} = \left\langle \dot{p}\right\rangle_S = \left\langle e\mathbf{E}\right\rangle_S = 0\,. \tag{12.2}$$

Yet the accelerating charge radiates energy at the rate

$$P = \frac{2}{3}\frac{Ke^2a^2}{c^2}\,, \tag{12.3}$$

with $K = Z_0/\left(4\pi\right) = \left(4\pi\varepsilon_0 c\right)^{-1} = \dot{3}0$ Ohms. Where does the energy of radiation come from?

In the time interval $d\tau$, the charge radiates fields with energy $Pd\tau$, but in the rest frame there is no kinetic energy to supply the radiation energy: $\dot{E}d\tau = 0$.

[1] And possibly angular momentum, as well.

There seems to be a problem here, since Poynting's theorem (see Chapt. 3) ensures energy conservation for any finite current distribution. What we have ignored is the interaction of the charge with the radiated field, or equivalently, the field energy arising from the interference of the radiation and transformed Coulomb fields.

One approach is to model the charge by a finite distribution. We may assume a spherical charge distribution in the charge rest frame. The shape will generally be different in different inertial frames. When the finite charge distribution is subjected to a field gradient, one side would be expected to accelerate faster than the other. What internal forces are required to keep the distribution spherical in its rest frame? Because of causality, one part of the charge distribution can communicate with another only at the speed of light. The model rapidly becomes rather messy for an "elementary" particle. To avoid the mess, we need to take a limit in which the charge distribution becomes a point.

A key point in any such model is the interaction of the charge with its own field. We look for the finite interaction of any extended model that approaches the interaction of a point charge with its own fields in an appropriate limit.

12.2 Self Interaction

We wish to find the field at positions x on the world line $r(\tau)$ of the charge. Of course if we simply put $x = r(\tau)$ and $R = 0$, the fields blow up. We follow Barut[2] in prescribing an interaction of the charge at $x = r(\tau)$ with the field produced by itself slightly earlier, at $\tau - s$, on its world line and by relaxing the light-cone condition. Instead of demanding that $R = x - r = r(\tau) - r(\tau - s)$ obey

$$R\bar{R} = 0\,, \tag{12.4}$$

we let R be timelike

$$R\bar{R} > 0\,. \tag{12.5}$$

After we derive an equation of motion, we will take the limit $s \to 0$ and hence $R\bar{R} \to 0$.

By expanding $r(\tau - s)$ about $r(\tau)$, we find

$$\begin{aligned} R &= r(\tau) - r(\tau - s) \\ &= c\left(us - \frac{1}{2}\dot{u}s^2 + \frac{1}{3!}\ddot{u}s^3\right) + \mathcal{O}(s^4)\,, \end{aligned} \tag{12.6}$$

[2]A. O. Barut, *Phys. Rev.* D**10**, 3335–3336 (1974).

where $\mathcal{O}(s^n)$ means terms of order s^n. Since u is unimodular,

$$\frac{d}{d\tau}(u\bar{u}) = \dot{u}\bar{u} + u\bar{\dot{u}} = 0\,, \tag{12.7}$$

and consequently

$$R\bar{R} = \bar{R}R = c^2s^2 + \mathcal{O}(s^4)\,. \tag{12.8}$$

If we revisit the derivation of the Liénard-Wiechert field of an accelerated charge in the last chapter, we find that with $R\bar{R} > 0$, there is one additional term:

$$\begin{aligned}\mathbf{F}(x) &= \frac{1}{2}Ke\left[\left(R\bar{\dot{u}} - c\right)\left(R\bar{u} + u\bar{R}\right) - \left(R\bar{\dot{u}} + \dot{u}\bar{R} - 2c\right)R\bar{u}\right] / \langle R\bar{u}\rangle_S^3 \\ &= \frac{1}{2}Ke\left[2c\langle R\bar{u}\rangle_V + R\bar{\dot{u}}u\bar{R} - \dot{u}\bar{R}R\bar{u}\right] / \langle R\bar{u}\rangle_S^3\,. \end{aligned} \tag{12.9}$$

All of the terms in $\mathbf{F}$ with u and its proper derivatives are to be evaluated at the retarded time $\tau - s$. We therefore need the expansions

$$2c\langle R\bar{u}\rangle_{V,ret} = c^2s^2\left(\dot{u}\bar{u} - \frac{2s}{3}\langle\ddot{u}\bar{u}\rangle_V\right) \tag{12.10}$$

$$\langle R\bar{u}\rangle_{S,ret} = cs\left(1 - \frac{s^2}{3!}\dot{u}\bar{\dot{u}}\right) \tag{12.11}$$

$$\left(R\bar{\dot{u}}u\bar{R}\right)_{ret} = -c^2s^2\left[\dot{u}\bar{u} - s\langle\ddot{u}\bar{u}\rangle_V\right] \tag{12.12}$$

$$\left(\dot{u}\bar{R}R\bar{u}\right)_{ret} = c^2s^2\left[\dot{u}\bar{u} - s\langle\ddot{u}\bar{u}\rangle_V\right]\,, \tag{12.13}$$

all of which are given up to terms $\mathcal{O}(s^4)$. Here we have again used the unimodularity of u, which implies $\ddot{u}\bar{u} + u\bar{\ddot{u}} + 2\dot{u}\bar{\dot{u}} = 0$ together with the expansion $u(\tau - s) = u - \dot{u}s + \frac{1}{2}\ddot{u}s^2 - \mathcal{O}(s^3)$. The expressions on the right-hand side of these expansions are evaluated at the instantaneous proper time τ of the charge.

Substitutions of these expansions into (12.9) gives the self field

$$\mathbf{F}_{\text{self}}[r(\tau), s] = \frac{Ke}{c}\left[-\frac{\dot{u}\bar{u}}{2s} + \frac{2\langle\ddot{u}\bar{u}\rangle_V}{3} + \mathcal{O}(s)\right]\,. \tag{12.14}$$

The Lorentz-force equation

$$m_0 c\dot{u}(\tau) = e\left\langle \mathbf{F}_{\text{self}} u(\tau)\right\rangle_{\Re} \tag{12.15}$$

for a charge e of bare mass m_0 at $r(\tau)$ interacting with this field is thus

$$\begin{aligned} m_0 c\dot{u} &= \frac{Ke^2}{c}\left\langle \left(-\frac{\dot{u}\bar{u}}{2s} + \frac{\ddot{u}\bar{u} - u\bar{\ddot{u}}}{3}\right) u \right\rangle_{\Re} + \mathcal{O}(s) \\ &= \frac{Ke^2}{c}\left(-\frac{\dot{u}}{2s} + \frac{\ddot{u} - u\bar{\ddot{u}}u}{3}\right) + \mathcal{O}(s)\ . \end{aligned} \tag{12.16}$$

The first term on the right is singular and blows up in the limit $s \to 0$.

12.2.1 Renormalized Mass

If we combine the singular term on the right with the bare-mass term on the left, we obtain

$$\left(m_0 c^2 + \frac{Ke^2}{2s}\right)\dot{u} = \frac{Ke^2}{3}\left(\ddot{u} - u\bar{\ddot{u}}u\right) + \mathcal{O}(s) \tag{12.17}$$

$$= \frac{2Ke^2}{3}\left(\ddot{u} + \dot{u}\bar{\dot{u}}u\right) + \mathcal{O}(s) \tag{12.18}$$

since $\ddot{u} + u\bar{\ddot{u}}u + 2\dot{u}\bar{\dot{u}}u = 0$. In the limit $s \to 0$, the singular term on the left is identified with the *renormalized mass*

$$m = m_0 + \frac{Ke^2}{2c^2 s} \tag{12.19}$$

which is taken to be a finite, physical quantity.

At this point, the derivation loses something of its mathematical rigor! It is sometimes argued that since the bare mass m_0 is not observable, we can let it be very large and negative (actually, unbounded from below) in order to cancel the infinity arising from the self-interaction term $Ke^2/(2s)$ as $s \to 0$. The idea of subtracting infinite terms to obtain a meaningful finite one is rather disconcerting, but it appears necessary not only here, but also in quantum electrodynamics (QED). Its main justification is that it works, and many of the resulting predictions of QED have been verified experimentally to unprecedented accuracy. It is rather less clear that the classical electrodynamic equation that results is as successful, as we discuss below.

In the limit $s \to 0$ we thus obtain

$$\begin{aligned} mc\dot{u} &= \frac{2Ke^2}{3c}\left(\ddot{u} + \dot{u}\bar{\dot{u}}u\right) \\ &= e\mathbf{f}_{\text{self}}u \end{aligned} \tag{12.20}$$

where

$$\begin{aligned} \mathbf{f}_{\text{self}} &= \frac{2Ke}{3c}\left\langle \ddot{u}\bar{u}\right\rangle_V \\ &= \frac{2Ke}{3c}\left(\ddot{u}\bar{u} + \dot{u}\bar{\dot{u}}\right) \\ &= \frac{2Ke}{3c}\frac{d}{d\tau}\left(\dot{u}\bar{u}\right) \end{aligned} \tag{12.21}$$

is the effective field of the self interaction and we have noted that the product $\mathbf{f}_{\text{self}}u$ is real:

$$\begin{aligned} \left[\frac{d\left(\dot{u}\bar{u}\right)}{d\tau}u\right]^{\dagger} &= u\frac{d}{d\tau}\left(\bar{u}\dot{u}\right) \\ &= u\bar{\dot{u}}\dot{u} + u\bar{u}\ddot{u} \\ &= \dot{u}\bar{\dot{u}}u + \ddot{u}\bar{u}u \\ &= \frac{d\left(\dot{u}\bar{u}\right)}{d\tau}u\,. \end{aligned} \tag{12.22}$$

At the third equality, we used the fact that $u\bar{u} = \bar{u}u$ and $\bar{\dot{u}}\dot{u} = \dot{u}\bar{\dot{u}}$ are scalars and thus commute with all other elements. Note also that $\dot{u}\bar{u}$ and hence its proper-time derivative is a spatial vector. Another way to state this is that the spacetime acceleration $\dot{u}$ is orthogonal to the proper velocity u: $\left\langle \dot{u}\bar{u}\right\rangle_S = 0$. Since u is a timelike paravector, the orthogonality implies that the acceleration $\dot{u}$ must be spacelike.

12.2.2 Avoiding Mass Renormalization

We can avoid the singular term and the need for renormalization by averaging the interaction of the charge at $r(\tau)$ with fields from times $\tau + s$ and $\tau - s$. From the expression (12.14) for the self field,

$$\frac{1}{2}\left\{\mathbf{F}_{\text{self}}[r(\tau)\,, s] + \mathbf{F}_{\text{self}}[r(\tau)\,, -s]\right\} = \frac{Ke}{c}\left[\frac{2\left\langle \ddot{u}\bar{u}\right\rangle_V}{3} + \mathcal{O}(s)\right], \tag{12.23}$$

which leads directly to the Lorentz-Dirac equation (12.20) for the particle of mass $m = m_0$.[3]

Although we have found a limiting procedure that gives a finite self interaction for a point charge, one may question the physical significance of the model used. The charge in our model has zero spatial extent, and we have thereby sidestepped difficulties in relativity with the shape of an accelerating charge distribution. It is a point charge that we let interact with its image at slightly later and earlier times on its trajectory. The interaction is therefore with retarded and advanced fields inside the light cone, that is, with fields that propagate at less than the speed of light. Of course, in the limit $s \to 0$, the interaction becomes instantaneous and possible problems with causality evidently disappear. We might not give the derivation a second thought if the result, namely the Lorentz-Dirac equation, were problem-free.

12.3 The Lorentz-Dirac Equation

The *Lorentz-Dirac equation* for the motion of a charge in an external field $\mathbf{F}$ is obtained from (12.20) by superimposing the effective self field with the external field:

$$mc\dot{u} = e\left\langle\left(\mathbf{F} + \mathbf{f}_{\text{self}}\right) u\right\rangle_{\Re} . \tag{12.24}$$

The scalar part times c gives the proper-time rate of change of the particle energy:

$$\begin{aligned} mc^2\dot{\gamma} &= e\gamma\mathbf{E}\cdot\mathbf{v} + \frac{2Ke^2}{3}\left\langle\ddot{u} + \dot{u}\bar{\dot{u}}u\right\rangle_S \\ &= e\gamma\mathbf{E}\cdot\mathbf{v} + \frac{2Ke^2}{3}\left(\langle\ddot{u}\rangle_S + \gamma\dot{u}\bar{\dot{u}}\right) \end{aligned} \tag{12.25}$$

and the change in particle energy is

$$mc^2\Delta\gamma = \int dt\,(e\mathbf{E}\cdot\mathbf{v} - P) + \frac{2Ke^2}{3}\langle\Delta\dot{u}\rangle_S , \tag{12.26}$$

[3] This is mathematically closely related to Dirac's use [P. A. M. Dirac, *Proc. R. Soc. Lond. A* **167**, 148 (1938)] of the antisymmetric field (half retarded minus advanced) and to the Wheeler-Feynman absorber theory [J. A. Wheeler and R. P. Feynman, *Rev. Mod. Phys.* **17**, 157–181 (1945)]. Unruh's use of the Penrose integral avoids the singularity in fields propagating on the forward and backward light cones [W. G. Unruh, *Proc. R. Soc. Lond. A* **348**, 447–465 (1976) and R. Penrose and W. Rindler (1984), p. 403.].

where

$$P = -\frac{2}{3} K e^2 \dot{u}\bar{\dot{u}} \tag{12.27}$$

is the Larmor power radiated. Between any two points on the particle world line at which the scalar part $\langle \dot{u} \rangle_S$ of the acceleration is the same, the contribution of the last term, known as the *Schott term*, in (12.26) vanishes. The sum of the particle kinetic energy and the radiated energy is then conserved. More generally, the Schott term represents energy stored in the interference of the bound Coulomb field $\mathbf{F}_c$ and the free radiated field $\mathbf{F}_r$ of the charge. Because radiation is emitted only during times of acceleration and the interference terms $\mathbf{F}_c\mathbf{F}_r^\dagger + \mathbf{F}_r\mathbf{F}_c^\dagger$ fall off as R_0^{-3}, the Schott energy must dissipate after long periods of unaccelerated motion.

12.4 Pathological Features

Through the effective field $\mathbf{f}_{\text{self}}$ (12.21), the Lorentz-Dirac equation is a third-order equation in derivatives of $r(\tau)$. This leads to acausal and unphysical behavior of some solutions. Since the third derivative $d^3 r / d\tau^3 = c\ddot{u}$ must remain finite, the acceleration $c\dot{u}$ itself must be continuous and the acceleration of a charge cannot jump discontinuously when an external field is suddenly turned on. In fact, physical solutions with bounded kinetic energies require *preacceleration* in most problems, so that the acceleration has reached its full value by the time the field reaches the charge.

The pathology of the Lorentz-Dirac equation is best illustrated if we use, as noted above, that $\mathbf{f}_{\text{self}} u$ is real. Then (12.24) can be rewritten

$$\begin{aligned} \left(1 - \tau_0 \frac{d}{d\tau}\right) (\dot{u}\bar{u}) &= \frac{e}{mc} \langle \mathbf{F} u \rangle_{\Re} \bar{u} \\ &= \frac{e}{2mc} \left(\mathbf{F} + u\mathbf{F}^\dagger \bar{u}\right), \end{aligned} \tag{12.28}$$

where

$$\tau_0 = \frac{2Ke^2}{3mc^2} = \frac{2r_e}{3c} = 6.266\,424 \times 10^{-24}\,\text{s} \tag{12.29}$$

is 2/3 of the time required for light to traverse the classical electron radius $r_e = Ke^2 / (mc) \simeq 2.81\,794\,1\,\text{fm}$. In the instantaneous rest frame of the charge, the equation (12.28) can be written

$$\mathbf{a} - \tau_0 \dot{\mathbf{a}} = \frac{e}{mc}\mathbf{E}\,. \tag{12.30}$$

This is the nonrelativistic limit of the Lorentz-force equation; it is known as the *Abraham-Lorentz equation.*

Since

$$\left(1 - \tau_0 \frac{d}{d\tau}\right) \dot{u}\bar{u} = -e^{\tau/\tau_0} \tau_0 \frac{d}{d\tau}\left(\dot{u}\bar{u}e^{-\tau/\tau_0}\right), \tag{12.31}$$

the formal solution to (12.28) is

$$\dot{u}\bar{u} = -e^{\tau/\tau_0} \frac{e}{2mc\tau_0} \int^{\tau} d\tau' \left(\mathbf{F} + u\mathbf{F}^{\dagger}\bar{u}\right) e^{-\tau'/\tau_0}\,. \tag{12.32}$$

Its nonrelativistic limit,

$$\mathbf{a} = -e^{\tau/\tau_0} \frac{e}{mc\tau_0} \int^{\tau} d\tau' \mathbf{E} e^{-\tau'/\tau_0}, \tag{12.33}$$

is easily seen satisfy the Abraham-Lorentz equation (12.30).

12.4.1 One-Dimensional Motion; Runaways

We can investigate some of the known pathological features of the Lorentz-Dirac equation by studying the special case of motion in a static electric field along one spatial direction. In this case, all algebraic quantities are real and commute. Let

$$\mathbf{F} = E(z)\, \mathbf{e}_3 \tag{12.34}$$

where $z := \mathbf{r} \cdot \mathbf{e}_3$ and express the proper velocity in terms of the rapidity w :

$$u = \exp(w\mathbf{e}_3)\ . \tag{12.35}$$

Then

$$\dot{u}\bar{u} = \dot{w}\mathbf{e}_3\,, \tag{12.36}$$

and the Lorentz-Dirac equation (12.28) in this case can be written

$$\left(1 - \tau_0 \frac{d}{d\tau}\right) \dot{w} = \frac{eE}{mc}\,. \tag{12.37}$$

Exercise 12.1. Show that

$$\dot{w} = c d\gamma / dz\,. \tag{12.38}$$

The formal solution (12.32) in the one-dimensional case becomes

$$\dot{w}(\tau) = -e^{\tau/\tau_0} \frac{e}{mc\tau_0} \int^{\tau} d\tau' E(\tau')\, e^{-\tau'/\tau_0}\,. \tag{12.39}$$

If we know $\dot{w}$ at any proper time τ_1 on its world line, we can replace the indefinite integral by a definite one:

$$\begin{aligned} \dot{w}(\tau) &= e^{\tau/\tau_0} \left[\dot{w}(\tau_1)\, e^{-\tau_1/\tau_0} - \frac{e}{mc\tau_0} \int_{\tau_1}^{\tau} d\tau' E(\tau')\, e^{-\tau'/\tau_0} \right] \\ &= \dot{w}(\tau_1)\, e^{(\tau-\tau_1)/\tau_0} - \frac{e}{mc\tau_0} \int_0^{\tau-\tau_1} ds E(\tau - s)\, e^{s/\tau_0}, \end{aligned} \tag{12.40}$$

where we changed the dummy integration variable from τ' to $s := \tau - \tau'$. The problem lies in the rapidly growing exponential factor. At one millisecond after τ_1, the factor

$$e^{(\tau-\tau_1)/\tau_0} \approx \exp\left(1.5 \times 10^{20}\right) \tag{12.41}$$

is unbelievably huge, something like $10^{65000000000000000000}$. It swamps everything and gives *runaway* solutions, even if the electric field vanishes, $E = 0$.

The severity of the runaways is actually *worse* than indicated above because the paravelocity u is the exponential of an exponential. In particular, with $E = 0$,

$$u = \exp(w\mathbf{e}_3) = \exp\left[\dot{w}_1 \tau_0 \left(e^{(\tau-\tau_1)/\tau_0}\right) \mathbf{e}_3\right]\,. \tag{12.42}$$

The runaway solutions are extreme examples of hypersensitivity of solutions to initial conditions. There may be one particular value of $\dot{w}_1$ that yields a physical solution, but other initial conditions give nonphysical solutions, and even the smallest (and usually unavoidable) error can change a well-behaved solution to a wild runaway in which w grows without limit.

12.4.2 Preacceleration and Nonuniqueness

Inspection of the Lorentz-Dirac equation (12.37) shows that its runaway behavior depends on the direction of time. If τ is reversed, unphysical solutions are damped. One way to ensure that the solution $\dot{w}$ vanishes at large proper times is to let $\tau_1 \to \infty$ and to put $\dot{w}_1 = 0$. Then, after changing the sign of the dummy integration variable, (12.40) gives

$$\dot{w}(\tau) = \frac{e}{mc\tau_0} \int_0^{\infty} ds E(\tau + s)\, e^{-s/\tau_0} \,. \tag{12.43}$$

This solution has eliminated the runaway behavior, and if changes in the electric field E are negligible over several τ_0, it can be approximately integrated to give the usual Lorentz-force result

$$\dot{w}(\tau) = \frac{eE(\tau)}{mc} \,. \tag{12.44}$$

The problem with the solution (12.43) is that the acceleration of the charge at proper time τ depends on the electric field at later times $\tau + s\,,\ \ s > 0$. This is the phenomenon of *preacceleration* mentioned earlier. Preacceleration violates our concept of causality, namely that the effect must *follow* the cause and not *vice versa.* It also admits multiple physical solutions to certain problems. For example, consider the case of a point charge at rest just outside the region of a strong constant field. One physical solution is for the charge to remain where it is. However, another possible physical solution is for the charge to be preaccelerated into the strong-field region (see Fig. 12.1).[4]

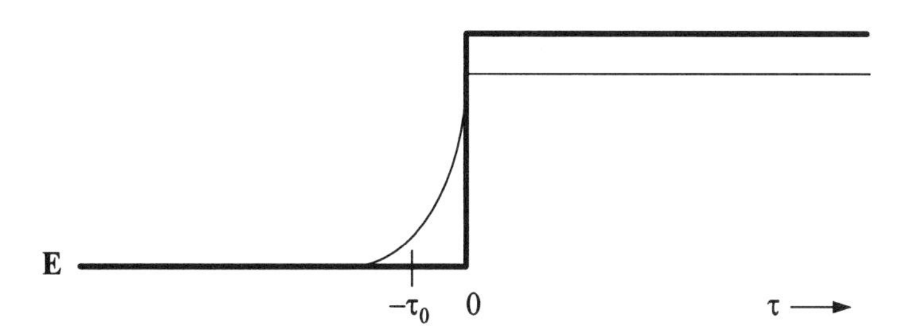

Fig. 12.1. Preacceleration occurs over small time scales of the order of τ_0. Here, the heavy line is the electric field, the light line, the acceleration $\dot{w}$.

[4] W. E. Baylis and J. Huschilt, "Nonuniqueness of physical solutions to the Lorentz-Dirac equation," *Phys. Rev.* D**13**, 3237–3239 (1976).

12.4.3 Potential Barrier

Another example of problems is provided by the potential-barrier problem. Consider the potential-energy step

$$V(x) = \begin{cases} 0, & x < 0 \\ V_0, & 0 < x < x_0 \\ 0, & x_0 < x \end{cases} \tag{12.45}$$

with the corresponding electric field

$$E(x) = -\frac{1}{e}\frac{dV}{dx} = \frac{V_0}{e}\left[\delta(x - x_0) - \delta(x)\right] . \tag{12.46}$$

Substitution of E into (12.43) gives

$$\dot{w}(x) = \frac{V_0}{mc\tau_0}\int_0^\infty \left[\delta(x' - x_0) - \delta(x')\right] e^{-s/\tau_0} ds \tag{12.47}$$

with $x = x(\tau)$ and $x' = x(\tau + s)$. Since $dx'/ds = c \sinh w(x')$, a change in variable from s to x' yields

$$\begin{aligned} \dot{w}(x) &= \frac{V_0}{mc^2\tau_0}\int_0^\infty \left[\delta(x' - x_0) - \delta(x')\right] e^{-s/\tau_0} \frac{dx'}{\sinh w(x')} \\ &= \frac{V_0 e^{\tau/\tau_0}}{mc^2\tau_0} \begin{cases} \left[\frac{\alpha}{\sinh w(x_0)} - \frac{1}{\sinh w(0)}\right], & x < 0 \\ \frac{\alpha}{\sinh w(x_0)}, & 0 < x < x_0 \\ 0, & x_0 < x, \end{cases} \end{aligned} \tag{12.48}$$

where

$$\alpha = \exp(-T/\tau_0) , \tag{12.49}$$

and $T > 0$ is the proper time at which the charge leaves the barrier: $x(T) = x_0$. Since w is constant for $x > x_0$,we can put $w(x_0) = w_f$ .We similarly define the incident rapidity $w_{in} = w(-\infty)$. A further integration over proper time gives

$$w(x) = \begin{cases} w_{in} + \frac{V_0}{mc^2} e^{\tau/\tau_0}\left[\frac{\alpha}{\sinh w_f} - \frac{1}{\sinh w(0)}\right], & x < 0 \\ w(0) + \frac{V_0}{mc^2}\left(e^{\tau/\tau_0} - 1\right)\frac{\alpha}{\sinh w_f}, & 0 < x < x_0 \\ w_f, & x_0 < x \end{cases} . \tag{12.50}$$

Even with the delta-function field, the third-order degree of the Lorentz-Dirac equation dictates that $\dot{w}$ should be finite and the rapidity w should therefore be continuous. By matching w at the potential steps, we obtain

$$w_{in} = w(0) + \frac{V_0}{mc^2}\left[\frac{1}{\sinh w(0)} - \frac{\alpha}{\sinh w_f}\right] \quad (12.51)$$

$$w(0) = w_f - \frac{V_0}{mc^2}\left[\frac{1-\alpha}{\sinh w_f}\right]. \quad (12.52)$$

These two equations together give w_{in} as a function of w_f.

For any barrier of physical thickness, the factor $\alpha = \exp(-T/\tau_0)$ is vanishingly small and can be ignored compared to 1 or to $\sinh w_f / \sinh w(0) > 1$. If we drop the terms α, we obtain a family of curves of w_{in} as a function of w_f for different relative heights V_0/mc^2, such as shown in the Maple plot below. We uncover that there are many values of w_{in} that give rise to two distinct values of w_f. In such cases, the physical solution of the Lorentz-Dirac equation for given initial values is not unique.

It is also interesting to look at the case of the unphysically thin barrier. In the limit of vanishing thickness, $\alpha \to 1$ and $w_{in} = w(0) = w_f$, independent of the barrier height. This is physically reasonable, since if the barrier has no thickness, it has no effect, no matter how "high." However, because of the continuity of the rapidity implied by the Lorentz-Dirac equation, it suggests that classical tunneling may occur through barriers whose widths are very thin but nonvanishing. To confirm this possibility, consider barriers so thin that

$$\alpha = 1 - x\left(mc^2/V_0\right)\sinh w_f. \quad (12.53)$$

Here, x is a small positive definite number in the range $0 < x < w_f$, but also small enough that $0 < \alpha < 1$. Then

$$w_f = w(0) + x \quad (12.54)$$

and

$$w_{in} = w_f + \frac{V_0}{mc^2}\left[\frac{1}{\sinh(w_f - x)} - \frac{1}{\sinh w_f}\right] > w_f. \quad (12.55)$$

For any given barrier height V_0 relative to mc^2, one can choose x sufficiently small that $w_{in} > w_f > w(0) > 0$ implies the tunneling condition:

$$\cosh w_{in} < 1 + V_0/mc^2. \quad (12.56)$$

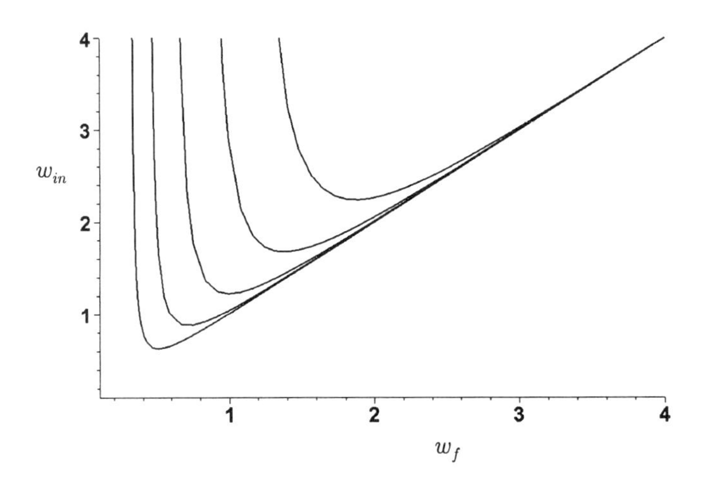

Fig. 12.2. A family of curves giving the incident rapidity w_{in} as a function of the final rapidity w_f for potential barriers of heights V_0/mc^2 from 0.1 to 1.6, increasing by factors of two from left to right at the top left. Since the curves show two values of w_f for most values of w_{in}, they demonstrate the nonuniqueness of physical solutions to the Lorentz-Dirac equation.

As an explicit example, let

$$x = w_f^3 mc^2/V_0 \ll 1\,. \tag{12.57}$$

Then

$$w_{in} = w_f \left(1 + \frac{w_f^2}{\sinh^2 w_f} \cosh w_f\right) \approx 2w_f\,, \tag{12.58}$$

where the last approximate equality holds for $w_f \ll 1$. The resultant w_{in} can easily make the incident kinetic energy below the barrier, but the requirement is that the barrier be unphysically thin. Indeed, substituting x (12.57) into $\alpha = \exp(-T/\tau_0)$ (12.53), we obtain a barrier-transition time of only

$$T = w_f^4 \tau_0 \tag{12.59}$$

for small w_f. Nevertheless, this does demonstrate that tunneling is possible in principle in the Lorentz-Dirac solutions.

12.4.4 Backward Time Integration

Similarly, runaway numerical solutions to the Lorentz-Dirac equation (12.28) in three spatial dimensions can be avoided by integrating the equation backward in time. This technique has been used to compute a number of physical solutions, although its application to most problems, in which the field at a charge depends on the retarded motion of other charges, requires special techniques, such as the self-consistent trajectory method.[5] The integro-differential equation analogous to (12.43) is

$$\dot{u}\bar{u} = \frac{e}{2mc\tau_0}\int_0^\infty ds\left[\mathbf{F}(\tau+s) + u\mathbf{F}(\tau+s)^\dagger\,\bar{u}\right]e^{-s/\tau_0}\,. \tag{12.60}$$

The value of the vector $\dot{u}\bar{u}$ at proper time τ depends on the field $\mathbf{F}$ at a slightly later time. As in the one-dimensional example above, solutions obtained by backward time integration lead invariably to preacceleration, and in some cases different final velocities used to start the integration lead to the same conditions at an earlier time. These are other examples of multiple physical solutions of the Lorentz-Dirac equation. Finally, there are some problems, in particular, the head-on collision of opposite charges, for which it has been proven that no physical solution to the Lorentz-Dirac equation exists.[6]

12.5 Eigenspinor Form

Solutions u of the Lorentz-Dirac equation (12.24) can be expressed in terms of the eigenspinor Λ by

$$u = \Lambda\Lambda^\dagger, \tag{12.61}$$

where

$$\dot{\Lambda} = \frac{e}{2mc}\left(\mathbf{F} + \mathbf{f}_{\text{self}}\right)\Lambda\,. \tag{12.62}$$

[5] J. Huschilt and W. E. Baylis, "Numerical solutions to two-body problems in classical electrodynamics: head-on collisions with retarded fields and radiation reaction. I. Repulsive case," *Phys. Rev. D* **13**, 3256–3261 (1976); J. Huschilt and W. E. Baylis, "Rutherford scattering with radiation reaction," *Phys. Rev. D* **17**, 985–993 (1978).

[6] C. J. Eliezer, *Proc. Camb. Philos. Soc.* **39**, 173 (1943); W. E. Baylis and J. Huschilt, "Numerical solutions to two-body problems in classical electrodynamics: head-on collisions with retarded fields and radiation reaction. II. Attractive case," *Phys. Rev. D* **13**, 3262–3268 (1976).

Exercise 12.2. Show solutions of the more general nonlinear eigenspinor equation

$$\dot{\Lambda} = \frac{e}{2mc}\left(\mathbf{F}^{(\alpha)} + \mathbf{f}_{\text{self}}^{(\beta)}\right)\Lambda\,, \tag{12.63}$$

where α and β are real parameters, and

$$\mathbf{F}^{(\alpha)} = (1-\alpha)\,\mathbf{F} + \alpha u \mathbf{F}_{\text{ext}}^{\dagger}\bar{u} \tag{12.64}$$

$$\mathbf{f}_{\text{self}}^{(\beta)} = (1-\beta)\,\mathbf{f}_{\text{self}} + \beta u \mathbf{f}_{\text{self}}^{\dagger}\bar{u}\,, \tag{12.65}$$

also satisfy the Lorentz-Dirac equation (12.24). Demonstrate further that $\mathbf{f}_{\text{self}}^{(\beta)}$ is independent of the value of β.

The eigenspinor can be factored into a "normal" spinor Λ_n that gives motion without radiation reaction:

$$\dot{\Lambda}_n = \frac{e}{2mc}\mathbf{F}\Lambda_n \tag{12.66}$$

and a "supplemental" spinor Λ_s that contributes the effects of radiation reaction:

$$\Lambda = \Lambda_n \Lambda_s\,. \tag{12.67}$$

An equation of motion for Λ_s is obtained by substituting $\Lambda = \Lambda_n\Lambda_s$ into (12.62):

$$\begin{aligned}
\dot{\Lambda} &= \dot{\Lambda}_n\Lambda_s + \Lambda_n\dot{\Lambda}_s && (12.68)\\
&= \frac{e}{2mc}\mathbf{F}\Lambda + \Lambda_n\dot{\Lambda}_s && (12.69)\\
&= \frac{e}{2mc}\left(\mathbf{F} + \mathbf{f}_{\text{self}}\right)\Lambda\,. && (12.70)
\end{aligned}$$

Thus,

$$\begin{aligned}
\dot{\Lambda}_s &= \frac{e}{2mc}\bar{\Lambda}_n\mathbf{f}_{\text{self}}\Lambda && (12.71)\\
&= \frac{e}{2mc}\Lambda_s\left(\bar{\Lambda}\mathbf{f}_{\text{self}}\Lambda\right) && (12.72)
\end{aligned}$$

and we recognize in $\bar{\Lambda}\mathbf{f}_{\text{self}}\Lambda$ the self field transformed to the rest frame of the charge:

$$\begin{aligned}
\bar{\Lambda}\mathbf{f}_{\text{self}}\Lambda &= \frac{2Ke}{3c}\bar{\Lambda}\left(\ddot{u}\bar{u} + \dot{u}\bar{\dot{u}}\right)\Lambda && (12.73)\\
&= \frac{2Ke}{3c}\left(\ddot{u}_{\text{rest}} - \mathbf{a}^2/c^2\right) && (12.74)\\
&= \frac{2Ke\dot{\mathbf{a}}}{3c^2} && (12.75)
\end{aligned}$$

where $\mathbf{a}$ is the rest-frame acceleration.

Exercise 12.3. Find $\ddot{u}$ and show that in the rest frame ($\mathbf{v} = 0$), it reduces to $\mathbf{a}^2/c^2 + \dot{\mathbf{a}}/c$.

The time-development equation for the supplemental spinor factor is therefore

$$\dot{\Lambda}_s = \frac{1}{2}\Lambda_s \frac{\dot{\mathbf{a}}\tau_0}{c}, \tag{12.76}$$

where as above, $\tau_0 = 2Ke^2/\left(3mc^2\right)$. One immediate consequence of this equation is that there is no effect of radiation reaction for a charge with constant rest-frame acceleration, for example a charge accelerated from rest in a constant electric field.

If the direction of $\dot{\mathbf{a}}$ remains constant, the solution to (12.76) is

$$\Lambda_s(\tau) = \Lambda_s(0) \exp\left[\frac{\mathbf{a}(\tau) - \mathbf{a}(0)}{2c}\tau_0\right]. \tag{12.77}$$

More generally, equations (12.66) and (12.76) for Λ_n and Λ_s are coupled because the rest-frame $\dot{\mathbf{a}}$ depends on Λ_n and the position of the charge at which $\mathbf{F}_{\text{ext}}$ is evaluated depends on Λ_s .

12.6 Conclusions

The standard current classical electrodynamics theory of an elementary charge and its interaction with external fields has been derived in many ways. We have presented what must rate as one of the simplest derivations, as the limit of the interaction of a point charge with its earlier image along the world line. It is fair to state that the basic theory is not in good shape. Practically all of the solutions to the third-order Lorentz-Dirac equation are unphysical runaway solutions in which the acceleration grows rapidly and without limit. The physical solutions can be extracted by integrating backward in time from a final vanishing acceleration, but such solutions display acausal behavior in the form of preacceleration. Furthermore, there are some problems for which no physical solution exists and others for which several exist.

Many alternatives to the Lorentz-Dirac equation have been proposed, but none has proved fully satisfactory. In some, energy is not conserved. In one proposal,[7] the effects of radiation reaction disappeared for head-on collisions of like

[7]T. C. Mo and C. H. Papas, *Phys. Rev. D*4, 3566 (1971).

charges,[8] even though this leads to large and unphysical gains in the kinetic energies.[9]

The quantum electrodynamics description of interactions between elementary charges and external fields seems in better shape in that it has provided predictions that have been experimentally confirmed to high precision. Nevertheless, its foundations require the subtraction of infinities in charge and mass renormalization, and most of the quantitative predictions are based on approximations of perturbation theory that are much less useful in the realm of quantum chromodynamics, which seeks to describe elementary interactions among quarks.

It is difficult to imagine any problems more fundamental to physics than the interaction of elementary charges, yet it is precisely here that basic problems exist. The problems cry for a clearer understanding and a better mathematical formulation of the underlying physics.

12.7 Problems

1. Consider the eigenspinor solution with radiation reaction when both electric and magnetic components of $\mathbf{F}_{\text{ext}}$ are aligned with the initial motion. Show in this case that the proper acceleration is $\mathbf{a} = \dot{\mathbf{w}}c$, where $\mathbf{w}$ is the rapidity, and that the equation of motion (12.37) is obtained. Also determine the rotation rate in the rest frame.

2. Show that the difference

$$\frac{1}{2}\bar{\Lambda}\left(\mathbf{F} - u\mathbf{F}^{\dagger}\bar{u}\right)\Lambda = ic\mathbf{B}_{\text{rest}} \tag{12.78}$$

gives the magnetic field $\mathbf{B}_{\text{rest}}$ in the rest frame of the charge. Interpret the condition

$$\mathbf{F} = u\mathbf{F}^{\dagger}\bar{u} \tag{12.79}$$

in terms of the rest-frame field.

---

[8] J. Huschilt and W. E. Baylis, "Solutions to the 'new' equation of motion for classical charged particles," *Phys. Rev. D*9, 2479-2480 (1974).

[9] J. Huschilt, W. E. Baylis, D. Leiter, and G. Szamosi, "Numerical solutions to two-body problems in classical electrodynamics: straight-line motion with retarded fields and no radiation reaction," *Phys. Rev. D*7, 2844-2850 (1973).

3. Let a charge e of mass m, initially at rest, be subjected to an electric field pulse

$$\mathbf{E}(\tau) = \begin{cases} 0\,, & \tau < 0 \\ \mathbf{E}_0\,, & 0 < \tau < T \\ 0, & \mathrm{T} < \tau \end{cases} \,. \tag{12.80}$$

Assume that the charge is initially at rest, $w\,(-\infty) = 0$,and find $\dot{w}(\tau)$ and $w(\tau)$. Sketch $\dot{w}(\tau)$ when $T \approx 5\tau_0$.

4. Show that in the previous problem, the particle energy at $\tau > T$ is independent of whether radiation reaction is included or not, but that with radiation reaction, the excess energy imparted to the particle by the field is just sufficient to supply the energy of the radiation in the nonrelativistic limit.

5. Consider a harmonically bound charge with radiation reaction, and show that at low velocities, the equation of motion is

$$\ddot{x} - \tau_0 \dddot{x} + \omega_0^2 x = 0\,, \tag{12.81}$$

where $\omega_0 = \sqrt{k/m}$ is the natural (angular) frequency of the oscillator. Demonstrate that at low velocities and with $\omega_0\tau_0 \ll 1$, the decay rate $-\dot{E}/E$ of the energy E is approximately

$$\Gamma = \omega_0^2\tau_0\,. \tag{12.82}$$

Let the radiation of angular frequency ω_0 have a wavelength in the visible range of about of $\frac{1}{2}\mu$m. Find the mean decay time $\Gamma^{-1}$and compare to that (roughly 16 ns) of the resonance transition of sodium to see whether the calculated value is in the right ball park.

Chapter 13

Rotations and Spherical Harmonics

In this chapter we study rotations, and spherical harmonics in angular coordinates. The reader is referred to Appendix A for a review of various coordinate systems in real three-dimensional Euclidean space, and of the gradient operator ∇, the Laplacian ∇^2, and the angular-momentum operator $\mathbf{L}$ in these coordinates. Here we introduce spherical harmonics and Legendre polynomials, and we study their properties with the help of ladder operators. Finally, in preparation for our study of radiation multipoles in the following chapter, we study rotations and their irreducible matrix representations.

13.1 Spherical Harmonics

The spherical harmonics

$$C^{lm}(\hat{\mathbf{r}}) := \sqrt{\frac{4\pi}{2l+1}}\, Y^{lm}(\hat{\mathbf{r}}) \tag{13.1}$$

are eigenfunctions of the angular momentum operators $\mathbf{L}^2$ and L_z (see Appendix A):

$$\begin{aligned} \mathbf{L}^2 C^{lm} &= l\,(l+1)\, C^{lm} && (13.2)\\ L_z C^{lm} &= m C^{lm} && (13.3) \end{aligned}$$

with the normalization and orthogonality

$$\int_{4\pi} d\hat{\mathbf{r}}\, C^{lm}(\hat{\mathbf{r}})^*\, C^{l'm'}(\hat{\mathbf{r}}) = \frac{4\pi}{2l+1}\delta^{ll'}\delta^{mm'}\,. \tag{13.4}$$

The orthogonality follows from the fact that the spherical harmonics are eigenfunctions of *hermitian operators*. An operator H is said to be hermitian on a space of functions if for every $\phi(x)\,,\psi(x)$ in the function space

$$\int dx\,\phi^*(x)\,[H\psi(x)] = \int dx\,[H\phi(x)]^*\,\psi(x)\;, \tag{13.5}$$

where the integration variable x can be more than one dimensional. In the case of the spherical harmonics,

$$\begin{aligned}
0 &= \int_{4\pi} d\hat{\mathbf{r}}\,C^{lm}(\hat{\mathbf{r}})^*\left(\mathbf{L}^2 - \mathbf{L}^2\right)C^{l'm'}(\hat{\mathbf{r}}) \\
&= \int_{4\pi} d\hat{\mathbf{r}}\,\left[\mathbf{L}^2 C^{lm}(\hat{\mathbf{r}})\right]^* C^{l'm'}(\hat{\mathbf{r}}) - \int_{4\pi} d\hat{\mathbf{r}}\,C^{lm}(\hat{\mathbf{r}})^*\left[\mathbf{L}^2 C^{l'm'}(\hat{\mathbf{r}})\right] && (13.6)\\
&= \int_{4\pi} d\hat{\mathbf{r}}\,C^{lm}(\hat{\mathbf{r}})^*\left[l\,(l+1) - l'\,(l'+1)\right]C^{l'm'}(\hat{\mathbf{r}})\;; && (13.7)\\
0 &= \int_{4\pi} d\hat{\mathbf{r}}\,C^{lm}(\hat{\mathbf{r}})^*\left(L_z - L_z\right)C^{l'm'}(\hat{\mathbf{r}}) = \\
&= \int_{4\pi} d\hat{\mathbf{r}}\,C^{lm}(\hat{\mathbf{r}})^*\,(m - m')\,C^{l'm'}(\hat{\mathbf{r}})\;. && (13.8)
\end{aligned}$$

The possible values of the eigenvalues $l\,(l+1)$ and m must be determined. From (13.3) we can easily determine the ϕ dependence of C^{lm}. If we write the argument as θ, ϕ instead of $\hat{\mathbf{r}}$, we can put

$$C^{lm}(\theta,\phi) = C^{lm}(\theta,0)\,e^{im\phi}\,. \tag{13.9}$$

We demand that C^{lm} be a single-valued function of θ and ϕ , and this implies

$$C^{lm}(\theta,2\pi) = C^{lm}(\theta,0) \tag{13.10}$$

and consequently that m be an integer. We will consider the possible values of l below.

The spherical harmonics form a complete basis in Hilbert space for any complex function of $\hat{\mathbf{r}}$. That is, any function $A(\hat{\mathbf{r}})$ can be expanded

$$A(\hat{\mathbf{r}}) = \sum_{lm} a_{lm} C^{lm}(\hat{\mathbf{r}}) \tag{13.11}$$

with the constant coefficients a_{lm} given by the orthogonality condition (13.4)

$$a_{lm} = \frac{2l+1}{4\pi} \int_{4\pi} d\hat{\mathbf{r}}\, C^{lm}(\hat{\mathbf{r}})^* \, A(\hat{\mathbf{r}}) \,. \tag{13.12}$$

In the special case that $A(\hat{\mathbf{r}})$ is the two-dimensional Dirac delta function

$$\delta(\hat{\mathbf{r}} - \hat{\mathbf{r}}') = \delta(\cos\theta - \cos\theta')\, \delta(\phi - \phi') \,, \tag{13.13}$$

we find

$$\delta(\hat{\mathbf{r}} - \hat{\mathbf{r}}') = \sum_{lm} \frac{2l+1}{4\pi} C^{lm}(\hat{\mathbf{r}}')^* \, C^{lm}(\hat{\mathbf{r}}) \,, \tag{13.14}$$

which is a common way to express the completeness relation.

13.2 Ladder Operators

Many properties of the spherical harmonics can be derived with the aid of the commutation relations [see the expressions for L_z and $L_\pm$ in Appendix A]

$$[L_z, L_\pm] = \pm L_\pm \tag{13.15}$$
$$\left[\mathbf{L}, \mathbf{L}^2\right] = 0 \,. \tag{13.16}$$

Exercise 13.1. Take the average of the two expressions $(\pm)$ in (13.15) to prove

$$[L_z, L_x] = iL_y \,. \tag{13.17}$$

Also prove by using cartesian coordinates to express the components L_x, L_y, and L_z. Generalize to cyclic permutations of (13.17) and more generally to

$$[\mathbf{L}\cdot\mathbf{a}, \mathbf{L}\cdot\mathbf{b}] = i\,(\mathbf{a}\times\mathbf{b})\cdot\mathbf{L} \tag{13.18}$$

for any fixed vectors $\mathbf{a}$ and $\mathbf{b}$.

Exercise 13.2. The commutation relations among components of $\mathbf{L}$ are sometimes written

$$\mathbf{L}\times\mathbf{L} = i\mathbf{L} \,. \tag{13.19}$$

Evaluate the z component of this relation to show its equivalence to a commutation relation for components L_x and L_y.

Exercise 13.3. Show explicitly that $\mathbf{L}$ commutes with $\mathbf{L}^2, \nabla^2$, and r^2.

Consider the new functions $L_\pm C^{lm}(\hat{\mathbf{r}})$. Are they eigenfunctions of $\mathbf{L}^2$ and L_z, and if so, what are their eigenvalues? From the commutation relations (13.15) and (13.16),

$$\begin{aligned} \mathbf{L}^2 L_\pm C^{lm}(\hat{\mathbf{r}}) &= L_\pm \mathbf{L}^2 C^{lm}(\hat{\mathbf{r}}) \\ &= l\,(l+1)\,L_\pm C^{lm}\,(\hat{\mathbf{r}}) \end{aligned} \tag{13.20}$$

and

$$L_z L_\pm C^{lm}(\hat{\mathbf{r}}) = L_\pm\,(L_z \pm 1)\,C^{lm}(\hat{\mathbf{r}}) \tag{13.21}$$

$$= (m \pm 1)\,L_\pm C^{lm}(\hat{\mathbf{r}})\ . \tag{13.22}$$

Consequently, $L_\pm C^{lm}(\hat{\mathbf{r}})$ must have the eigenvalues $l\,(l+1)$ of $\mathbf{L}^2$ and $m \pm 1$ of L_z. If we expand $L_\pm C^{lm}(\hat{\mathbf{r}})$ in the $\left\{C^{lm}(\hat{\mathbf{r}})\right\}$ functional basis, only one term contributes:

$$L_\pm C^{lm}(\hat{\mathbf{r}}) = a C^{lm\pm 1}(\hat{\mathbf{r}})\ , \tag{13.23}$$

and the coefficient a can be found from the normalization (13.4) when we observe that

$$L_- L_+ = L_x^2 + L_y^2 + i\,[L_x, L_y] \tag{13.24}$$

$$= \mathbf{L}^2 - L_z^2 - L_z \tag{13.25}$$

and

$$L_+ = L_-^\dagger\ . \tag{13.26}$$

Thus,

$$\frac{4\pi}{2l+1}\,|a|^2 = \int_{4\pi} d\hat{\mathbf{r}}\ \left[L_+ C^{lm}(\hat{\mathbf{r}})\right]^* \left[L_+ C^{lm}(\hat{\mathbf{r}})\right] \tag{13.27}$$

$$\begin{aligned} &= \int_{4\pi} d\hat{\mathbf{r}}\, C^{lm}(\hat{\mathbf{r}})^*\ (L_- L_+)\, C^{lm}(\hat{\mathbf{r}}) \\ &= \int_{4\pi} d\hat{\mathbf{r}}\, C^{lm}(\hat{\mathbf{r}})^*\ \left(\mathbf{L}^2 - L_z^2 - L_z\right) C^{lm}(\hat{\mathbf{r}}) \\ &= \frac{4\pi}{2l+1}\,[l\,(l+1) - m\,(m+1)]\ . \end{aligned} \tag{13.28}$$

With the standard choice of phase, we can therefore put

$$a = \sqrt{l\,(l+1) - m\,(m+1)} \tag{13.29}$$

and thus

$$L_+ C^{lm} = \sqrt{l\,(l+1) - m\,(m+1)} C^{lm+1}\,. \tag{13.30}$$

Similarly,

$$L_- C^{lm} = \sqrt{l\,(l+1) - m\,(m-1)} C^{lm-1}\,. \tag{13.31}$$

The operators L_+ and L_- are called *ladder operators*, or more specifically, the *raising* and *lowering operators*. For any eigenfunction C^{lm}, we can, by applying $L_\pm$, generate new eigenfunctions with the same value of l but with m replaced by $m \pm 1$. This will continue indefinitely unless there is a maximum value m_{max} and a minimum value m_{min} of m where the coefficients vanish:

$$\begin{aligned} l\,(l+1) &= m_{\text{max}}\,(m_{\text{max}}+1) & (13.32) \\ &= m_{\text{min}}\,(m_{\text{min}}-1)\,. & (13.33) \end{aligned}$$

Evidently $l = m_{\text{max}} = -m_{\text{min}}$. Thus l is a positive integer and m ranges over the $2l+1$ values $-l \leq m \leq l$.

It is convenient to write the eigenfunctions in terms of the variables

$$\begin{aligned} x_\pm &= x \pm iy = r\sin\theta e^{\pm i\phi} & (13.34) \\ z &= r\cos\theta\,. & (13.35) \end{aligned}$$

Evidently

$$r^2 = x_+ x_- + z^2\,. \tag{13.36}$$

In terms of these variables, we find from the chain rule

$$\begin{aligned} L_z &= -i\left(x\frac{\partial}{\partial y} - y\frac{\partial}{\partial x}\right) & (13.37) \\ &= -ix\left(\frac{\partial x_+}{\partial y}\frac{\partial}{\partial x_+} + \frac{\partial x_-}{\partial y}\frac{\partial}{\partial x_-}\right) + iy\left(\frac{\partial x_+}{\partial x}\frac{\partial}{\partial x_+} + \frac{\partial x_-}{\partial x}\frac{\partial}{\partial x_-}\right) \\ &= x\left(\frac{\partial}{\partial x_+} - \frac{\partial}{\partial x_-}\right) + iy\left(\frac{\partial}{\partial x_+} + \frac{\partial}{\partial x_-}\right) \\ &= x_+\frac{\partial}{\partial x_+} - x_-\frac{\partial}{\partial x_-}\,. & (13.38) \end{aligned}$$

Similarly,

$$L_{\pm} = \pm\left(2z\frac{\partial}{\partial x_{\mp}} - x_{\pm}\frac{\partial}{\partial z}\right). \tag{13.39}$$

13.2.1 Generating the Spherical Harmonics

Consider the function x_-^l . Note

$$\begin{aligned} L_z x_-^l &= -l x_-^l \\ L_- x_-^l &= 0 \\ \mathbf{L}^2 x_-^l &= \left(L_+L_- + L_z^2 - L_z\right) x_-^l = l\,(l+1)\, x_-^l\,, \end{aligned} \tag{13.40}$$

which shows that x_-^l is proportional to the spherical harmonic $C^{l,-l}$. Because $x_-x_+ = r^2\sin^2\theta$, then putting $\xi = \cos\theta$ and noting

$$\begin{aligned} \int_{4\pi} d\hat{\mathbf{r}}\, \left(x_-x_+\right)^l &= 2\pi r^{2l}\int_{-1}^{+1} d\xi\,\left(1-\xi^2\right)^l \\ &= 4\pi\frac{2^{2l}\,(l!)^2}{(2l+1)!}\,, \end{aligned} \tag{13.41}$$

we find that the normalization condition (13.4) gives

$$r^l C^{l,-l}(\hat{\mathbf{r}}) = \frac{\sqrt{(2l)!}}{l!}\left(\frac{x_-}{2}\right)^l. \tag{13.42}$$

All other C^{lm} can be found by successive applications of the ladder operator L_+ [see (13.39)]

$$L_+C^{lm} = \sqrt{l\,(l+1) - m\,(m+1)}C^{l,m+1}. \tag{13.43}$$

For example,

$$L_+r^lC^{l,-l} = \frac{\sqrt{(2l)!}}{(l-1)!}z\left(\frac{x_-}{2}\right)^{l-1} = \sqrt{2l}r^lC^{l,1-l} \tag{13.44}$$

which gives

$$r^lC^{l,1-l} = \frac{\sqrt{(2l-1)!}}{(l-1)!}z\left(\frac{x_-}{2}\right)^{l-1}. \tag{13.45}$$

$l=$	0	1	2	3	4
$m=0$	1	z	$\frac{1}{2}\left(3z^2-r^2\right)$	$\frac{1}{2}\left(5z^2-3r^2\right)z$	$\frac{1}{8}\left(35z^4-30r^2z^2+3r^4\right)$
-1		$\frac{x_-}{\sqrt{2}}$	$\sqrt{\frac{3}{2}}zx_-$	$\frac{\sqrt{3}}{4}\left(5z^2-r^2\right)x_-$	$\frac{\sqrt{5}}{4}\left(7z^2-3r^2\right)zx_-$
-2			$\frac{1}{2}\sqrt{\frac{3}{2}}x_-^2$	$\frac{\sqrt{30}}{4}zx_-^2$	$\frac{\sqrt{10}}{8}\left(7z^2-r^2\right)x_-^2$
-3				$\frac{\sqrt{5}}{4}x_-^3$	$\frac{\sqrt{35}}{4}zx_-^3$
-4					$\frac{\sqrt{70}}{16}x_-^4$

Table 13.1. The spherical harmonics. A table of $r^l C^{lm}\left(\hat{\mathbf{r}}\right)$.

In this manner, a table of spherical harmonics can be generated. Note the parity property

$$C^{lm}(-\hat{\mathbf{r}}) = (-)^l\, C^{lm}(\hat{\mathbf{r}}) \tag{13.46}$$

and the special value

$$C^{lm}(\mathbf{e}_3) = \delta^{m0} . \tag{13.47}$$

Only spherical harmonics with $m \leq 0$ are shown. Those with $m > 0$ are given by

$$C^{l,-m}(\hat{\mathbf{r}}) = (-)^m\, C^{lm*}(\hat{\mathbf{r}}) . \tag{13.48}$$

The harmonics with $m = 0$ are evidently real. They are called *Legendre polynomials*

$$C^{l0}(\hat{\mathbf{r}}) =: P_l(\mathbf{e}_3 \cdot \hat{\mathbf{r}}) = P_l(\cos\theta) . \tag{13.49}$$

The last relation is a special case of the more general sum

$$\sum_{m=-l}^{l} C^{lm}\left(\hat{\mathbf{k}}\right)^* C^{lm}(\hat{\mathbf{r}}) = P_l\left(\hat{\mathbf{k}} \cdot \hat{\mathbf{r}}\right) , \tag{13.50}$$

which is known as the *addition theorem*, which can be proved from (13.49) and the invariance of (13.50) under rotations of $\mathbf{k}$ and $\mathbf{r}$ together. The set of $2l+1$ components $\left\{C^{lm}\right\}$ for a given l value is sometimes referred to as a *spherical* *tensor of rank l.* The addition theorem may be thought of as a generalization of the dot product from vectors to spherical tensors of rank l.

Exercise 13.4. Show that in the common case with $l = 1$, (13.50) becomes

$$\cos\theta_1 \cos\theta_2 + \sin\theta_1 \sin\theta_2 \cos(\phi_1 - \phi_2) = \cos\theta_{12}\,, \tag{13.51}$$

where $\mathbf{k}$ and $\mathbf{r}$ point in the directions (θ_1, ϕ_1) and (θ_2, ϕ_2), respectively, and θ_{12} is the angle between them. Show further that this is the trigonometric form of

$$\begin{aligned} \mathbf{k}\cdot\mathbf{e}^1\,\mathbf{e}_1\cdot\mathbf{r} + \mathbf{k}\cdot\mathbf{e}^2\,\mathbf{e}_2\cdot\mathbf{r} + \mathbf{k}\cdot\mathbf{e}^3\,\mathbf{e}_3\cdot\mathbf{r} &= \mathbf{k}\cdot\mathbf{e}^k\,\mathbf{e}_k\cdot\mathbf{r} \\ &= \mathbf{k}\cdot\mathbf{r} \end{aligned} \tag{13.52}$$

divided by the magnitudes of $\mathbf{k}$ and $\mathbf{r}$.

Exercise 13.5. Use the completeness relation (13.14) to prove

$$\sum_l (2l+1)\, P_l\left(\hat{\mathbf{k}}\cdot\hat{\mathbf{r}}\right) = 4\pi\delta\left(\hat{\mathbf{k}} - \hat{\mathbf{r}}\right). \tag{13.53}$$

Exercise 13.6. Show that with the identification (13.49), the orthogonality condition (13.4) on the spherical harmonics implies

$$\left(l + \frac{1}{2}\right)\int_{-1}^{1} d\xi\, P_l(\xi)\, P_{l'}(\xi) = \delta_{ll'}. \tag{13.54}$$

From the eigenvalue equation (13.2) and the operator relation for $\mathbf{L}^2$ in spherical coordinates (see Appendix A), the Legendre polynomials satisfy

$$\frac{d}{d\xi}\left(1 - \xi^2\right)\frac{dP_l(\xi)}{d\xi} + l\,(l+1)\,P_l(\xi) = 0\,. \tag{13.55}$$

This differential equation together with the value $P_l(1) = 1$ imply the Rodrigues formula

$$P_l(\xi) = \frac{1}{2^l l!}\frac{d^l}{d\xi^l}\left(\xi^2 - 1\right)^l, \tag{13.56}$$

from which a number of recursion relations follow:

$$P'_{l+1} - P'_{l-1} = (2l+1)\,P_l \tag{13.57}$$

$$P'_{l+1} - \xi P'_l = (l+1)\,P_l \tag{13.58}$$

$$\xi P'_l - P'_{l-1} = lP_l \tag{13.59}$$

$$(l+1)\,P_{l+1} + lP_{l-1} = (2l+1)\,\xi P_l \tag{13.60}$$

$$\left(1 - \xi^2\right)P'_l + l\xi P_l = lP_{l-1}\,. \tag{13.61}$$

Example 13.7. The Rodrigues formula gives

$$\begin{aligned} P_{l+1}(\xi) &= \frac{1}{2^{l+1}(l+1)!}\frac{d^l}{d\xi^l}\left[\frac{d}{d\xi}\left(\xi^2-1\right)^{l+1}\right] \\ &= \frac{1}{2^{l+1}(l+1)!}\frac{d^l}{d\xi^l}\left[2(l+1)\xi\left(\xi^2-1\right)^l\right] \\ &= \frac{1}{2^l l!}\frac{d^l}{d\xi^l}\left[\xi\left(\xi^2-1\right)^l\right]. \end{aligned} \tag{13.62}$$

Taking a derivative, we obtain the first of the recursion relations:

$$\begin{aligned} P'_{l+1}(\xi) &= \frac{1}{2^l l!}\frac{d^l}{d\xi^l}\frac{d}{d\xi}\left[\xi\left(\xi^2-1\right)^l\right] \\ &= \frac{1}{2^l l!}\frac{d^l}{d\xi^l}\left[\left(\xi^2-1\right)^l+2l\xi^2\left(\xi^2-1\right)^{l-1}\right] \\ &= \frac{1}{2^l l!}\frac{d^l}{d\xi^l}\left[(2l+1)\left(\xi^2-1\right)^l+2l\left(\xi^2-1\right)^{l-1}\right] \\ &= (2l+1)P_l+P'_{l-1}. \end{aligned} \tag{13.63}$$

On the other hand, we can also note

$$\begin{aligned} P'_{l+1}(\xi) &= \frac{1}{2^l l!}\frac{d^{l+1}}{d\xi^{l+1}}\left[\xi\left(\xi^2-1\right)^l\right] \\ &= \frac{(l+1)}{2^l l!}\frac{d^l}{d\xi^l}\left(\xi^2-1\right)^l+\frac{\xi}{2^l l!}\frac{d^{l+1}}{d\xi^{l+1}}\left(\xi^2-1\right)^l \\ &= (l+1)P_l+\xi P'_l, \end{aligned} \tag{13.64}$$

which is the second recursion relation. The difference of the first and second relations gives the third recursion formula.

Exercise 13.8. Demonstrate that the first, second, and third recursion relations imply the fourth.

Exercise 13.9. Show how the second and third recursion relations can be combined to give the fifth one. Differentiate the fifth relation and combine with the third to prove that the differential equation (13.55) is satisfied.

Consider the difference vector $\mathbf{r}=\mathbf{x}-\mathbf{x}'$, and let

$$x_> = \text{greater of } \{|\mathbf{x}|, |\mathbf{x}'|\} \tag{13.65}$$

$$x_< = \text{smaller of } \{|\mathbf{x}|, |\mathbf{x}'|\}. \tag{13.66}$$

Thus,

$$x_> + x_< = |\mathbf{x}| + |\mathbf{x}'| \tag{13.67}$$

$$x_> - x_< = ||\mathbf{x}| - |\mathbf{x}'|| . \tag{13.68}$$

Any analytical function of $r = \sqrt{x_>^2 + x_<^2 - 2x_> x_< \hat{\mathbf{x}} \cdot \hat{\mathbf{x}}'}$ can be expanded in the Legendre polynomials $P_l(\hat{\mathbf{x}} \cdot \hat{\mathbf{x}}')$. The expansion coefficients are found by applying the orthogonality and recursion relations for the $P_l(\xi)$. In particular,

$$r = x_> \sum_{l=0}^{\infty} \left(\frac{x_<}{x_>}\right)^l \left[\frac{(x_</x_>)^2}{2l+3} - \frac{1}{2l-1}\right] P_l(\hat{\mathbf{x}} \cdot \hat{\mathbf{x}}') \tag{13.69}$$

$$r^{-1} = x_>^{-1} \sum_{l=0}^{\infty} \left(\frac{x_<}{x_>}\right)^l P_l(\hat{\mathbf{x}} \cdot \hat{\mathbf{x}}') \tag{13.70}$$

$$r^{-3} = \left[x_> \left(x_>^2 - x_<^2\right)\right]^{-1} \sum_l (2l+1) \left(\frac{x_<}{x_>}\right)^l P_l(\hat{\mathbf{x}} \cdot \hat{\mathbf{x}}') \tag{13.71}$$

The completeness relation (13.14) for spherical harmonics implies that any function of $\hat{\mathbf{r}}$ can be expanded in the $C^{lm}(\hat{\mathbf{r}})$. A particularly useful expansion is the *composition relation* for the product of two spherical harmonics:

$$C^{lm}(\hat{\mathbf{r}})\, C^{kq}(\hat{\mathbf{r}}) = \sum_{jn} \langle lk00|j0\rangle \, \langle lkmq|jn\rangle \, C^{jn}(\hat{\mathbf{r}}) \tag{13.72}$$

where $\langle lkmq|jn\rangle$ is a Clebsch-Gordan coefficient (see below, section 13.3.2).[1] Common examples are

$$\cos\theta\, C^{lm} = a_0(l,m)\, C^{l-1,m} + b_0(l,m)\, C^{l+1,m} \tag{13.73}$$

$$e^{\pm i\phi} \sin\theta\, C^{lm} = a_\pm(l,m)\, C^{l-1,m\pm1} + b_\pm(l,m)\, C^{l+1,m\pm1} , \tag{13.74}$$

where the coefficients are

$$a_0(l,m) = \frac{1}{2l+1}\sqrt{l^2 - m^2} \tag{13.75}$$

$$b_0(l,m) = \frac{1}{2l+1}\sqrt{(l+1)^2 - m^2} \tag{13.76}$$

$$a_\pm(l,m) = \frac{1}{2l+1}\sqrt{(l \mp m)(l \mp m - 1)} \tag{13.77}$$

$$b_\pm(l,m) = -\frac{1}{2l+1}\sqrt{(l \pm m + 2)(l \pm m + 1)} . \tag{13.78}$$

[1] These are discussed and tabulated in many quantum mechanics texts. See, for example, D. M. Brink and G. R. Satchler (1968). See also Section 13.3.2 below.

13.3 Rotations and Matrix Representations

The origin of the composition relation (13.72) is most easily seen if we view the spherical harmonics as elements of rotation matrices. The simplest faithful representation of SU(2), the two-fold covering group of the rotation group $SO_+(3)$ in three dimensions, is the standard matrix representation of the rotation element $R = \exp(-i\boldsymbol{\theta}/2)$ in the Pauli algebra. The matrix structure depends on the rotation axis, but since any rotation can be expressed in Euler-angle rotations about $\mathbf{e}_3$, then about $\mathbf{e}_2$, and finally about $\mathbf{e}_3$ again, we can concentrate on the two rotations with matrix representations

$$\mathbf{D}^{(1/2)}(\theta\mathbf{e}_2) := R(\theta\mathbf{e}_2) = \exp(-i\mathbf{e}_2\theta/2) = \begin{pmatrix} \cos\theta/2 & -\sin\theta/2 \\ \sin\theta/2 & \cos\theta/2 \end{pmatrix} \tag{13.79}$$

and

$$\mathbf{D}^{(1/2)}(\phi\mathbf{e}_3) := R(\phi\mathbf{e}_3) = \exp(-i\mathbf{e}_3\phi/2) = \begin{pmatrix} \exp(-i\phi/2) & 0 \\ 0 & \exp(i\phi/2) \end{pmatrix}. \tag{13.80}$$

The rotation matrices are unitary, and in the standard representation, the matrix for rotations about $\mathbf{e}_2$ is always real whereas that for rotations about $\mathbf{e}_3$ is always diagonal.

The matrices (13.79) and (13.80) form a representation of the group SU(2). There is a matrix for every rotation of the group, and it can be found from products of (13.79) and (13.80). The representation is said to be *irreducible* because, in the case of 2×2 matrices, there is no single similarity transformation that diagonalizes both $\mathbf{D}^{(1/2)}(\theta\mathbf{e}_2)$ and $\mathbf{D}^{(1/2)}(\phi\mathbf{e}_3)$.[2]

The rotation of a vector $\mathbf{V} = V^k\mathbf{n}_k$ is obtained by rotating the basis vectors:

$$\begin{aligned} \mathbf{n}_k \to R\mathbf{n}_k R^\dagger = R\mathbf{n}_k\bar{R} &= \mathbf{n}_j \left\langle \mathbf{n}^j R\mathbf{n}_k\bar{R}\right\rangle_S \\ &= \mathbf{n}_j\mathbf{D}^j{}_k\,, \end{aligned} \tag{13.81}$$

where $\mathbf{D}$ is a matrix with elements

$$\mathbf{D}^j{}_k = \left\langle \mathbf{n}^j R\mathbf{n}_k\bar{R}\right\rangle_S . \tag{13.82}$$

[2]The concept of irreducible representations is important in group theory. It will be extended to representations larger than 2×2 below.

Thus,

$$\begin{aligned}\mathbf{V} &\rightarrow V'^j\mathbf{n}_j = V^k R\mathbf{n}_k\bar{R} \\ &= \mathbf{n}_j\mathbf{D}^j{}_k V^k \qquad (13.83)\end{aligned}$$

and the rotated components V'^j are given on the original basis by:

$$V'^j = \mathbf{D}^j{}_k V^k\,. \qquad (13.84)$$

In *matrix form*, one often writes

$$\mathbf{V}' = \mathbf{D}\mathbf{V} \qquad (13.85)$$

where here

$$\mathbf{V}' \equiv \begin{pmatrix} V'^1 \\ V'^2 \\ V'^3 \end{pmatrix}, \; \mathbf{V} \equiv \begin{pmatrix} V^1 \\ V^2 \\ V^3 \end{pmatrix}, \qquad (13.86)$$

In a cartesian basis, with $\mathbf{n}^k = \mathbf{e}^k = \mathbf{e}_k$,

$$\begin{aligned} R(\phi\mathbf{e}_3)\,\mathbf{e}_1\bar{R}(\phi\mathbf{e}_3) &= \mathbf{e}_1\cos\phi + \mathbf{e}_2\sin\phi \qquad (13.87)\\ R(\phi\mathbf{e}_3)\,\mathbf{e}_2\bar{R}(\phi\mathbf{e}_3) &= \mathbf{e}_2\cos\phi - \mathbf{e}_1\sin\phi \qquad (13.88)\\ R(\phi\mathbf{e}_3)\,\mathbf{e}_3\bar{R}(\phi\mathbf{e}_3) &= \mathbf{e}_3\,, \qquad (13.89)\end{aligned}$$

which can be combined to a single matrix equation in the form of (13.81):

$$R(\phi\mathbf{e}_3)\,(\mathbf{e}_1,\mathbf{e}_2,\mathbf{e}_3)\,\bar{R}(\phi\mathbf{e}_3) = (\mathbf{e}_1,\mathbf{e}_2,\mathbf{e}_3)\,\mathbf{D}(\phi\mathbf{e}_3) \qquad (13.90)$$

with

$$\mathbf{D}(\phi\mathbf{e}_3) = \begin{pmatrix} \cos\phi & -\sin\phi & 0 \\ \sin\phi & \cos\phi & 0 \\ 0 & 0 & 1 \end{pmatrix}. \qquad (13.91)$$

Cyclic permutation (rows and columns) gives

$$\mathbf{D}(\theta\mathbf{e}_2) = \begin{pmatrix} \cos\theta & 0 & \sin\theta \\ 0 & 1 & 0 \\ -\sin\theta & 0 & \cos\theta \end{pmatrix}. \qquad (13.92)$$

A linear transformation of the basis can diagonalize one but not both of these matrices. As shown in problem 1, each of the basis vectors $\{\mathbf{n}_+, \mathbf{n}_0, \mathbf{n}_-\}$ where $\mathbf{n}_\pm = \mathbf{e}_1 \pm i\mathbf{e}_2$ and $\mathbf{n}_0 = \mathbf{e}_3$ is an eigenvector of rotations about $\mathbf{e}_3$. We need normalized versions of these basis vectors, which we define with phases like those of the spherical harmonics C^{1m} by

$$\hat{\mathbf{n}}_{\pm 1} = \mp \mathbf{n}_\pm / \sqrt{2} = \mp (\mathbf{e}_1 \pm i\mathbf{e}_2) / \sqrt{2} \tag{13.93}$$

$$\hat{\mathbf{n}}_0 = \mathbf{n}_0 = \mathbf{e}_3 \,. \tag{13.94}$$

The reciprocal basis (A.5) comprises $\hat{\mathbf{n}}^0 = \hat{\mathbf{n}}_0$ and $\hat{\mathbf{n}}^{\pm 1} = (\hat{\mathbf{n}}_{\pm 1})^\dagger = -\hat{\mathbf{n}}_{\mp 1}$. In this basis, the rotation matrix (13.82) has elements usually written

$$\mathbf{D}^{(1)}_{mn} = \left\langle \hat{\mathbf{n}}^m R \hat{\mathbf{n}}_n \bar{R} \right\rangle_S = \left\langle \hat{\mathbf{n}}^\dagger_m R \hat{\mathbf{n}}_n \bar{R} \right\rangle_S , \tag{13.95}$$

with $m, n = +1, 0, -1$, which gives

$$\mathbf{D}^{(1)}(\phi \mathbf{e}_3) = \begin{pmatrix} e^{-i\phi} & 0 & 0 \\ 0 & 1 & 0 \\ 0 & 0 & e^{i\phi} \end{pmatrix} . \tag{13.96}$$

The matrix (13.92) for rotations about $\mathbf{e}_2$ is changed to

$$\mathbf{D}^{(1)}(\theta \mathbf{e}_2) = \frac{1}{2} \begin{pmatrix} 1 + \cos\theta & -\sqrt{2}\sin\theta & 1 - \cos\theta \\ \sqrt{2}\sin\theta & 2\cos\theta & -\sqrt{2}\sin\theta \\ 1 - \cos\theta & \sqrt{2}\sin\theta & 1 + \cos\theta \end{pmatrix} \tag{13.97}$$

by the transformation to the $\{\hat{\mathbf{n}}_{+1}, \hat{\mathbf{n}}_0, \hat{\mathbf{n}}_{-1}\}$basis. In accordance with (13.81), the rotation of the basis vectors is given by

$$\hat{\mathbf{n}}_n \to \hat{\mathbf{n}}'_n = R \hat{\mathbf{n}}_n R^\dagger \tag{13.98}$$

$$= \hat{\mathbf{n}}_m \left\langle \hat{\mathbf{n}}^m R \hat{\mathbf{n}}_n R^\dagger \right\rangle_S \tag{13.99}$$

$$\equiv \sum_m \hat{\mathbf{n}}_m D^{(1)}_{mn} \,. \tag{13.100}$$

(The sum over m is inserted explicitly because in the standard notation for the rotation matrix elements, j appears as a lower rather than upper index.)

The $\mathbf{D}^{(1)}$ matrices also form an irreducible representation of SU(2). It is an irreducible representation because there is no single similarity transformation that even block-diagonalizes both $\mathbf{D}^{(1)}(\phi \mathbf{e}_3)$ and $\mathbf{D}^{(1)}(\theta \mathbf{e}_2)$. The matrices $\mathbf{D}^{(1)}$ of the representation act on vectors in $\mathbb{E}^3$, and their irreducibility means that there are no subspaces (directions or oriented planes) that are invariant under all rotations.

13.3.1 Product Representations and Their Reduction

Another way to generate the matrices $\mathbf{D}^{(1)}$ is from direct products of the $\mathbf{D}^{(1/2)}$ matrices. Such products can arise when we rotate products of two vectors, say $\mathbf{r}_1$ and $\mathbf{r}_2$, which can be associated either with two different particles, for example, or with two properties of the same particle. The most general product is the tensor product $\mathbf{r}_1 \otimes \mathbf{r}_2$ which transforms under rotation according to

$$\mathbf{r}_1 \otimes \mathbf{r}_2 \rightarrow (R \otimes R)(\mathbf{r}_1 \otimes \mathbf{r}_2)\left(R^\dagger \otimes R^\dagger\right) \tag{13.101}$$

$$= \left(R\mathbf{r}_1 R^\dagger\right) \otimes \left(R\mathbf{r}_2 R^\dagger\right) . \tag{13.102}$$

If we represent the rotations R by matrices $\mathbf{D}$,then the matrices $(R \otimes R)$ are represented by direct products $\mathbf{D} \otimes \mathbf{D}$, whose elements include products of every element of the matrix on the left with every element of the one on the right.

Starting with matrices from two irreducible representations $\mathbf{D}^{(k)}$ and $\mathbf{D}^{(l)}$, where k, l are whole or half integers, we obtain matrices of a reducible product representation

$$\mathbf{D}^{(k \otimes l)} = \mathbf{D}^{(k)} \otimes \mathbf{D}^{(l)} . \tag{13.103}$$

Thus from (13.79) and (13.80), we find

$$\mathbf{D}^{\left(\frac{1}{2}\otimes\frac{1}{2}\right)}(\mathbf{e}_3\phi) = \begin{pmatrix} \exp(-\frac{i}{2}\phi) & 0 \\ 0 & \exp(\frac{i}{2}\phi) \end{pmatrix} \otimes \begin{pmatrix} \exp(-\frac{i}{2}\phi) & 0 \\ 0 & \exp(\frac{i}{2}\phi) \end{pmatrix}$$

$$= \begin{pmatrix} \exp(-i\phi) & 0 & 0 & 0 \\ 0 & 1 & 0 & 0 \\ 0 & 0 & 1 & 0 \\ 0 & 0 & 0 & \exp(i\phi) \end{pmatrix} \tag{13.104}$$

and

$$\mathbf{D}^{\left(\frac{1}{2}\otimes\frac{1}{2}\right)}(\mathbf{e}_2\theta) = \begin{pmatrix} \cos\theta/2 & -\sin\theta/2 \\ \sin\theta/2 & \cos\theta/2 \end{pmatrix} \otimes \begin{pmatrix} \cos\theta/2 & -\sin\theta/2 \\ \sin\theta/2 & \cos\theta/2 \end{pmatrix}$$

$$= \begin{pmatrix} c\begin{pmatrix} c & -s \\ s & c \end{pmatrix} & -s\begin{pmatrix} c & -s \\ s & c \end{pmatrix} \\ s\begin{pmatrix} c & -s \\ s & c \end{pmatrix} & c\begin{pmatrix} c & -s \\ s & c \end{pmatrix} \end{pmatrix}$$

$$= \begin{pmatrix} c^2 & -cs & -cs & s^2 \\ cs & c^2 & -s^2 & -cs \\ cs & -s^2 & c^2 & -cs \\ s^2 & cs & cs & c^2 \end{pmatrix} , \tag{13.105}$$

where we have applied the short-hand notation $c = \cos\theta/2$ and $s = \sin\theta/2$. A transformation that block diagonalizes $\mathbf{D}^{\left(\frac{1}{2}\otimes\frac{1}{2}\right)}$ for all rotations can be found from inspection. First note that the two middle elements of the diagonal matrix $\mathbf{D}^{\left(\frac{1}{2}\otimes\frac{1}{2}\right)}(\mathbf{e}_3\phi)$ are the same; the 2×2 matrix in the center of $\mathbf{D}^{\left(\frac{1}{2}\otimes\frac{1}{2}\right)}(\mathbf{e}_3\phi)$ is the identity and is invariant under any 2×2 unitary transformation. Next observe that the transformation

$$\mathbf{T} = \begin{pmatrix} 1 & 0 & 0 & 0 \\ 0 & 1/\sqrt{2} & -1/\sqrt{2} & 0 \\ 0 & 1/\sqrt{2} & 1/\sqrt{2} & 0 \\ 0 & 0 & 0 & 1 \end{pmatrix} \tag{13.106}$$

reduces the central part of $\mathbf{D}^{\left(\frac{1}{2}\otimes\frac{1}{2}\right)}(\mathbf{e}_2\theta)$ and a final transformation by

$$\mathbf{S} = \begin{pmatrix} 0 & 1 & 0 & 0 \\ 0 & 0 & 1 & 0 \\ 1 & 0 & 0 & 0 \\ 0 & 0 & 0 & 1 \end{pmatrix} \tag{13.107}$$

cyclicly permutes the first three rows and columns to give

$$\begin{aligned} (\mathbf{TS})^\dagger\, \mathbf{D}^{\left(\frac{1}{2}\otimes\frac{1}{2}\right)}(\mathbf{e}_2\theta)\, \mathbf{TS} &= \begin{pmatrix} 1 & 0 & 0 & 0 \\ 0 & c^2 & -\sqrt{2}cs & s^2 \\ 0 & \sqrt{2}cs & c^2 - s^2 & -\sqrt{2}cs \\ 0 & s^2 & \sqrt{2}cs & c^2 \end{pmatrix} \\ &= \mathbf{D}^{(0)} \oplus \mathbf{D}^{(1)}(\mathbf{e}_2\theta)\,, \end{aligned} \tag{13.108}$$

where in the second line, we have expressed the block-diagonal matrix as a direct sum of the trivial representation $\mathbf{D}^{(0)} = 1$ and the irreducible three-dimensional representation $\mathbf{D}^{(1)}$ (13.97) for rotations about $\mathbf{e}_2$. Since (a) the same similarity transformation with $\mathbf{TS}$ puts rotations $\mathbf{D}^{\left(\frac{1}{2}\otimes\frac{1}{2}\right)}(\mathbf{e}_3\phi)$ in the same block diagonal form, (b) any rotation can be expressed as a product of rotations about $\mathbf{e}_2$ and $\mathbf{e}_3$, and (c) any product of matrices in a given block-diagonal form has the same form, all rotations in the product representation $\frac{1}{2}\otimes\frac{1}{2}$ are reduced by the transformation $\mathbf{TS}$ and we can write the equivalence relation

$$\mathbf{D}^{\left(\frac{1}{2}\otimes\frac{1}{2}\right)} \equiv \mathbf{D}^{\left(\frac{1}{2}\right)} \otimes \mathbf{D}^{\left(\frac{1}{2}\right)} \simeq \mathbf{D}^{(0)} \oplus \mathbf{D}^{(1)}\,. \tag{13.109}$$

In a similar way, the direct product of irreducible representations $\mathbf{D}^{(k)}$ and $\mathbf{D}^{(l)}$ is generally reducible and can be reduced to a direct sum of representations $\mathbf{D}^{(J)}$ with $|k-l| \leq J \leq k+l$:

$$\mathbf{D}^{(k)} \otimes \mathbf{D}^{(l)} \simeq \bigoplus_{J=|k-l|}^{k+l} \mathbf{D}^{(J)}, \tag{13.110}$$

and the total dimension of the product representation is $(2k+1)(2l+1)$.

Exercise 13.10. Use dimension counting to verify that the J values of the irreducible representations in the direct sum (13.110) are just

$$J = |k-l|, |k-l|+1, \cdots, k+l, \tag{13.111}$$

with each appearing just once, when the product representation $k \otimes l$ is reduced. In other words, prove that

$$(2k+1)(2l+1) = \sum_{J=|k-l|}^{k+l} (2J+1) . \tag{13.112}$$

13.3.2 Clebsch-Gordan Coefficients

The transformation matrix corresponding to $\mathbf{TS}$, needed to perform the reduction, usually cannot be guessed by inspection, but its matrix elements are real and have been tabulated in many texts on quantum mechanics. They are called Clebsch-Gordan coefficients[3], denoted $\langle klqm|JM\rangle = \langle JM|klqm\rangle$:

$$D_{qq'}^{(k)}(\boldsymbol{\theta})\, D_{mm'}^{(l)}(\boldsymbol{\theta}) = \sum_{JMM'} \langle klqm|JM\rangle\, D_{MM'}^{(J)}(\boldsymbol{\theta})\, \langle JM'|klq'm'\rangle . \tag{13.113}$$

Only terms $M = q+m$, $M' = q'+m'$ contribute. The inverse transformation is

$$D_{MM'}^{(J)}(\boldsymbol{\theta}) = \sum_{mm'qq'} \langle JM|klqm\rangle\, D_{qq'}^{(k)}(\boldsymbol{\theta})\, D_{mm'}^{(l)}(\boldsymbol{\theta})\, \langle klq'm'|JM'\rangle . \tag{13.114}$$

[3]These are often denoted $\langle klqm|klJM\rangle$ .We use the common Wigner sign convention. See, for example, A. E. Edmonds, *Angular Momentum in Quantum Mechanics* (Princeton U. Press, 1957).

Explicit equations for the matrix elements $D^{(J)}_{MM'}$ can be derived with these formulas. For rotations about $\mathbf{e}_3$, they are quite simple:

$$D^{(J)}_{MM'}(\phi\mathbf{e}_3) = \exp(-iM\phi)\,\delta_{MM'}\,. \tag{13.115}$$

For rotations about $\mathbf{e}_2$, they are given by the Wigner formula:

$$\begin{aligned} D^{(J)}_{MM'}(\theta\mathbf{e}_2) &= (\cos\theta/2)^{2J}\sum_k (-)^k\,[\cdots]^{1/2}\,(\tan\theta/2)^{2k+M'-M} \\ [\cdots] &= \binom{J+M}{k}\binom{J-M'}{k}\binom{J-M}{k+M'-M}\binom{J+M'}{k+M'-M}, \end{aligned} \tag{13.116}$$

where k ranges over all integer values for which the coefficient $[\cdots]$ does not vanish and

$$\binom{n}{m} = \frac{n!}{m!\,(n-m)!} \tag{13.117}$$

is a binomial coefficient.

The Clebsch-Gordan coefficients are conveniently tabulated in terms of the related but more symmetric $3-j$ symbols $\begin{pmatrix} j_1 & j_2 & j_3 \\ m_1 & m_2 & m_3 \end{pmatrix}$, in terms of which

$$\langle klqm|JM\rangle = (-)^{k-l+M}\,(2J+1)^{1/2}\begin{pmatrix} k & l & J \\ q & m & -M \end{pmatrix}. \tag{13.118}$$

Under the interchange of any pair of columns or under the change in sign of the lower row (all three m values), the $3-j$ symbol changes by $(-1)^{k+l+J}$. The symbols vanish unless the lower row sums to zero, unless each m-value is no larger in magnitude than the j value in the same column, and unless the three values in the top row satisfy the triangular relationships: $|J-k| \le l \le J+k$, $|J-l| \le k \le J+l$, and $|k-l| \le J \le k+l$. All 6 entries in the table must be integer or half-integer. The few values that follow allow all nonvanishing $3-j$

symbols to be determined for which the smallest j value is 0,1/2, or 1:

$$\begin{pmatrix} j & j & 0 \\ m & -m & 0 \end{pmatrix} = (-)^{j-m}(2j+1)^{-1/2} \tag{13.119}$$

$$\begin{pmatrix} k & k+\frac{1}{2} & \frac{1}{2} \\ q & -q-\frac{1}{2} & \frac{1}{2} \end{pmatrix} = (-)^{k-q-1}\left[\frac{k+q+1}{2(j+1)(2j+1)}\right]^{1/2} \tag{13.120}$$

$$\begin{pmatrix} k & k & 1 \\ q & -q-1 & 1 \end{pmatrix} = (-)^{k-q}\left[\frac{(k-q)(k+q+1)}{2k(k+1)(2k+1)}\right]^{1/2} \tag{13.121}$$

$$\begin{pmatrix} k & k & 1 \\ q & -q & 0 \end{pmatrix} = (-)^{k-q}\frac{q}{[k(k+1)(2k+1)]^{1/2}} \tag{13.122}$$

$$\begin{pmatrix} k & k+1 & 1 \\ q & -q-1 & 1 \end{pmatrix} = (-)^{k-q}\left[\frac{(k+q+1)(k+q+2)}{2(k+1)(2k+1)(2k+3)}\right]^{1/2} \tag{13.123}$$

$$\begin{pmatrix} k & k+1 & 1 \\ q & -q & 0 \end{pmatrix} = (-)^{k-q-1}\left[\frac{(k-q+1)(k+q+1)}{(k+1)(2k+1)(2k+3)}\right]^{1/2}. \tag{13.124}$$

A number of computer programs are available to calculate the $3-j$ symbols.[4]

Exercise 13.11. Use the symmetry of the $3-j$ symbol and its relation (13.118) to Clebsch-Gordan coefficients to prove

$$\langle kl, -q, -m|J, -M\rangle = (-)^{k+q-J}\langle klqm|JM\rangle \tag{13.125}$$

$$\langle lkmq|JM\rangle = (-)^{k+q-J}\langle klqm|JM\rangle\,. \tag{13.126}$$

An important relation for rotation matrices follows from the closure of the rotation group: the product of rotations is a rotation:

$$\sum_n D^{(l)}_{mn}(\boldsymbol{\theta}_2)\,D^{(l)}_{nm'}(\boldsymbol{\theta}_1) = D^{(l)}_{mm'}(\Theta) \tag{13.127}$$

where Θ is the result of successive rotations by $\boldsymbol{\theta}_1$ and then by $\boldsymbol{\theta}_2$. In particular, if the two rotations are inverses of each other,

$$\sum_n D^{(l)}_{mn}(-\boldsymbol{\theta})\,D^{(l)}_{nm'}(\boldsymbol{\theta}) = \sum_n D^{(l)\dagger}_{mn}(\boldsymbol{\theta})\,D^{(l)}_{nm'}(\boldsymbol{\theta}) = \delta_{mm'}\,. \tag{13.128}$$

From Table 13.1, the spherical harmonics are sums of products of $x_\pm/r$ and z/r, and hence products of the spherical harmonics C^{1m}. The behavior of the C^{lm}

[4]See, for example, the share package "Wigner" in the professional version of Maple V rel. 5.

under rotations are in principle determined by the rotational behavior of the C^{1m}, and the linear combinations are chosen so as to make this behavior particularly simple. The C^{1m} are the hermitian conjugates of the spherical vector components of $\hat{\mathbf{r}}$,

$$\begin{aligned}\hat{\mathbf{r}} &= \left(\frac{x-iy}{\sqrt{2}r}\right)\left(\frac{\mathbf{e}_1+i\mathbf{e}_2}{\sqrt{2}}\right)+\frac{z}{r}\mathbf{e}_3+\left(\frac{x+iy}{\sqrt{2}r}\right)\left(\frac{\mathbf{e}_1-i\mathbf{e}_2}{\sqrt{2}}\right)\\ &= -C^{1,-1}(\hat{\mathbf{r}})\,\mathbf{e}_+ + C^{1,0}(\hat{\mathbf{r}})\,\mathbf{e}_0 - C^{1,1}(\hat{\mathbf{r}})\,\mathbf{e}_-\\ &= C^{1m*}(\hat{\mathbf{r}})\,\mathbf{e}_m \end{aligned} \tag{13.129}$$

and transform under rotations as the hermitian conjugate of (13.84):

$$C^{1m} \to C^{1n} D^{(1)\dagger}_{nm} . \tag{13.130}$$

More generally, the spherical harmonics transform as

$$C^{lm}(\hat{\mathbf{r}}_2) = \sum_n C^{ln}(\hat{\mathbf{r}}_1)\, D^{(l)\dagger}_{nm}(\boldsymbol{\theta}_{21}) = \sum_n D^{(l)*}_{mn}(\boldsymbol{\theta}_{21})\, C^{ln}(\hat{\mathbf{r}}_1) , \tag{13.131}$$

where $\boldsymbol{\theta}_{21}$ is the rotation that takes $\mathbf{r}_1$ into $\mathbf{r}_2$. The spherical harmonic C^{lm} is said to be an element of the carrier space of the representation $\mathbf{D}^{(l)\dagger}$. If $\hat{\mathbf{r}}_1 = \mathbf{e}_3$, then since $C^{lm}(\mathbf{e}_3) = \delta^{m0}$, one finds that the spherical harmonics are also just special rotation matrix elements:

$$C^{lm}(\hat{\mathbf{r}}) = D^{(l)\dagger}_{0m}(\boldsymbol{\theta}) = D^{(l)*}_{m0}(\boldsymbol{\theta}) \tag{13.132}$$

where the rotation by $\boldsymbol{\theta}$ rotates $\mathbf{e}_3$ into $\hat{\mathbf{r}}$.

Exercise 13.12. Show that the reduction formula (13.113) implies the composition relation (13.72) for spherical harmonics.

13.4 Infinitesimal Generators

The differential operator $\mathbf{L} = -i\mathbf{r}\times\boldsymbol{\nabla}$ plays an important role as a generator of infinitesimal rotations. Consider a scalar field $\chi(\mathbf{r})$ and apply an infinitesimal translation of the field by an amount $d\mathbf{r}$:

$$\chi(\mathbf{r}) \to \chi'(\mathbf{r}) . \tag{13.133}$$

The effect of the translation is to take the value of the original field at $\mathbf{r}$ and make it the value of the transformed field at $\mathbf{r} + d\mathbf{r}$, and the value of the transformed field at $\mathbf{r}$ is that of the untransformed field χ at $\mathbf{r} - d\mathbf{r}$:

$$\begin{aligned} \chi'(\mathbf{r}) &= \chi(\mathbf{r} - d\mathbf{r}) \\ &= \chi(\mathbf{r}) - d\mathbf{r} \cdot \nabla\chi(\mathbf{r}) \\ &= (1 - d\mathbf{r} \cdot \nabla)\,\chi(\mathbf{r}) \ . \end{aligned} \tag{13.134}$$

If the infinitesimal transformation is a rotation described by the angle $d\boldsymbol{\theta}$, the transformation can be written

$$\begin{aligned} \chi(\mathbf{r}) &\rightarrow \chi\,(\mathbf{r} - d\boldsymbol{\theta} \times \mathbf{r}) \\ &= (1 - d\boldsymbol{\theta} \times \mathbf{r} \cdot \nabla)\,\chi(\mathbf{r}) \\ &= (1 - d\boldsymbol{\theta} \cdot \mathbf{r} \times \nabla)\,\chi(\mathbf{r}) \\ &= (1 - i d\boldsymbol{\theta} \cdot \mathbf{L})\,\chi(\mathbf{r}) \ . \end{aligned} \tag{13.135}$$

The operator $\mathbf{L}$ is called the generator of infinitesimal rotations for scalar functions. It is associated with the orbital angular momentum of the system in quantum theory. Finite rotations can be built up from many small ones, and for a rotation by $\boldsymbol{\theta}$ we find

$$\begin{aligned} \chi(\mathbf{r}) &\rightarrow \lim_{N\to\infty} \left(1 - \frac{i\boldsymbol{\theta} \cdot \mathbf{L}}{N}\right)^N \chi(\mathbf{r}) \\ &= \exp(-i\boldsymbol{\theta} \cdot \mathbf{L})\,\chi(\mathbf{r}) \ . \end{aligned} \tag{13.136}$$

If the field is a vector field $\mathbf{F}$ rather than a scalar one, rotation will also mix the vector components:

$$\begin{aligned} \mathbf{F}(\mathbf{r}) &\rightarrow (1 + d\boldsymbol{\theta}\times)\,(1 - i d\boldsymbol{\theta} \cdot \mathbf{L})\,\mathbf{F}(\mathbf{r}) \\ &\rightarrow (1 + d\boldsymbol{\theta}\times - i d\boldsymbol{\theta} \cdot \mathbf{L})\,\mathbf{F}(\mathbf{r}) \ . \end{aligned} \tag{13.137}$$

If we define a "spin operator" $\mathbf{S}$ for vectors by

$$\mathbf{a} \cdot \mathbf{S}\,\mathbf{F} := i\mathbf{a} \times \mathbf{F}\,, \tag{13.138}$$

the infinitesimal rotation (13.137) becomes

$$\begin{aligned} \mathbf{F}(\mathbf{r}) &\rightarrow (1 - i d\boldsymbol{\theta} \cdot \mathbf{S} - i d\boldsymbol{\theta} \cdot \mathbf{L})\,\mathbf{F}(\mathbf{r}) \\ &= (1 - i d\boldsymbol{\theta} \cdot \mathbf{J})\,\mathbf{F}(\mathbf{r}) \end{aligned} \tag{13.139}$$

where the operator for the infinitesimal rotation is now

$$\mathbf{J} := \mathbf{L} + \mathbf{S} \tag{13.140}$$

and is associated with the total angular momentum.

13.5 Multipoles of Charge Distributions

As an important application of spherical harmonics, consider the interaction of isolated static charge densities. We start with the simple expression for the electrostatic potential at $\mathbf{x}$ of a point charge Q at $\mathbf{R}$, namely

$$\phi(\mathbf{x}) = \frac{1}{4\pi\varepsilon_0}\frac{Q}{|\mathbf{R}-\mathbf{x}|} \tag{13.141}$$

and apply the expansion (13.70) for the case $R > x$, where here $R = |\mathbf{R}|$ and $x = |\mathbf{x}|$:

$$\frac{1}{|\mathbf{R}-\mathbf{x}|} = \frac{1}{R}\sum_{L\geq 0}\left(\frac{x}{R}\right)^L P_L\left(\hat{\mathbf{x}}\cdot\hat{\mathbf{R}}\right) \tag{13.142}$$

$$= \sum_{L\geq 0}\frac{1}{R^{2L+1}}\sum_{M=-L}^{L} R^{L,M*}x^{L,M}, \tag{13.143}$$

where we have applied the addition theorem (13.50) and identified the multipoles

$$R^{L,M} = R^L C^{LM}\left(\hat{\mathbf{R}}\right),\ x^{L,M} = x^L C^{LM}(\hat{\mathbf{x}})\,. \tag{13.144}$$

The potential energy of the interacting system when a charge q is placed at $\mathbf{x}$ is thus

$$V = q\phi(\mathbf{x}) = \frac{qQ}{4\pi\varepsilon_0}\sum_{L\geq 0}\frac{1}{R^{2L+1}}\sum_{M=-L}^{L} R^{L,M*}x^{L,M}. \tag{13.145}$$

Each term in the multipole expansion (13.145) has the dependence on $\mathbf{x}$ and $\mathbf{R}$ factored into a multipole dependent on $\mathbf{x}$ times one dependent on $\mathbf{R}$. This factorization is particularly useful when instead of a single point charge q we have a distribution $\rho(\mathbf{x})$ with

$$q = \int d^3\mathbf{x}\,\rho(\mathbf{x})\,. \tag{13.146}$$

Then the interaction energy is

$$V = \frac{Q}{4\pi\varepsilon_0}\int d^3\mathbf{x}\,\frac{\rho(\mathbf{x})}{|\mathbf{R}-\mathbf{x}|} \tag{13.147}$$

$$= \frac{qQ}{4\pi\varepsilon_0}\sum_{L\geq 0}\frac{1}{R^{2L+1}}\sum_{M=-L}^{L} R^{L,M*}\left\langle x^{L,M}\right\rangle_\rho, \tag{13.148}$$

where

$$q\left\langle x^{L,M}\right\rangle_{\rho} = \int d^3\mathbf{x}\,\rho(\mathbf{x})\,x^L C^{LM}(\hat{\mathbf{x}}) \tag{13.149}$$

is the L, Mth multipole moment of the charge distribution. Of course, the interaction energy V is also Q times the electrostatic potential at $\mathbf{R}$ of the charge distribution $\rho(\mathbf{x})$: $V = Q\Phi(\mathbf{R})$. If $\rho(\mathbf{x})$ is spherically symmetric, only the monopole moment is nonvanishing ($\left\langle x^{L,M}\right\rangle_{\rho} = \delta^{L0}\delta^{M0}$ and the potential outside the distribution is identical to that for a point charge q at $\mathbf{x} = 0$. More generally, the moments $\left\langle x^{L,M}\right\rangle_{\rho}$ of the distribution can be determined from the angular dependence of the potential $\Phi(\mathbf{R})$ (see problem 6).

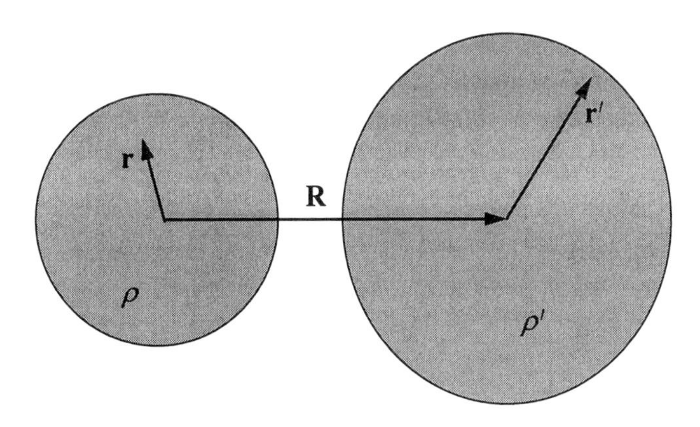

Fig. 13.1. The interaction of two nonoverlapping charge distributions can be expressed in terms of multipoles.

Suppose now that in place of the single charge Q at $\mathbf{R}$, we have a second charge distribution. More precisely, suppose we have two isolated charge distributions $\rho(\mathbf{r})$ and $\rho'(\mathbf{r}')$ whose centers are separated by $\mathbf{R}$ (see Fig. 13.1). We want to expand the interaction energy

$$V = \frac{1}{4\pi\varepsilon_0}\int d^3\mathbf{r}\int d^3\mathbf{r}'\frac{\rho(\mathbf{r})\,\rho'(\mathbf{r}')}{|\mathbf{R}-\mathbf{r}+\mathbf{r}'|} \tag{13.150}$$

in multipole moments of the two charge distributions. We need to extend the multipole expansion (13.143) by a decomposition theorem that relates moments $x^{L,M}$ to $r^{l,m}$ and $r'^{l'm'}$, where $\mathbf{x} = \mathbf{r} - \mathbf{r}'$.

Theorem 13.1 *(Multipole Decomposition) Let* $\mathbf{x} = \mathbf{r} - \mathbf{r}'$. *Then*

$$x^{L,M} = \sum_{l=0}^{L} (-)^{l'} \binom{2L}{2l}^{1/2} \left[r^{(l)} \otimes r'^{(l')} \right]^{L,M} , \tag{13.151}$$

where $l' = L - l$. *Here,* $\binom{2L}{2l}$ *is the binomial coefficient*

$$\binom{2L}{2l} = \frac{(2L)!}{(2l)!\,(2l')!} \tag{13.152}$$

and

$$\left[r^{(l)} \otimes r'^{(l')} \right]^{L,M} = \sum_{m,m'} r^{l,m} r'^{l',m'} \langle ll'mm'|LM\rangle \ . \tag{13.153}$$

Proof: The multipole $x^{L,M}$ must have the rotational behavior of C^{LM}, and the only combination of $r^{l,m}$ and $r'^{l',m'}$ with this behavior is $\left[r^{(l)} \otimes r'^{(l')} \right]^{L,M}$. This follows from the relation (13.114) for the rotation matrices in the special case $m' = q' = M' = 0$. Therefore,

$$x^{L,M} = \sum_{ll'} a_{ll'}^{(L)} \left[r^{(l)} \otimes r'^{(l')} \right]^{L,M} \tag{13.154}$$

and we only need to find the coefficients $a_{ll'}^{(L)}$. Since these coefficients are independent of M, m, and m', we can take $M = -L$ and use the relation (13.42) together with the binomial expansion of $x_- = r_- - r'_-$:

$$x^{L,-L} = \frac{\sqrt{(2L)!}}{2^L L!} \left(r_- - r'_- \right)^L \tag{13.155}$$

$$= \sum_{l=0}^{L} (-)^{l'} \binom{2L}{2l}^{1/2} r^{l,-l} r'^{l',-l'}, \ l' = L - l \, . \tag{13.156}$$

Since the Clebsch-Gordan coefficient $\langle ll', -l, -l'|L, -L\rangle = 1$ when $l + l' = L$, the coefficient $a_{ll'}^{(L)}$ is

$$a_{ll'}^{(L)} = (-)^{l'} \delta_{l+l',L} \binom{2L}{2l}^{1/2} . \tag{13.157}$$

Substitution into $x^{L,M}$ gives directly the multipole decomposition theorem. ■

When combined with the multipole expansion (13.143), the theorem gives

$$\frac{1}{|\mathbf{R}-\mathbf{r}+\mathbf{r}'|} = \sum_{L\geq 0} \frac{1}{R^{2L+1}} \sum_{M=-L}^{L} R^{L,M*} \sum_{ll'} a_{ll'}^{(L)} \left[r^{(l)} \otimes r'^{(l')}\right]^{L,M} , \qquad (13.158)$$

and the interaction energy of the two distributions is

$$\begin{aligned} V &= \frac{1}{4\pi\varepsilon_0} \sum_{LM} \frac{R^{L,M*}}{R^{2L+1}} \sum_{ll'} (-)^{l'} \delta_{l+l',L} \binom{2L}{2l}^{1/2} \\ &\quad \times \sum_{m,m'} \left\langle q r^{l,m} \right\rangle_{\rho} \left\langle q' r'^{l',m'} \right\rangle_{\rho'} \langle ll'mm'|LM\rangle , \qquad (13.159) \end{aligned}$$

where the multipole moments $\left\langle q r^{l,m} \right\rangle_{\rho}$, $\left\langle q' r'^{l',m'} \right\rangle_{\rho'}$ are given by

$$\left\langle q r^{l,m} \right\rangle_{\rho} = \int d^3\mathbf{r}\, \rho(\mathbf{r})\, r^l C^{lm}(\hat{\mathbf{r}}) . \qquad (13.160)$$

The expression is simplified if the $\mathbf{e}_3$ axis is chosen along $\mathbf{R}$. Then $R^{L,M} = \delta^{M,0} R^L$ and

$$V = \frac{qq'}{4\pi\varepsilon_0} \sum_{Lll'} \frac{\mathcal{V}_{ll'}}{R^{L+1}} \qquad (13.161)$$

with

$$\mathcal{V}_{ll'} = \sum_{ll'} (-)^{l'} \delta_{l+l',L} \binom{2L}{2l}^{1/2} \sum_{m,m'} \left\langle r^{l,m} \right\rangle_{\rho} \left\langle r'^{l',m'} \right\rangle_{\rho'} \langle ll'mm'|L0\rangle \qquad (13.162)$$

determining the contribution of the interaction of the $2l$-multipole of ρ with the $2l'$ multipole of ρ'. For example, for the interaction of the $2l$-multipole with a monopole, the interaction is

$$\frac{qq'}{4\pi\varepsilon_0} \frac{\mathcal{V}_{l0}}{R^{l+1}} \qquad (13.163)$$

with $\mathcal{V}_{l0} = \left\langle r^{l,0} \right\rangle_{\rho}$, and for the interaction of the $2l$-multipole with a dipole, the interaction is

$$\frac{qq'}{4\pi\varepsilon_0} \frac{\mathcal{V}_{l1}}{R^{l+2}} , \qquad (13.164)$$

with

$$\mathcal{V}_{l1}=\sqrt{\frac{l\,(l+1)}{2}}\left[(l+1)\left\langle r^{l,0}\right\rangle_{\rho}\left\langle r'^{1,0}\right\rangle_{\rho'}-\left\langle r^{l,1}\right\rangle_{\rho}\left\langle r'^{1,-1}\right\rangle_{\rho'}-\left\langle r^{l,-1}\right\rangle_{\rho}\left\langle r'^{1,1}\right\rangle_{\rho'}\right].$$

13.6 Problems

1. Consider the complex basis $\{\mathbf{n}_+,\mathbf{n}_0,\mathbf{n}_-\}$ where $\mathbf{n}_\pm=\mathbf{e}_1\pm i\mathbf{e}_2$ and $\mathbf{n}_0=\mathbf{e}_3$.

   (a) Show that the reciprocal basis (A.5) comprises $\mathbf{n}^0=\mathbf{n}_0$ and $\mathbf{n}^\pm=\frac{1}{2}\left(\mathbf{n}_\pm\right)^\dagger=\frac{1}{2}\mathbf{n}_\mp$.

   (b) Demonstrate that $\mathbf{n}^\pm$ are null cliffors and write them as products of the projector $P=\frac{1}{2}\left(1+\mathbf{e}_3\right)$ and $\mathbf{e}_1$.

   (c) Rotate the basis elements $\mathbf{n}^\pm$ and $\mathbf{n}^0$ by an angle ϕ about $\mathbf{e}_3$ and show that they change by at most a phase factor. (The fact that $\mathbf{n}^\pm$ are eigenelements of $\exp(-i\mathbf{e}_3\phi/2)$ is a source of their appeal.)

   (d) Expand the position vector $\mathbf{r}$ in the reciprocal basis

   $$\mathbf{r}=r_m\mathbf{n}^m\,,\quad m=0,\pm \tag{13.165}$$

   and express the resulting complex coordinates in both cartesian and spherical coordinates.

   (e) Verify by the differentiation chain rule and the real cartesian expression of $\mathbf{r}$ that the complex tangent vectors $\partial\mathbf{r}/\partial r_m$ are the reciprocal basis vectors.

2. Show how the nilpotent complex vectors $\mathbf{n}_\pm=\mathbf{e}_1\pm i\mathbf{e}_2$ can be interpreted as the biparavector swept out when $\mathbf{e}_1$ moves along the light-cone direction $\mathbf{e}_0+\mathbf{e}_3$.

3. Use the rotational behavior of the spherical harmonics to prove that

   $$\sum_m C^{lm*}\left(\hat{\mathbf{k}}\right)C^{lm}\left(\hat{\mathbf{r}}\right) \tag{13.166}$$

   is invariant under rotations. Then derive the addition theorem (13.50) from the definition of Legendre polynomials (13.49).

4. Use the symmetry of the $3-j$ symbols to relate the values of $\langle klqm|JM\rangle$ and $\langle kJ,q,-M|l,-m\rangle$.

5. Calculate the explicit Clebsch-Gordan coefficients $\left\langle \frac{1}{2}\frac{1}{2}qm|JM\right\rangle$ and use these to derive the unitary matrix that reduces the direct-product representation $\mathbf{D}^{\left(\frac{1}{2}\otimes\frac{1}{2}\right)}$.

6. Let $\Phi(\mathbf{R})$ be the electrostatic potential of a charge distribution $\rho(\mathbf{x})$ confined to a volume of radius $x<R$. Use the orthogonality of the spherical harmonics to give the L,Mth multipole moment $q\left\langle x^{L,M}\right\rangle_\rho$ of the distribution as an integral over directions $\hat{\mathbf{R}}$ of $\Phi(\mathbf{R})\,C^{LM}\left(\hat{\mathbf{R}}\right)$.

7. Show that the multipole expansion (13.161) has the expected form for the interaction of two dipoles, namely

$$V=\frac{qq'}{4\pi\varepsilon_0}\frac{\mathcal{V}_{11}}{R^3}, \tag{13.167}$$

where $\mathcal{V}_{11}=\langle\mathbf{r}\rangle_\rho\cdot\langle\mathbf{r}'\rangle_{\rho'}-3\,\langle\mathbf{r}\cdot\mathbf{e}_3\rangle_\rho\,\langle\mathbf{r}'\cdot\mathbf{e}_3\rangle_{\rho'}$.

Chapter 14

Radiation Multipoles

In this final chapter we study multipole radiation. Such radiation can be conveniently expressed in terms of vector spherical harmonics. As with the scalar spherical harmonics of the previous chapter, the vector version exploits the isotropy of space to separate the wave equation into radial and angular components. First we discuss the vector wave equation; next the vector spherical harmonics and spherical Bessel functions are introduced; finally we consider the radiation multipoles themselves.

14.1 Wave Equation

As we have seen in earlier chapters, Maxwell's equation in source-free space is

$$\bar{\partial}\mathbf{F} = 0 \tag{14.1}$$

and with the Lorenz gauge condition ($\langle \partial \bar{A} \rangle_S = 0$),

$$\mathbf{F} = c\partial \bar{A}\,, \tag{14.2}$$

so that Maxwell's equation becomes the wave equation for the spacetime potential A :

$$\Box A = 0\,, \tag{14.3}$$

where $\Box = c^{-2}\partial_t^2 - \nabla^2$. Application of ∂ to Maxwell's equation also gives the wave equation for each component of $\mathbf{F}$:

$$\Box \mathbf{F} = 0\,. \tag{14.4}$$

In this section we want to investigate solutions to the wave equation in spherical coordinates, which are appropriate to scattering theory.

It is convenient mathematically to assume solutions undergoing duality rotations

$$\mathbf{F}(\mathbf{r},t) = \mathbf{F}(\mathbf{r},0)\, e^{-i\omega t}, \tag{14.5}$$

in which the electric and magnetic fields exchange place every quarter of a cycle at a fixed angular frequency ω. Solutions with other time dependencies can be formed from linear superpositions of these harmonic ones. The wave equation (14.4) then becomes the Helmholtz equation

$$\left(\nabla^2 + k^2\right) \mathbf{F}(\mathbf{r},0) = 0 \tag{14.6}$$

where $k := \omega/c$. In spherical coordinates,

$$\left(\frac{1}{r}\frac{\partial^2}{\partial r^2}r - \frac{\mathbf{L}^2}{r^2} + k^2\right) \mathbf{F}(\mathbf{r},0) = 0\,, \tag{14.7}$$

where $r = \sqrt{\mathbf{r}^2}$.

14.2 Vector Spherical Harmonics

When $\mathbf{F}$ is expanded in spherical harmonics, each component in the complex spherical basis $\{\hat{\mathbf{n}}_{+1}, \hat{\mathbf{n}}_0, \hat{\mathbf{n}}_{-1}\}$ can be written

$$\hat{\mathbf{n}}^q \cdot \mathbf{F}(\mathbf{r},0) = \sum_{lm} F^q_{lm}(r)\, C^{lm}(\hat{\mathbf{r}}) \tag{14.8}$$

and the radial parts satisfy

$$\left(\frac{1}{r}\frac{\partial^2}{\partial r^2}r - \frac{l\,(l+1)}{r^2} + k^2\right) F^q_{lm}(r) = 0\,. \tag{14.9}$$

The vector-field solution is thus

$$\mathbf{F}(\mathbf{r},0) = \sum_{qlm} F^q_{lm} \hat{\mathbf{n}}_q C^{lm}(\hat{\mathbf{r}})\ . \tag{14.10}$$

The components of $\mathbf{F}$ are labeled with q, and the dependence of $\mathbf{F}$ on its argument $\mathbf{r}$ is given by the expansion in spherical harmonics. Under rotations of the vector field $\mathbf{F}$, the vector $\mathbf{F}$ itself should be rotated forward while its argument $\mathbf{r}$ is rotated backward. As a result, the terms $\hat{\mathbf{n}}_q C^{lm}(\hat{\mathbf{r}})$ transform according to the representation

$$\mathbf{D}^{(1)}(\boldsymbol{\theta}) \otimes \mathbf{D}^{(l)\dagger}(-\boldsymbol{\theta}) = \mathbf{D}^{(1)}(\boldsymbol{\theta}) \otimes \mathbf{D}^{(l)}(\boldsymbol{\theta}) \ , \tag{14.11}$$

which is a reducible representation. However, we can find linear combinations that transform under the irreducible representations $\mathbf{D}^{(J)}$:

$$\mathbf{C}^{JM}_{l1}(\hat{\mathbf{r}}) \quad : \; = \sum_{mn} \langle JM|lm1n\rangle \, C^{lm}(\hat{\mathbf{r}}) \, \hat{\mathbf{n}}_n \tag{14.12}$$

$$= \; : \sqrt{\frac{4\pi}{2l+1}} \mathbf{Y}^{JM}_{l1}(\hat{\mathbf{r}}) \tag{14.13}$$

These are called *vector spherical harmonics.* The angular dependence of any vector field can be expressed in terms of the $\mathbf{C}^{JM}_{l1}(\hat{\mathbf{r}})$. They rotate under the representation $\mathbf{D}^{(J)}$ of the rotation group:

$$\mathbf{C}^{JM}_{l1}(\hat{\mathbf{r}}) \to \mathbf{C}^{JN}_{l1}(\hat{\mathbf{r}}) \, D^{(J)}_{NM}(\boldsymbol{\theta}) \ . \tag{14.14}$$

The Clebsch-Gordan coefficients $\langle JM|lm1n\rangle$ are (see Chapt. 13)

$$\langle Jm\pm 1|lm1\pm 1\rangle = \begin{cases} \left[\frac{(l\mp m-1)(l\mp m)}{2l(2l+1)}\right]^{1/2}, & J=l-1 \\ \mp\left[\frac{(l\pm m+1)(l\mp m)}{2l(l+1)}\right]^{1/2}, & J=l \\ \left[\frac{(l\pm m+1)(l\pm m+2)}{2(2l+1)(l+1)}\right]^{1/2}, & J=l+1 \end{cases} \tag{14.15}$$

and

$$\langle Jm|lm10\rangle = \begin{cases} -\left[\frac{l^2-m^2}{l(2l+1)}\right]^{1/2}, & J=l-1 \\ \frac{m}{\sqrt{l(l+1)}}, & J=l \\ \left[\frac{(l+1)^2-m^2}{(2l+1)(l+1)}\right]^{1/2}, & J=l+1 \end{cases} \ . \tag{14.16}$$

As elements of a unitary (also orthogonal) transformation, they obey the orthonormality conditions

$$\sum_{mn} \langle JM|lmkn\rangle \, \langle lmkn|J'M'\rangle \; = \; \delta_{JJ'}\delta_{MM'} \tag{14.17}$$

$$\sum_{JM} \langle lmkn|JM\rangle \, \langle JM|lm'kn'\rangle \; = \; \delta_{mm'}\delta_{nn'} \, . \tag{14.18}$$

They also obey symmetry relations

$$\langle lmkn|JM\rangle = (-)^{l+k-J} \langle knlm|JM\rangle \tag{14.19}$$

$$= \sqrt{\frac{2J+1}{2l+1}} (-)^{l-J+n} \langle lm|JMk-n\rangle \ . \tag{14.20}$$

Exercise 14.1. Prove the orthonormality relation

$$\int_{4\pi} d\hat{\mathbf{r}}\, \mathbf{C}_{l1}^{JM*}(\hat{\mathbf{r}}) \cdot \mathbf{C}_{l'1}^{J'M'}(\hat{\mathbf{r}}) = \frac{4\pi}{2l+1} \delta^{JJ'} \delta^{MM'} \delta_{ll'} \ . \tag{14.21}$$

When viewed as a vector field, $\mathbf{r}$ is invariant under rotations because the forward rotation of its components is canceled by the backward rotation of its argument [see also problems (1) and (2)]. Consequently, $\mathbf{r}C^{lm}$ transforms under rotations according to the representation $\mathbf{D}^{(l)}$:

$$\mathbf{r}C^{lm}(\hat{\mathbf{r}}) \to \mathbf{r} \sum_n C^{ln}(\hat{\mathbf{r}})\, D_{nm}^{(l)}(\boldsymbol{\theta}) \ . \tag{14.22}$$

This is just the way the vector harmonic $\mathbf{C}_{k1}^{lm}$ behaves under rotations. Because the $\left\{\mathbf{C}_{k1}^{LM}\right\}$ is a complete orthogonal basis for the rotational behavior of vector fields, we expect $\mathbf{r}C^{lm}(\hat{\mathbf{r}})$ to be a linear combination of $r\mathbf{C}_{k1}^{lm}$ with $k = l, l \pm 1$. Indeed, from the expansion of $\hat{\mathbf{r}}$ (13.129) and the composition relation (13.72) for spherical harmonics given in Chapt. 13,

$$\begin{aligned}
\mathbf{r}C^{lm}(\hat{\mathbf{r}}) &= rC^{lm}(\hat{\mathbf{r}})\, C^{1n*}(\hat{\mathbf{r}})\, \hat{\mathbf{n}}_n \\
&= rC^{lm}(\hat{\mathbf{r}})\, C^{1n}(\hat{\mathbf{r}})\, \hat{\mathbf{n}}_n^* \\
&= r \sum_{LMn} \langle l010|L0\rangle \langle lm1n|LM\rangle\, C^{LM}(\hat{\mathbf{r}})\, \hat{\mathbf{n}}_{-n} (-)^n \\
&= r \sum_{LMn} \sqrt{\frac{2L+1}{2l+1}} \langle l0|L010\rangle \langle lm|LM1-n\rangle\, C^{LM}(\hat{\mathbf{r}})\, \hat{\mathbf{n}}_{-n} \\
&= r \sum_L \sqrt{\frac{2L+1}{2l+1}} \langle l0|L010\rangle\, \mathbf{C}_{L1}^{lm} \ .
\end{aligned} \tag{14.23}$$

Furthermore, since rotations mix only terms of the same l value, the same rotational behavior will be observed for products of the form $\mathbf{r} f_l(r)\, C^{lm}(\hat{\mathbf{r}})$.

The rotational behavior of $\mathbf{r}C^{lm}(\hat{\mathbf{r}})$ is unchanged when $\mathbf{r}$ is replaced by any other system vector, even if this vector is an operator, in particular when $\mathbf{r}$ is replaced by the vector operator ∇, by their vector product $\mathbf{L} = -i\mathbf{r} \times \nabla$, or by $\nabla \times \mathbf{L}$.

Exercise 14.2. Determine the vector components of $\mathbf{L}C^{lm}(\hat{\mathbf{r}})$ and show that they are the components of the vector spherical harmonic

$$\mathbf{L}C^{lm}(\hat{\mathbf{r}}) = \sqrt{l(l+1)}\mathbf{C}^{lm}_{l1}(\hat{\mathbf{r}}) \,. \tag{14.24}$$

14.3 Spherical Bessel Functions

The solutions to the radial Helmholtz equation (14.9) are the spherical Bessel functions $j_l(kr)$, $n_l(kr)$, $h_l^{(1)}(kr)$, and $h_l^{(2)}(kr)$, which are related to the standard cylindrical Bessel functions J_ν, N_ν, $H_\nu^{(1)}$, and $H_\nu^{(2)}$ as given in Table 14.1. In the third row of this table, the $+$ and $-$ signs correspond to $h_l^{(1)}$ and $h_l^{(2)}$, respectively. Note that $h_l^{(2)}(z) = h_l^{(1)*}(z^*)$.

$j_l(z)$	$\sqrt{\frac{\pi}{2z}}J_{l+\frac{1}{2}}(z)$	$\frac{1}{2}\left[h_l^{(1)}(z) + h_l^{(2)}(z)\right]$	$(-z)^l\left(\frac{1}{z}\frac{d}{dz}\right)^l\left(\frac{\sin z}{z}\right)$
$n_l(z)$	$\sqrt{\frac{\pi}{2z}}N_{l+\frac{1}{2}}(z)$	$\frac{1}{2i}\left[h_l^{(1)}(z) - h_l^{(2)}(z)\right]$	$(-z)^l\left(\frac{1}{z}\frac{d}{dz}\right)^l\left(-\frac{\cos z}{z}\right)$
$h_l^{(1,2)}(z)$	$\sqrt{\frac{\pi}{2z}}H^{(1,2)}_{l+\frac{1}{2}}(z)$	$j_l(z) \pm in_l(z)$	$(-z)^l\left(\frac{1}{z}\frac{d}{dz}\right)^l\left(\frac{\exp \pm iz}{\pm iz}\right)$

Table 14.1. Spherical Bessel functions can be expressed in several ways in terms of cylindrical Bessel functions, in terms of each other, or in terms of a generating formula.

The last column of Table 14.1 gives expressions for generating the spherical Bessel functions explicitly in terms of trigonometric functions. The first three are given in Table 14.2: The asymptotic forms are shown in Table 14.3. Note that j_l

	$l=0$	$l=1$	$l=2$
$j_l(z)$	$\frac{\sin z}{z}$	$\frac{\sin z}{z^2} - \frac{\cos z}{z}$	$\left(\frac{3}{z^3} - \frac{1}{z}\right)\sin z - \frac{3}{z^2}\cos z$
$n_l(z)$	$-\frac{\cos z}{z}$	$-\frac{\cos z}{z^2} - \frac{\sin z}{z}$	$-\left(\frac{3}{z^3} - \frac{1}{z}\right)\cos z - \frac{3}{z^2}\sin z$
$h_l^{(1)}$	$\frac{\exp iz}{iz}$	$-\frac{\exp iz}{z}\left(1 + \frac{i}{z}\right)$	$\frac{i\exp iz}{iz}\left(1 + \frac{3i}{z} - \frac{3}{z^2}\right)$.

Table 14.2. Explicit values of spherical Bessel functions.

		$z \longrightarrow 0$	$z \longrightarrow \infty$
$j_l(z)$	$\longrightarrow$	$\frac{z^l}{(2l+1)!!}$	$\frac{1}{z}\sin\left(z - l\pi/2\right)$
$n_l(z)$	$\longrightarrow$	$-\frac{(2l-1)!!}{z^{l+1}}$	$-\frac{1}{z}\cos\left(z - l\pi/2\right)$
$h_l^{(1)}$	$\longrightarrow$	$-\frac{\exp iz}{z}\left(1 + \frac{i}{z}\right)$	$\frac{1}{iz}\exp\left[i\left(z - \pi l/2\right)\right]$.

Table 14.3. Asymptotic forms of spherical Bessel functions.

is regular at the origin but n_l is irregular. There are several recurrence relations that can be derived from the generating formulas:

$$u_{l+1}(z) = -z^l \frac{d}{dz} z^{-l} u_l(z) \tag{14.25}$$

$$\frac{2l+1}{z} u_l(z) = u_{l+1}(z) + u_{l-1}(z) \tag{14.26}$$

$$(2l+1)\, u_l'(z) = l u_{l-1}(z) - (l+1)\, u_{l+1}(z) \tag{14.27}$$

where $u = j, n$, or $h^{(1,2)}$.

With the help of the spherical Bessel functions, we can extend the list of functions of r that can be expanded. As in the last chapter, with $\mathbf{r} = \mathbf{x} - \mathbf{x}'$ we have $r = \sqrt{\mathbf{x}^2 + \mathbf{x}'^2 - 2x_> x_< \hat{\mathbf{x}} \cdot \hat{\mathbf{x}}'}$ and

$$x_> = \text{greater of } \{|\mathbf{x}|, |\mathbf{x}'|\} \tag{14.28}$$

$$x_< = \text{smaller of } \{|\mathbf{x}|, |\mathbf{x}'|\} . \tag{14.29}$$

Monochromatic spherical waves can be expressed

$$\frac{\exp ikr}{ikr} = \sum_{l=0}^{\infty} (2l+1)\; j_l(kx_<)\, h^{(1)}(kx_>)\, P_l(\hat{\mathbf{x}} \cdot \hat{\mathbf{x}}') \tag{14.30}$$

$$\frac{\cos kr}{kr} = -\sum_{l=0}^{\infty} (2l+1)\, j_l(kx_<)\, n_l(kx_>)\, P_l(\hat{\mathbf{x}} \cdot \hat{\mathbf{x}}') \tag{14.31}$$

$$\frac{\sin kr}{kr} = \sum_{l=0}^{\infty} (2l+1)\, j_l(kx_<)\, j_l(kx_>)\, P_l(\hat{\mathbf{x}} \cdot \hat{\mathbf{x}}') . \tag{14.32}$$

Finally, the important plane wave $\exp\left(i\mathbf{k} \cdot \mathbf{r}\right)$ solution of the Helmholtz equation (14.6) can also be expanded in Legendre polynomials:

$$e^{i\mathbf{k}\cdot\mathbf{r}} = \sum_{l=0}^{\infty} i^l\, (2l+1)\; j_l(kr)\, P_l\Big(\hat{\mathbf{k}} \cdot \hat{\mathbf{r}}\Big) . \tag{14.33}$$

By virtue of the addition theorem,

$$e^{i\mathbf{k}\cdot\mathbf{r}} = \sum_{lm} \phi_{lm}(\mathbf{r})\, C^{lm*}\left(\hat{\mathbf{k}}\right) \tag{14.34}$$

where the spherical wave is $\phi_{lm}(\mathbf{r}) = i^l\,(2l+1)\,j_l(kr)\,C^{lm}(\hat{\mathbf{r}})$.

14.4 Electromagnetic Multipoles

For vector multipole fields with wave number k and the angular behavior of $C^{lm}(\hat{\mathbf{r}})$, the set of three fields

$$\begin{aligned} \mathbf{F}_{lm}^{\text{long}} &= (ik)^{-1}\,\nabla\phi_{lm} \\ \mathbf{F}_{lm}^{\text{el}} &= k^{-1}\left[l\,(l+1)\right]^{-1/2}\nabla\times\mathbf{L}\phi_{lm} \\ \mathbf{F}_{lm}^{\text{mag}} &= \left[l\,(l+1)\right]^{-1/2}\mathbf{L}\phi_{lm}\,, \end{aligned} \tag{14.35}$$

where $\phi_{lm} = f_l(r)\,C^{lm}(\hat{\mathbf{r}})$, form a complete vector basis. All three have the same rotational behavior and the same normalization as ϕ_{lm}, but they differ in their differential properties and their parity: the fields $\mathbf{F}_{lm}^{\text{long}}$ are irrotational ($\nabla\times\mathbf{F}_{lm} = 0$) and have parity $(-)^{l+1}$ whereas $\mathbf{F}_{lm}^{\text{el}}$ and $\mathbf{F}_{lm}^{\text{mag}}$ are divergence-free (solenoidal) and have parity $(-)^{l+1}$ and $(-)^l$, as expected of electric and magnetic fields in source-free space, respectively.

Consider a plane wave directed in the $\mathbf{e}_3$ direction with polarization vector $\hat{\mathbf{n}}_m$. It can be expanded in vector spherical harmonics:

$$\hat{\mathbf{n}}_m e^{ikz} = \sum_l i^l\,(2l+1)\,j_l(kr)\,C^{l0}(\hat{\mathbf{r}})\,\hat{\mathbf{n}}_m \tag{14.36}$$

$$= \sum_{lL} i^l\,(2l+1)\,j_l(kr)\,\mathbf{C}_{l1}^{Lm}(\hat{\mathbf{r}})\,\langle Lm|l01m\rangle\,. \tag{14.37}$$

Alternatively, it can be expressed in terms of the fields (14.35). The longitudinal component can be written

$$\hat{\mathbf{n}}_0 e^{ikz} = (ik)^{-1}\,\nabla e^{ikz} \tag{14.38}$$

$$= (ik)^{-1}\,\nabla\sum_l \phi_{l0} \tag{14.39}$$

$$= \sum_l \mathbf{F}_{l0}^{\text{long}}\,. \tag{14.40}$$

The transverse components are

$$\hat{\mathbf{n}}_{\pm 1} e^{ikz} = -\frac{1}{\sqrt{2}} \sum_l \left(\mathbf{F}^{\text{el}}_{l,\pm 1} \pm \mathbf{F}^{\text{mag}}_{l,\pm 1} \right) . \tag{14.41}$$

Since the rotational behavior of the spherical multipoles is simple, we can readily rotate the wave to be directed along $\mathbf{k}$:

$$\hat{\mathbf{n}}'_{\pm 1} e^{i\mathbf{k}\cdot\mathbf{x}} = -\frac{1}{\sqrt{2}} \sum_{lm} \left(\mathbf{F}^{\text{el}}_{lm} \pm \mathbf{F}^{\text{mag}}_{lm} \right) D^{(l)}_{m,\pm 1}(\boldsymbol{\theta}) \tag{14.42}$$

where $\boldsymbol{\theta}$ is the angle of rotation from $\mathbf{e}_3$ to $\hat{\mathbf{k}}$.

Any electromagnetic field in free space can be expanded in electric and magnetic multipoles of the form

$$\mathbf{F}(\hat{\mathbf{r}}) = \sum_{lm} \left(a^{\text{el}}_{lm} \mathbf{F}^{\text{el}}_{lm} + a^{\text{mag}}_{lm} \mathbf{F}^{\text{mag}}_{lm} \right) \tag{14.43}$$

with $\mathbf{F}^{\text{el}}_{lm}$ and $\mathbf{F}^{\text{mag}}_{lm}$ given by (14.35) and $f_l(r)$ a linear combination of two linearly independent spherical Bessel functions. The advantage of this form is that each radiation multipole coefficient a^{el}_{lm} or a^{mag}_{lm} can be related directly to a single source multipole. These have the forms

$$\begin{aligned} a^{\text{el}}_{lm} &= \frac{ik(2l+1)}{c\sqrt{l(l+1)}} \int d^3\mathbf{r}\, \mathbf{F}^{\text{mag}}_{lm}(\mathbf{r})^* \cdot [\nabla \times \mathbf{j} + c\nabla \times (\nabla \times \mathbf{M})] \qquad (14.44) \\ a^{\text{mag}}_{lm} &= -\frac{k^2(2l+1)}{\sqrt{l(l+1)}} \int d^3\mathbf{r}\, \mathbf{F}^{\text{mag}}_{lm}(\mathbf{r})^* \cdot \left[\nabla \times \mathbf{M} + \frac{1}{ck^2} \nabla \times (\nabla \times \mathbf{j}) \right] \end{aligned}$$

where $\mathbf{j}$ is the vector current density and $\mathbf{M}$ the magnetization of the source. In the long wavelength limit where $\mathbf{M} = 0$, a^{el}_{lm} reduces to

$$a^{\text{el}}_{lm} \simeq \frac{k^{2+l}}{i(2l-1)!!} \sqrt{\frac{l+1}{l}} \int d^3\mathbf{r}\, r^l C^{lm*}(\hat{\mathbf{r}})\, \rho(\mathbf{r}) . \tag{14.45}$$

14.5 Problems

1. From the definition (14.12) of vector spherical harmonics [see also equation (13.129) from the previous chapter], show directly that the unit vector in the radial direction is

$$\hat{\mathbf{r}} = -\sqrt{3}\mathbf{C}^{00}_{11}(\hat{\mathbf{r}}) . \tag{14.46}$$

   Verify that this is a special case of (14.23).

2. One of the simplest vector fields is the electric field of a stationary point charge. It is clearly spherically symmetric. Show that it is indeed invariant under rotations in three ways:

   (a) Rotate separately the vector components of the field and the position vector at which the field is evaluated; then combine the results.

   (b) Express the field in terms of the vector spherical harmonics and show that only the $(JM) = (00)$ term contributes.

   (c) Show that the field is an eigenfunction with eigenvalue zero of the total angular momentum operator $\mathbf{J}$ with scalar products $\mathbf{J} \cdot \mathbf{a} = \mathbf{L} \cdot \mathbf{a} + \mathbf{S} \cdot \mathbf{a}$ where $\mathbf{S} \cdot \mathbf{a}$ is the operator $\imath \mathbf{a} \times$ for any constant vector $\mathbf{a}$.

Appendix A

Other Coordinate Systems

In most of this text, references to explicit coordinate systems have been minimized. Since Clifford's geometric algebra allows most problems to be cast in component-free notation, we have largely been able to avoid mention of any coordinate system, and where expansions in basis sets were needed we could use the common Cartesian system, with its fixed, orthonormal system of basis vectors.

However, other coordinate systems are frequently used. In particular, orthogonal curvilinear systems such as spherical and cylindrical coordinates often conform better to the symmetries of the problem than Cartesian coordinates. The geometric algebra is readily applied to these and also to more general, nonorthogonal coordinate systems. This appendix reviews the geometry and formalism behind such applications. For readers familiar with general curvilinear coordinates, this appendix can serve to illustrate a simple application of the geometric algebra $\mathcal{C}\ell_3$.

A.1 Coordinate Systems in Physical Space

Let the spatial position be $\mathbf{r}$. In previous work we have expanded $\mathbf{r}$ in an orthonormal basis $\{\mathbf{e}_1, \mathbf{e}_2, \mathbf{e}_3\}$ that we associated with the Cartesian directions:

$$\begin{aligned} \mathbf{r} &= r^k \mathbf{e}_k \\ &= x\mathbf{e}_1 + y\mathbf{e}_2 + z\mathbf{e}_3 , \end{aligned} \tag{A.1}$$

where

$$\mathbf{e}_j \mathbf{e}_k + \mathbf{e}_k \mathbf{e}_j = 2\mathbf{e}_j \cdot \mathbf{e}_k = 2\delta_{jk} . \tag{A.2}$$

If we wish to use other coordinates (ξ^1, ξ^2, ξ^3) that may not be Cartesian, we must first define the tangent vectors

$$\mathbf{n}_k = \frac{\partial \mathbf{r}}{\partial \xi^k}, \tag{A.3}$$

which are not necessarily of unit length. The position vector can be expanded in terms of these tangent vectors:

$$d\mathbf{r} = \frac{\partial \mathbf{r}}{\partial \xi^k} d\xi^k = \mathbf{n}_k d\xi^k. \tag{A.4}$$

Any set $\{\mathbf{n}_1, \mathbf{n}_2, \mathbf{n}_3\}$ of three linearly independent directions forms a basis of physical space $\mathbb{E}^3$. Every vector is fully defined by its components on such a set, as in (A.4). To determine the components $d\xi^k$, we need to invert the expansion (A.4). The inversion is easily accomplished with the help of the reciprocal basis $\{\mathbf{n}^1, \mathbf{n}^2, \mathbf{n}^3\}$, defined by the scalar products

$$\left\langle \mathbf{n}^j \mathbf{n}_k \right\rangle_s = \mathbf{n}^j \cdot \mathbf{n}_k = \delta^j_k\,. \tag{A.5}$$

The scalar product of $d\mathbf{r}$ (A.4) with $\mathbf{n}^j$ gives immediately

$$d\xi^j = \mathbf{n}^j \cdot d\mathbf{r}\,. \tag{A.6}$$

Exercise A.1. Verify that the reciprocal basis to the set $\{\mathbf{n}_1, \mathbf{n}_2, \mathbf{n}_3\}$ in $\mathbb{E}^3$can be written

$$\mathbf{n}^1 = \frac{\langle \mathbf{n}_2 \mathbf{n}_3 \rangle_V}{\langle \mathbf{n}_1 \mathbf{n}_2 \mathbf{n}_3 \rangle_S} \tag{A.7}$$

and cyclic permutations $(1 \to 2 \to 3 \to 1)$ thereof.
[Hint: since $\langle \langle \mathbf{n}_2 \mathbf{n}_3 \rangle_S \, \mathbf{n}_k \rangle_S = 0$, $\langle \langle \mathbf{n}_2 \mathbf{n}_3 \rangle_V \, \mathbf{n}_k \rangle_S = \langle \mathbf{n}_2 \mathbf{n}_3 \mathbf{n}_k \rangle_S$.]

Exercise A.2. Show that

$$\langle \mathbf{n}_1 \mathbf{n}_2 \mathbf{n}_3 \rangle_S \left\langle \mathbf{n}^1 \mathbf{n}^2 \mathbf{n}^3 \right\rangle_S = -1\,. \tag{A.8}$$

[Hint: it may be useful as an intermediate step to prove the relation

$$\langle \langle \mathbf{n}_2 \mathbf{n}_3 \rangle_V \langle \mathbf{n}_3 \mathbf{n}_1 \rangle_V \rangle_V = - \langle \mathbf{n}_1 \mathbf{n}_2 \mathbf{n}_3 \rangle_S \, \mathbf{n}_3 \,]\,. \tag{A.9}$$

The reciprocal vector $\mathbf{n}^1$ is the vector dual to the plane $\langle \mathbf{n}_2 \mathbf{n}_3 \rangle_V$. It is not necessarily aligned with $\mathbf{n}_1$. Of course the bases $\{\mathbf{n}_1, \mathbf{n}_2, \mathbf{n}_3\}$ and $\{\mathbf{n}^1, \mathbf{n}^2, \mathbf{n}^3\}$

span the same three-dimensional space. Consequently, each $\mathbf{n}^j$ can be expanded in the $\mathbf{n}_k$ and *vice versa:*

$$\mathbf{n}^j = \left(\mathbf{n}^j\right)^k \mathbf{n}_k \,. \tag{A.10}$$

Here, the symbol $(\mathbf{n}^j)^k$ denotes the kth component of the vector $\mathbf{n}^j$. Dotting both sides into $\mathbf{n}^l$ inverts the equation to give

$$\left(\mathbf{n}^j\right)^l = \mathbf{n}^j \cdot \mathbf{n}^l. \tag{A.11}$$

If we define the metric tensors

$$\eta^{kl} = \left\langle \mathbf{n}^k \mathbf{n}^l \right\rangle_s = \mathbf{n}^k \cdot \mathbf{n}^l \tag{A.12}$$

we obtain

$$\mathbf{n}^j = \eta^{jk} \mathbf{n}_k \,. \tag{A.13}$$

Similarly,

$$\mathbf{n}_j = \eta_{jk} \mathbf{n}^k \tag{A.14}$$

with

$$\eta_{kl} := \left\langle \mathbf{n}_k \mathbf{n}_l \right\rangle_s = \mathbf{n}_k \cdot \mathbf{n}_l \,. \tag{A.15}$$

The equations (A.13) and (A.14) are sometimes referred to as index-raising and -lowering equations, respectively.

From (A.4) and (A.6) we see that any vector $d\mathbf{r}$ can be expanded in the form[1]

$$d\mathbf{r} = \mathbf{n}_k \, \mathbf{n}^k \cdot d\mathbf{r} \,. \tag{A.16}$$

It follows that the metric tensors η_{jk} and η^{jk} are inverses of each other:

$$\eta_{jk} \eta^{kl} = \mathbf{n}_j \cdot \mathbf{n}_k \, \mathbf{n}^k \cdot \mathbf{n}^l = \mathbf{n}_j \cdot \mathbf{n}^l = \delta_j^l \,. \tag{A.17}$$

In the special case of orthogonal coordinates, the metric tensors are diagonal

$$\eta_{jk} = \delta_j^k \eta_{kk} \tag{A.18}$$

---

[1] In terms of dyads, $\mathbf{n}_k \mathbf{n}^k = \overleftrightarrow{\mathbf{1}}$ is the unit element.

(no sum), and the basis vectors $\mathbf{n}_k$ are just the inverses of $\mathbf{n}^k$:

$$\mathbf{n}_k = \eta_{kj}\mathbf{n}^j = \eta_{kk}\mathbf{n}^k = \left(\mathbf{n}^k\right)^{-1} . \tag{A.19}$$

More generally, the determinant of the matrix tensor is proportional to the volume element formed by the basis vectors:

$$\begin{aligned} \left\langle \mathbf{n}^1\mathbf{n}^2\mathbf{n}^3 \right\rangle_S &= \eta^{1j}\eta^{2k}\eta^{3l} \left\langle \mathbf{n}_j\mathbf{n}_k\mathbf{n}_l \right\rangle_S \\ &= \det\eta \left\langle \mathbf{n}_1\mathbf{n}_2\mathbf{n}_3 \right\rangle_S \\ &= i\sqrt{\det\eta} \end{aligned} \tag{A.20}$$

where η represents the 3×3 matrix with elements η^{jk} .

A.2 Gradient Operator

The gradient operator in nonorthonormal coordinates is best defined in terms of a directional derivative. Let ξ^k be a coordinate, then ∇ is defined by

$$\frac{\partial}{\partial \xi^k} = \frac{\partial \mathbf{r}}{\partial \xi^k} \cdot \nabla = \mathbf{n}_k \cdot \nabla , \tag{A.21}$$

which can be inverted to give

$$\nabla = \mathbf{n}^k \frac{\partial}{\partial \xi^k} . \tag{A.22}$$

If the basis is orthogonal, we can write

$$\nabla = \left(\frac{\partial \mathbf{r}}{\partial \xi^k}\right)^{-1} \frac{\partial}{\partial \xi^k} . \tag{A.23}$$

Otherwise, we can use [apply results (A.7) and (A.20) to equation (A.23)]

$$\nabla = \frac{\varepsilon_{jkl}}{\sqrt{\det\eta}} \frac{\partial \mathbf{r}}{\partial \xi^k} \times \frac{\partial \mathbf{r}}{\partial \xi^l} \frac{\partial}{\partial \xi^j} , \tag{A.24}$$

where ε_{jkl} is the fully antisymmetric third-rank tensor, with $\varepsilon_{123} = 1$.

A.3 Spherical Coordinates

The spherical coordinates θ, ϕ, r are orthogonal curvilinear coordinates, defined by the rotation of $r\mathbf{e}_3$ into the position $\mathbf{r}$:

$$\mathbf{r} = rR\mathbf{e}_3R^\dagger, \tag{A.25}$$

where R rotates $\mathbf{e}_3$ to $\hat{\mathbf{r}}$ with polar angle θ and azimuthal angle ϕ :

$$R = \exp\left(-i\mathbf{e}_3\phi/2\right)\exp\left(-i\mathbf{e}_2\theta/2\right) . \tag{A.26}$$

Note the partial derivatives

$$\begin{aligned} \frac{\partial R}{\partial \theta} &= -R\frac{i\mathbf{e}_2}{2} \\ \frac{\partial R}{\partial \phi} &= -\frac{i\mathbf{e}_3}{2}R \\ \frac{\partial R}{\partial r} &= 0 . \end{aligned} \tag{A.27}$$

These give the orthogonal (but not normalized) tangent vectors that form a basis of $\mathbb{R}^3$

$$\begin{aligned} \frac{\partial \mathbf{r}}{\partial \theta} &= -irR\frac{\mathbf{e}_2\mathbf{e}_3-\mathbf{e}_3\mathbf{e}_2}{2}R^\dagger = rR\mathbf{e}_1R^\dagger \\ \frac{\partial \mathbf{r}}{\partial \phi} &= -ir\frac{\mathbf{e}_3\hat{\mathbf{r}}-\hat{\mathbf{r}}\mathbf{e}_3}{2} = \mathbf{e}_3\times\mathbf{r} = r\sin\theta R\mathbf{e}_2R^\dagger \\ \frac{\partial \mathbf{r}}{\partial r} &= R\mathbf{e}_3R^\dagger . \end{aligned} \tag{A.28}$$

It is convenient to define orthogonal unit vectors parallel to the tangent vectors

$$\begin{aligned} \hat{\boldsymbol{\theta}} &\equiv R\mathbf{e}_1R^\dagger = (\mathbf{e}_1\cos\phi + \mathbf{e}_2\sin\phi)\cos\theta - \mathbf{e}_3\sin\theta \\ \hat{\boldsymbol{\phi}} &\equiv R\mathbf{e}_2R^\dagger = -\mathbf{e}_1\sin\phi + \mathbf{e}_2\cos\phi \\ \hat{\mathbf{r}} &\equiv R\mathbf{e}_3R^\dagger = (\mathbf{e}_1\cos\phi + \mathbf{e}_2\sin\phi)\sin\theta + \mathbf{e}_3\cos\theta . \end{aligned} \tag{A.29}$$

From (A.23), we then find

$$\nabla = \frac{\hat{\boldsymbol{\theta}}}{r}\frac{\partial}{\partial\theta} + \frac{\hat{\boldsymbol{\phi}}}{r\sin\theta}\frac{\partial}{\partial\phi} + \hat{\mathbf{r}}\frac{\partial}{\partial r} . \tag{A.30}$$

What makes spherical coordinates challenging is the dependence of the unit vectors on position. This means, for example, that $\hat{\boldsymbol{\theta}}$ and $\partial/\partial\theta$ do not commute. Instead,

$$\frac{\partial}{\partial\theta}\hat{\boldsymbol{\theta}} = \hat{\boldsymbol{\theta}}\frac{\partial}{\partial\theta} + \frac{\partial\hat{\boldsymbol{\theta}}}{\partial\theta} \tag{A.31}$$

and similarly for other differentials and unit vectors. From (A.27) and (A.29) one finds directly

$$\begin{aligned}
\frac{\partial\hat{\mathbf{r}}}{\partial r} &= \frac{\partial\hat{\boldsymbol{\theta}}}{\partial r} = \frac{\partial\hat{\boldsymbol{\phi}}}{\partial r} = 0 && \text{(A.32)}\\
\frac{\partial\hat{\mathbf{r}}}{\partial\theta} &= \hat{\boldsymbol{\theta}},\ \frac{\partial\hat{\boldsymbol{\theta}}}{\partial\theta} = -\hat{\mathbf{r}},\ \frac{\partial\hat{\boldsymbol{\phi}}}{\partial\theta} = 0 && \text{(A.33)}\\
\frac{\partial\hat{\mathbf{r}}}{\partial\phi} &= \hat{\boldsymbol{\phi}}\sin\theta,\ \frac{\partial\hat{\boldsymbol{\theta}}}{\partial\phi} = \hat{\boldsymbol{\phi}}\cos\theta,\ \frac{\partial\hat{\boldsymbol{\phi}}}{\partial\phi} = -\hat{\boldsymbol{\theta}}\cos\theta - \hat{\mathbf{r}}\sin\theta\,. && \text{(A.34)}
\end{aligned}$$

With these relations, we can also put

$$\begin{aligned}
\nabla &= \frac{1}{r}\left(\frac{\partial}{\partial\theta}\hat{\boldsymbol{\theta}} - \frac{\partial\hat{\boldsymbol{\theta}}}{\partial\theta}\right) + \frac{1}{r\sin\theta}\left(\frac{\partial}{\partial\phi}\hat{\boldsymbol{\phi}} - \frac{\partial\hat{\boldsymbol{\phi}}}{\partial\phi}\right) + \frac{\partial}{\partial r}\hat{\mathbf{r}} - \frac{\partial\hat{\mathbf{r}}}{\partial r}\\
&= \frac{1}{r}\left(\frac{\partial}{\partial\theta}\hat{\boldsymbol{\theta}} + \hat{\mathbf{r}}\right) + \frac{1}{r\sin\theta}\left(\frac{\partial}{\partial\phi}\hat{\boldsymbol{\phi}} + \hat{\boldsymbol{\theta}}\cos\theta + \hat{\mathbf{r}}\sin\theta\right) + \frac{\partial}{\partial r}\hat{\mathbf{r}}\\
&= \frac{1}{r}\left(\frac{\partial}{\partial\theta} + \frac{\cos\theta}{\sin\theta}\right)\hat{\boldsymbol{\theta}} + \frac{1}{r\sin\theta}\frac{\partial}{\partial\phi}\hat{\boldsymbol{\phi}} + \left(\frac{\partial}{\partial r} + \frac{2}{r}\right)\hat{\mathbf{r}}\\
&= \frac{1}{r\sin\theta}\left(\frac{\partial}{\partial\theta}\sin\theta\,\hat{\boldsymbol{\theta}} + \frac{\partial}{\partial\phi}\hat{\boldsymbol{\phi}}\right) + \frac{1}{r^2}\frac{\partial}{\partial r}r^2\hat{\mathbf{r}}\,. && \text{(A.35)}
\end{aligned}$$

Note that (A.30) and (A.35) are expressions of the same operator. They differ only in which side of the differentials the unit vectors appear on. It is common to use expression (A.30) for the gradient operator (∇ applied to a scalar function of $\mathbf{r}$) and expression (A.35) for the divergence (the scalar part of ∇ applied to a vector field) since in this way partial derivatives of the curvilinear unit vectors can be avoided.

The Laplacian in spherical coordinates is found by dotting (A.35) into (A.30):

$$\begin{aligned}
\nabla^2 &= \left[\frac{1}{r\sin\theta}\left(\frac{\partial}{\partial\theta}\sin\theta\,\hat{\boldsymbol{\theta}}+\frac{\partial}{\partial\phi}\hat{\boldsymbol{\phi}}\right)+\frac{1}{r^2}\frac{\partial}{\partial r}r^2\hat{\mathbf{r}}\right]\cdot\left[\frac{\hat{\boldsymbol{\theta}}}{r}\frac{\partial}{\partial\theta}+\frac{\hat{\boldsymbol{\phi}}}{r\sin\theta}\frac{\partial}{\partial\phi}+\hat{\mathbf{r}}\frac{\partial}{\partial r}\right] \\
&= \frac{1}{r^2\sin\theta}\frac{\partial}{\partial\theta}\sin\theta\frac{\partial}{\partial\theta}+\frac{1}{r^2\sin^2\theta}\frac{\partial^2}{\partial\phi^2}+\frac{1}{r}\frac{\partial^2}{\partial r^2}r \\
&= \frac{1}{r}\frac{\partial^2}{\partial r^2}r-\frac{\mathbf{L}^2}{r^2} \qquad \text{(A.36)}
\end{aligned}$$

where

$$\begin{aligned}
\mathbf{L} &= -\langle\mathbf{r}\nabla\rangle_V = -i\mathbf{r}\times\boldsymbol{\nabla} \\
&= -ir\hat{\mathbf{r}}\times\left(\frac{\hat{\boldsymbol{\theta}}}{r}\frac{\partial}{\partial\theta}+\frac{\hat{\boldsymbol{\phi}}}{r\sin\theta}\frac{\partial}{\partial\phi}\right) \\
&= -i\left(\hat{\boldsymbol{\phi}}\frac{\partial}{\partial\theta}-\hat{\boldsymbol{\theta}}\frac{1}{\sin\theta}\frac{\partial}{\partial\phi}\right). \qquad \text{(A.37)}
\end{aligned}$$

The expression $\hbar\mathbf{L}$ is the orbital angular momentum operator in quantum theory.

Exercise A.3. Verify from the last expression and from (A.33) and (A.34) that

$$\mathbf{L}^2 = -\frac{1}{\sin\theta}\frac{\partial}{\partial\theta}\sin\theta\frac{\partial}{\partial\theta}-\frac{1}{\sin^2\theta}\frac{\partial^2}{\partial\phi^2}. \qquad \text{(A.38)}$$

In chapter 12, we need the components

$$\begin{aligned}
L_z &= \mathbf{L}\cdot\mathbf{e}_3 = -i\left(x\frac{\partial}{\partial y}-y\frac{\partial}{\partial x}\right) \qquad \text{(A.39)} \\
&= -i\left(-\hat{\boldsymbol{\theta}}\cdot\mathbf{e}_3\frac{1}{\sin\theta}\frac{\partial}{\partial\phi}\right) \\
&= -i\frac{\partial}{\partial\phi} \qquad \text{(A.40)}
\end{aligned}$$

and

$$\begin{aligned}
L_\pm &= \mathbf{L}\cdot(\mathbf{e}_1\pm i\mathbf{e}_2) \qquad \text{(A.41)} \\
&= -i\left(\pm ie^{\pm i\phi}\frac{\partial}{\partial\theta}-e^{\pm i\phi}\cos\theta\frac{1}{\sin\theta}\frac{\partial}{\partial\phi}\right) \\
&= \pm e^{\pm i\phi}\left(\frac{\partial}{\partial\theta}\pm i\cot\theta\frac{\partial}{\partial\phi}\right), \qquad \text{(A.42)}
\end{aligned}$$

where from (A.29) we noted the complex components

$$\begin{aligned} \hat{\boldsymbol{\theta}} \cdot (\mathbf{e}_1 \pm i\mathbf{e}_2) &= e^{\pm i\phi} \cos\theta \\ \hat{\boldsymbol{\phi}} \cdot (\mathbf{e}_1 \pm i\mathbf{e}_2) &= \pm i e^{\pm i\phi} \\ \hat{\mathbf{r}} \cdot (\mathbf{e}_1 \pm i\mathbf{e}_2) &= e^{\pm i\phi} \sin\theta \,. \end{aligned} \tag{A.43}$$

A.4 Integrals

It is convenient to use $d\hat{\mathbf{r}}$ for the two-dimensional element that specifies the direction of the unit spatial $\hat{\mathbf{r}}$. In terms of spherical coordinates,

$$d\hat{\mathbf{r}} = d\cos\theta \, d\phi \,. \tag{A.44}$$

Thus

$$\int_{4\pi} d\hat{\mathbf{r}} = \int_0^{2\pi} d\phi \int_{-1}^{1} d\cos\theta \,. \tag{A.45}$$

A.5 Cylindrical Coordinates

The cylindrical coordinates are ρ, ϕ, z, in terms of which the position vector is

$$\mathbf{r} = \rho \left(\mathbf{e}_1 \cos\phi + \mathbf{e}_2 \sin\phi \right) + z\mathbf{e}_3 \,. \tag{A.46}$$

We leave it as an exercise to work out the detailed relationships.

Exercise A.4. (a) Find the tangent vectors for the cylindrical coordinates and define unit vectors in their directions.
(b) Expand $\mathbf{r}$ in terms of the cylindrical unit vectors.
(b) Expand ∇ in terms of the cylindrical unit vectors and give expressions both with the unit vectors on the left of the differential operators and on the right. [Partial ans.:

$$\nabla = \hat{\boldsymbol{\rho}} \frac{\partial}{\partial \rho} + \hat{\boldsymbol{\phi}} \frac{1}{\rho} \frac{\partial}{\partial \phi} + \hat{\mathbf{z}} \frac{\partial}{\partial z} \,.] \tag{A.47}$$

(c) Expand $\mathbf{L}$ in the cylindrical unit vectors.

Appendix B

Maple Worksheet rad3.mws

This Maple V worksheet computes and displays the animated electric field of a moving charge, using the virtual-photon stream method. First we set some constants. It is presented here to demonstrate the simplicity of the calculation. T is the time interval for which motion is displayed. The field lines are shown as streams of virtual photons as they are sprayed from the charge over a period of J time intervals. The calculation tracks virtual photons emitted at time intervals of dt along 2N rays that are isotropically distributed in the instantaneous rest frame of the charge.

This and other relevant worksheets for recent releases of Maple can be downloaded from the site http://www.birkhauser.com. Search for this text.

Set initial values:

```
> restart;
> Digits:=5: with(plots): nplt:=0: j:=0: J:=64: N:=6:
> v:=Pi*.25: dt:=.0625*Pi/v: TJ:=J*dt: T:=2*Pi/v: L:=.7*TJ: a:=v/(4*dt):
> for n from 0 to N do cs.n:=cos(Pi*n/N): sn.n:=sin(Pi*n/N) od:
```

Begin the main time loop:

```
> for t from dt-TJ by dt to T-dt do
```

Select one of three possible motions of the charge by adding and removing comment signs (#), or program any other linear motion:

1. A relativistic oscillator:

```
>  #X:=sin(v*t): V:=v*cos(v*t):
```

2. A short impulse during $0<t<4dt$:

```
>  if t<0 then X:=0: V:=0:
>  elif t<4*dt then X:=.5*a*t^2: V:=a*t:
```

```
> else X:=v*(t-2*dt): V:=v fi;
```
3. Uniform acceleration, instantaneously at rest at t=0:
```
> #s:=t/4: sr:=sqrt(s^2+1): X:=4*(sr-1): V:=s/sr:
```
Begin computation of the virtual photons released:
```
> j:=j+1: if j <= J then imax := j fi;
> for n from 0 to N do
>   for i from imax by -1 to 2 do # shift photon positions x,y.
>   dx.n[i]:=dx.n[i-1]: x.n[i]:=x.n[i-1]+dx.n[i]:
>   dy.n[i]:=dy.n[i-1]: y.n[i]:=y.n[i-1]+dy.n[i]
>   od:
>   dx.n[1]:=dt*(V+cs.n)/(1+V*cs.n): x.n[1]:=X: # transform the angles.
>   dy.n[1]:=dt*sqrt(1-V^2)*sn.n/(1+V*cs.n): y.n[1]:=0
> od:
```
When enough points have been computed, collect them and store the plots:
```
> if j >= J and j mod 2 = 0 then
>   pointseqtop:=seq([seq([x.n[i],y.n[i]],i=1..imax)],n=0..N):
>   pointseqbot:=seq([seq([x.n[i],-y.n[i]],i=1..imax)],n=1..N-1):
>   plt[nplt]:=plot({pointseqtop,pointseqbot},x=-L..L,y=-L..L,axes=boxed,
>         color=black); # add "style=point,symbol=circle" to show photons.
>   nplt:=nplt+1 fi:
> od:
```
Now animate the fields!
```
> display([seq(plt[n],n=0..nplt-1)],insequence=true,scaling=constrained);
```

Bibliography

[1] A. O. Barut. *Electrodynamics and Classical Theory of Fields and Particles*. Dover Publications, Inc., New York, 1980.

[2] W. E. Baylis, editor. *Clifford (Geometric) Algebra with Applications to Physics, Mathematics, and Engineering*, Birkhäuser, Boston, 1996.

[3] M. Born and E. Wolf. *Principles of Optics*. Pergamon, Oxford, U.K., 1965.

[4] D. M. Brink and G. R. Satchler. *Angular Momentum*. Clarendon Press, Oxford, 2nd edition, 1968.

[5] W. K. Clifford. *Mathematical Papers*. Chelsea, Bronx, N.Y., 1968. A 1968 reprint of the 1882 edition.

[6] N. H. Fletcher and T. D. Rossing. *The Physics of Musical Instruments*. Springer-Verlag, New York, 1991.

[7] G. R. Fowles. *Introduction to Modern Optics*. Holt, Rinehart, and Winston, New York, 2nd edition, 1975.

[8] D. J. Griffiths. *Introduction to Electrodynamics*. Prentice-Hall, Inc., Englewood Cliffs, NJ, 2nd edition, 1989.

[9] W. R. Hamilton. *Elements of Quaternions*, volume I and II. Chelsea, New York, 3rd edition, 1969. A reprint of the 1866 edition published by Longmans Green (London) with corrections by C. J. Jolly.

[10] D. Hestenes. *Space-Time Algebra*. Gordon and Breach, New York, 1966.

[11] D. Hestenes and G. Sobczyk. *Clifford Algebra to Geometric Calculus*. Kluwer Academic, Dordrecht, 1984.

[12] B. J. Hunt. *The Maxwellians*. Cornell University Press, Ithaca, NY, 1991.

[13] J. D. Jackson. *Classical Electrodynamics*. John Wiley and Sons, Inc., New York, 3rd edition, 1998.

[14] B. Jancewicz. *Multivectors and Clifford Algebras in Electrodynamics*. World Scientific, Singapore, 1989.

[15] M. Klein. *Mathematical Thought from Ancient to Modern Times*, volume 2, chapter 32. Oxford University Press, Oxford, U. K., 1972.

[16] L. D. Landau and E. M. Lifshitz. *The Classical Theory of Fields*. Pergamon Press, Inc., Oxford, U. K., 4th edition, 1975.

[17] P. Lounesto. *Clifford Algebras and Spinors*. Cambridge University Press, Cambridge, U. K., 1997.

[18] J. B. Marion and M. A. Heald. *Classical Electromagnetic Radiatiion*. Harcourt Brace Jovanovich, Inc., Orlando, Florida, second edition, 1980.

[19] J. C. Maxwell. *Treatise on Electricity and Magnetism*, volume 1 and 2. Clarendon Press, Oxford, U. K., 1873. 3rd edition (1891) reprinted by Dover Publications, New York, 1954.

[20] C. W. Misner, K. S. Thorne, and J. A. Wheeler. *Gravitation*. W. H. Freeman & Co., San Francisco, 1973.

[21] H. C. Ohanian. *Classical Electrodynamics*. Allyn and Bacon, Inc., Newton, Massachusetts, 1988.

[22] W. K. H. Panofsky and M. Phillips. *Classical Electricity and Magnetism*. Addison-Wesley, Inc., Reading, Massachusetts, second edition, 1962.

[23] R. Penrose and W. Rindler. *Spinors and Spacetime*. Cambridge University Press, Cambridge, U. K., Vol. 1: 1984 and Vol. 2: 1986.

[24] M. Schwartz. *Principles of Electrodynamics*. McGraw-Hill, Inc., New York, 1972.

[25] J. Snygg. *Clifford Algebra, a Computational Tool for Physicists*. Oxford University Press, Oxford, U. K., 1997.

[26] E. F. Taylor and J. A. Wheeler. *Spacetime Physics.* W. H. Freeman & Co., San Francisco, 2nd edition, 1992.

[27] J. B. Westgard. *Electrodynamics: A Concise Introduction.* Springer-Verlag, New York, 1995.

[28] E. T. Whittaker. *The History of the Theories of Aether and Electricity.* Longman, London, 1910.

Index

Abraham-Lorentz equation, 311
addition theorem, 328
adjoint, 23
Aharonov-Bohm effect, 97
algebra
    center of, 22
    Clifford, xiii, 1
        existence of, 10
    Dirac, 20
    geometric, xiii, 1
    Pauli, xiii, 11
    quaternion, 7
    vector, 8
Ampere's law, 80
angular momentum
    of charge-monopole pair, 100
    of electromagnetic field, 107
angular momentum operator, 363
angular momentum operators, 322
antenna
    linear, 299
antisymmetric products, 30
atom
    modeled by oscillator, 286
atomic decay time
    of sodium resonance transition, 321
atomic lifetimes, 295, 303
atomic units, 2
autocorrelation function, 155

Babinet's principle, 299
basis
    reciprocal, 358
basis bivectors, 13
basis vectors, 20
beats, 162
Biot-Savart law, 80, 104
    generalized, 276
biparavectors, 46
    basis of, 46
    elliptic, 50
    handedness of, 50
    hyperbolic, 50
    lightlike, 50
    null, 50
    orthonormal, 47
    simple and compound, 49
    spacelike, 50
    timelike, 50
birefringent material, 288
bivector, 13, 14
    as generator of rotations, 15
boost parameter, 43
boosts, 41
    limiting, 140
    product of, 137
Bremsstrahlung, 263, 281
Brewster angle, 229

canonical element, 21
canonical momentum, 183
capacitance per unit length

of transmission line, 156
capture solutions
of charge in Coulomb field, 141
causality, 248, 270
breakdown in Lorentz-Dirac equation, 310
in model of elementary charge, 305
center of an algebra, 22
characteristic impedance
of transmission line, 158
characteristic impedance of vacuum, 82
circular polarization, 196
$\mathcal{C}\ell_{0,1}$, 20
$\mathcal{C}\ell_{0,2}$, 20
$\mathcal{C}\ell_{1,3}$, xiii
$\mathcal{C}\ell_{2}$, 11
$\mathcal{C}\ell_{3}$, 20
$\mathcal{C}\ell_{3}$
basis of, 15
$\mathcal{C}\ell_{3}$, xiii, 11
$\mathcal{C}\ell_{3,1}$, xiii
$\mathcal{C}\ell_{4}$, 20
$\mathcal{C}\ell_{n}$, 25
$\mathcal{C}\ell_{p,q}$, 20
classical electron radius, 103, 267
Clausius-Mossotti equation, 301
Clebsch-Gordan coefficients, 331, 350
cliffor, 23
Clifford algebra, 1
Clifford analysis, 17
Clifford dual, 21
Clifford number, 23
Clifford, William Kingdon, 1, 8
clocks
synchronization of, 68
coaxial line, 164
coherency density, 198
complete set of solutions, 147
completeness relation, 153, 331
complex analysis, 17
complex field, 20
composition relation, 331
compound fields, 127
Compton effect, 37
Compton scattering, 37
inverse, 77
conducting screen, 231
conductivity, 212
conductors
reflection from, 234
conjugate momentum, 137
conjugation
Clifford, 23
complex, 23
hermitian, 24
conservation of charge, 91
conservation of energy, 98
constitutive equations, 80
continuity conditions, 222
continuity equation, 91
contragradient transformation, 63
contravariant indices, 40
convective derivative, 138
corner cube, 28
coulomb
large as electrostatic unit, 4
Coulomb field
motion of charge in, 130
Coulomb force law, 3
Coulomb gauge, 102, 271
Coulomb's law, 79
Coulomb-Faraday law
generalized, 276
covariant and invariant, 63

covariant description, 33
covariant indices, 40
critical angle, 225
cross products, 22
current density, 27
curvilinear coordinates, 357
cutoff frequency, 243
cyclotron frequency
    proper, 126
cylindrical coordinates, 364

D'Alembertian, 91
delta function
    Dirac, 151
density fluctuations
    and blue sky, 293
dielectric constant, 207
dielectric dyadic (tensor), 288
differential forms, xiv
diffraction, 297
diffusion equation, 214
dipole polarizability, 287
dipole radiation, 272
Dirac delta function, 151, 163
    derivatives of, 163
Dirac matrices, xiii, 20
dispersion, 153, 209
Doppler shift, 75
Doppler width, 76
Doppler-reduced spectroscopy, 76
drift frame, 126, 141
drift velocity, 126
Drude model, 303
    of Van der Waals forces, 295
dual
    Clifford, 21
    Hodge, 21
dual field tensor, 105
dual subspaces, 15
duality rotation, 176
dyadic expression, 104
dynamic polarizability, 289

eigenprojectors, 173, 189
eigenspinor, 110
    Lorentz transformation of, 113
    time evolution of, 120
eigenspinor form
    of Lorentz-Dirac equation, 317
eigenvalues
    of projectors, 173
Einstein's principle of equivalence, 261
Einstein, Albert, 1
electric dipole
    induced, 286
electric field, 69
    motional, 75
electromagnetic field, 69
    angular momentum of, 107
    energy density of, 98
    energy flow of, 98
    gives spacetime rotation rate, 122
    interpretations of, 91
    invariants of, 71
    momentum conservation in, 133
    momentum density of, 98
    of accelerating charge, 253
    of charge in uniform gravitational
        field, 261
    of uniformaly accelerated charge,
        259
electromagnetic field tensor, 105
electron
    classical radius of, 103, 267
    g factor, 122
    model of, 305

rest energy of, 40
self-energy of, 103
electron flow
in copper wire, 160
elliptical polarization, 196
energy
internal, 39
rest
of electrons and positrons, 40
threshold, 39
in different frames, 41
energy conservation
for accelerating charge, 305
Euler relation, 17
evanescent wave, 225
evanescent waves, 216
even part, 25
events
spacetime, 33

Faraday, 70, 108
Faraday's law, 79
Fermat's principle, 245
Fibonacci sequence, 32
field lines
of a static electric charge, 72
of moving charge, 76
field tensor, 105
dual, 105
fine-structure constant, 2
flag, 175
flagpole, 175
Fourier integrals, 150
Fourier series, 149
Fourier transform, 150
frame
center-of-momentum, 39, 41
target, 41
frame space, 118
Fraunhofer diffraction, 297
Fresnel diffraction, 297
Fresnel equations, 231
fringes
diffraction, 297
Frobenius, F. Georg, 10
functions of paravectors, 30
fundamental axiom of Clifford algebra, 8

g factor of electron, 122
Galilean transformations, 28
conflict with Maxwell theory, 1
gauge
Coulomb, 170, 271
Lorenz, 170
transverse, 272
gauge invariance, 97
gauge transformations, 98
Lorenz condition, 102
gauss
magnetic field unit, 5
Gauss' law, 79
Gaussian shape, 161
Gaussian units, 3
generator
of rotation, 19
of spacetime rotation, 52
geometric algebra, 8
Gibbs, Josiah Willard, 7
Gibbs-Heaviside vector analysis, 8
golden ratio, 31
grade automorphism, 25
gradient operator
in nonorthonormal coordinates, 360
Grassmann, Hermann G., 8
Green function, 269

retarded, 270
group
of Lorentz transformations, 117
group velocity, 154
in three dimensions, 239

Hamilton, William Rowan, 6
Hamiltonian
of charge in external field, 138
handedness
of biparavector, 50
of coordinates, 22
Hanle effect, 303
harmonic oscillator
as model of atom, 286
harmonic time dependence, 147
headlight effect, 57
Heaviside step function, 152
Heaviside, Oliver, 6
Heaviside-Lorentz units, 3
helicity, 177, 196
helicity basis, 197
Helmholtz equation, 349
Helmholtz motion, 160
hermitian operators, 323
Hertz, Heinrich, 6
Hestenes' space-time algebra, xiii
Hodge dual, xiv, 21
of electromagnetic field, 84
Huygen's principle, 299
hyperbolic motion, 125, 138
hypercomplex numbers, 6
hysteresis curve, 207

idempotent, 171
idempotent decomposition, 189
imaginary part, 24
impedance, 158
input, 159
impedance matching, 158
optical, 245
index of refraction, 208, 287
of glass, 218
of water, 218
inductance per unit length
of transmission line, 157
interference
destructive, 148
inverse
of paravector, 25
inverse Compton scattering, 77
inversion, 67
involution, 23
involutions, 29

Jones vector, 198
Joule heat, 106

Kronecker delta, 26

ladder operators, 324
Lagrangian
of charge in paravector potential, 137
Laplacian operator
in spherical coordinates, 363
Larmor power formula, 263
Legendre polynomials, 328
recursion relations for, 329
Levi-Cevita symbol, 31
Liénard-Wiechart potential, 251
lifetimes of atomic levels, 303
light
speed of, 2
light-cone condition, 249
lightlike (null) paravectors, 27
linear polarization, 198

Lorentz contraction, 74
Lorentz force, 75
Lorentz transformation, 40, 112
  active and passive, 60
  linearity of, 44
  matrix representation, 61
  of eigenspinor, 113
  orthochronous, 43
  restricted, 43
    as boost times rotation, 44
  simple, 51
  spinorial form of, 42
  unrestricted homogeneous group, 43
Lorentz-Dirac equation, 304, 309
  backward integration of, 317
  eigenspinor form of, 317
  nonuniqueness of physical solutions, 313
  pathological features of, 311
  preacceleration in physical solutions, 310
  runaway solutions of, 312
  tunneling solutions of, 316
Lorentz-force equation
  covariant form, 108
  solution for constant fields, 123
  solution in Coulomb field, 130
  solution in directed plane wave, 180
  solution in plane wave plus electric field, 192
  solution in plane wave plus magnetic field, 193
  spinorial, 180
  spinorial form, 123
  vector form, 108
Lorenz condition, 97, 102
magnetic brake, 139
magnetic field, 69
magnetic memory, 207
magnetic monopoles, 79, 93
  angular momentum associated with, 100
  Dirac's model, 95
MapleV worksheet, 278
mass
  and internal energy, 39
mass renormalization, 307
  avoiding, 308
mass-shell condition, 183
matching conditions
  at non-conducting boundaries, 222
matrix representations, 10
  irreducible, 335
  reducible, 335
Maxwell stress dyad, 135
Maxwell stress tensor, 103
Maxwell's equation(s)
  and vector concept, 6
  component form, 86
  covariance of, 83
  covariant form of, 81
  homogeneous, 86
  inhomogeneous, 86
  linearity of, 83
  symmetries of, 85
  vector form, 79
Maxwell's macroscopic equations, 206
Maxwell, James Clerk, 1
metric
  negative definite, 19
metric space, xiv
metric tensor, 9, 15, 26, 40, 359
Minkowski spacetime, 20, 34
Minkowski spacetime metric, 26

model atom, 286
modes
  normal, 148
  TE and TM, 243
momentum
  spacetime, 37
momentum conservation
  in electromagnetic fields, 133
momentum paravector, 27
multipole expansion, 345
multipole radiation, 348
multivector, 14

natural modes, 148
neutron
  mean lifetime of, 58
normal modes, 148
null paravectors, 27, 29
null spacetime planes, 102

odd part, 25
Ohm's law, 212
operators
  raising and lowering, 326
optical theorem, 302

pacwoman property
  inverse, 179
  of projectors, 173
paravector metric, 26
paravector potential, 27
paravector space, 23, 25
  planes in, 45
paravectors, 23
  as spacetime vectors, 34
  classification of, 27
  functions of, 30
  orthogonal, 27, 51
  orthogonality of, 26
  parallel, 47
parity inversion, 85
parity operator, 67
particle accelerators
  synchrotron radiation in, 281
particle creation, 41
Pauli algebra, xiii, 11
Pauli spin matrices, 11
penetration depth, 216
periscope, 19
permeability, 207
permittivity, 207, 287
  frequency dependent, 289
phase shifters, 201
phase velocity, 167
  in three dimensions, 239
photon streams, 276
Planck's constant, 37
plane wave
  directed, 167
  motion of charge in, 180
plane waves
  circularly polarized, 177
  monochromatic, 177
Poincaré sphere, 200
Poincaré spinor, 197
Poisson's equation, 269
polarizability
  dipole, 287
  dynamic, 289
polarizability dyadic (tensor), 288
polarization, 194
  circular, 177, 196
  elliptical, 178
  elliptical polarization, 196
  linear, 179, 198
  of dilute gas, 287
  of media, 206

of reflected waves, 229
polarizers, 201
ponderomotive momentum, 185
positron
rest energy of, 40
Poynting vector, 98
Poynting's theorem, 98, 305
preacceleration, 313
in Lorentz-Dirac equation, 310
projector, 29, 171
proper time, 35
retarded, 249
proper velocity, 27, 35, 54
pseudo-Euclidean space, 15

quarter-wave line, 165
quarter-wave plate, 204
quaternion algebra, 25
quaternions, 6
unit, 13

radiation field
energy-momentum density of, 262
radiation pattern
from linear antenna, 303
radiation reaction, 304
of harmonically bound charge, 321
on charged oscillator, 293
one-dimensional motion with, 311
rapidity, 43
Rayleigh scattering, 291
real part, 24
reciprocal basis, 358
reciprocal basis paravectors, 27, 64
reflection
from conductors, 234
in spacetime, 65
of wave at mirror, 191
of wave in one dimension, 148
spatial, 66
total internal, 225
reflection coefficient, 227
reflection of waves
at conducting surfaces, 217
reflection symmetry, 85
reflections, 16
in intersecting planes, 19
in spacetime planes, 103
reflections in mirrors, 19
refraction
index of, 208
renormalization
of mass, 307
representation
irreducible, 332
resistor
Joule heat of, 106
reversion, 24
rifle transformation, 62
Rodrigues formula
for Legendre polynomials, 329
rotating frames, 30
rotation matrices, 332
rotations
equivalent to two reflections, 19
in n dimensions, 18
in plane, 15
linear combinations of, 17
spacetime, 41
runaway solutions
to Lorentz-Dirac equation, 312

scalar part, 24
scalar product, 41
scattering
Rayleigh, 291

resonant, 290
Thomson, 291
scattering amplitude, 301
scattering of radiation
in constant magnetic field, 293
Schott energy, 310
self-interaction
of electron, 305
separation of variables, 146
SI units, 3
simple fields, 127
skin depth, 216
sky
reason for blue color, 291
SL(2,C), 110, 112, 117
structure equations, 78
SL(2,C) diagrams, 118
Snell's law, 224
snowflake effect, 58
$SO_+(1,3)$, 61
$SO_+(3)$, 332
solenoid
current of, 163
magnetic field of, 104
space
pseudo-Euclidean, 15
spacelike paravectors, 27
spacetime diagrams, 67
of planes, 69
spacetime manifold, 33
spacetime momentum, 37
spacetime plane, 45
spacetime planes
orthogonal, 51
spacetime reflection, 65
spacetime rotation, 41, 51, 110
generators of, 52
spacetime vectors, 27, 34
classification of, 27
orthogonality of, 26
spatial inversion, 25
spatial reversal, 23
spectral decomposition, 30, 173, 188
spectral distribution, 155
spectral lines
Doppler width of, 76
spectrum, 155
speed of light, 2
spherical Bessel functions, 352
spherical coordinates, 361
spherical harmonics, 322
table of, 328
spherical tensor, 329
spherical waves, 353
spinning particles, 137
spinorial form
of Lorentz transformation, 42
spinors, xiv
standing wave, 191
standing waves, 185
parallel electric and magnetic fields in, 187
statcoulomb, 4
statvolt, 28
stellar aberration, 57
step function
Heaviside, 152
Stokes parameters, 199
Stokes subspace, 199
storage rings
synchrotron radiation in, 281
stress energy-momentum tensor
electromagnetic, 135
string
motion of bowed, 160
structure equations

of SL(2,C), 78
structure factor, 292
SU(2), 332
subalgebras, 25
superposition
coherent and incoherent, 202
superposition principle, 83, 144, 231
surface charge, 232
surface current, 232
surfing
of charges in plane waves, 185
susceptibility
electric and magnetic, 207
synchrotron radiation, 281

tangent space, 34
tangent vectors, 34
TEM waves, 240
tensor
of biparavector elements, 48
tensor notation, xv
termination impedance, 165
tesla
SI unit of magnetic field, 5
tetrad, 59
Thomas precession, 121
Thomson scattering, 267, 291
threshold energy, 39
in different frames, 41
time
in Newtonian physics, 35
in relativity, 35
proper, 35
time dilation, 55, 74
time evolution
of eignespinor, 120
time reversal, 66, 85
timelike paravectors, 27
total internal reflection, 225
transformation
active and passive, 42, 63
antiunimodular, 65
contragradient, 63
Lorentz, *see* Lorentz transformation
nonorthochronous, 66
orthochronous, 66
proper and improper, 43
rifle, 62
velocity, 41
transmission coefficient, 227
transmission line, 156
triparavector, 89
trivector, 14
tunneling through barriers
in Lorentz-Dirac solutions, 316
twin lead, 164
twin paradox, 75

uncertainty relation, 161
unimodular elements, 26, 44
unit paravector, 55
unit quaternions, 13
unitary elements, 44
units
electromagnetic
lack of conversion factors for, 5
Gaussian, 3
Heaviside-Lorentz, 2, 3
MKSA, 2
SI, 2, 3

van der Waals forces
Drude model for, 295
van der Waals interaction

higher-order corrections to, 303
vector algebra, 8
vector concept
in electromagnetic theory, 6
vector part, 24
vector products, 10
vector space
of nondefinite metric, 20
vector spherical harmonics, 350
vectors
of grade m, 14
velocity
of electrons in wire, 160
proper, 35, 54
velocity composition, 74
versors, 14
vibration recipe, 149
virtual "photon streams", 276
volume element, 21

wave equation, 167, 348
for dielectric media, 209
for the electromagnetic field, 93
in conducting media, 214
one-dimensional, 143
wave guides, 238
cutoff frequency of, 243
wave number, 147
wave packet, 150, 153, 161
Gaussian, 161
group velocity of, 154
spreading of, 154, 161
wave speed, 153
wave velocity, 167
waves
at boundaries, 220
evanescent, 225
finite train of, 162
group velocity of, 239
in string, 145
phase velocity of, 239
polarization of reflected, 229
standing, 147
transverse electromagnetic (TEM), 240
wedge products, 12
Wien velocity filter, 138
wire loop
magnetic field of, 104

zero-divisors, 10, 172
zero-point oscillations, 296

Progress in Physics

—A collection of research-oriented monographs, reports, and notes arising from lectures or seminars.
—Quickly published; concurrent with research.
—Easily accessible through international distribution facilities.
—Reasonably priced.
—Reporting research development combining original results with an expository treatment of particular subject area.
—A contribution to the international scientific community: for colleagues and for graduate students who are seeking current information and directions in their graduate and postgraduate work.

Manuscripts should be no less than 100 and preferably no more than 500 pages in length. We encourage preparation of manuscripts in such forms as LaTeX or AMS TeX for delivery in camera-ready copy which leads to rapid publication, or in electronic form for interfacing with laser printers or typesetters. However, we continue to accept other forms of preparation as well.

Proposals should be sent to:
Birkhäuser Boston, Inc., 675 Massachusetts Avenue, Cambridge, Massachusetts, 02139, or Birkhäuser Verlag AG., P.O. Box 133, CH-4010 Basel / Switzerland

1 COLLET/ECKMANN. Iterated Maps on the Interval as Dynamical Systems
ISBN 3-7643-3510-6
2 JAFFE/TAUBES. Vortices and Monopoles, Structure of Static Gauge Theories
ISBN 3-7643-3025-2
3 MANIN. Mathematics and Physics
ISBN 3-7643-3027-9
4 ATWOOD/BJORKEN/BRODSKY/STROYNOWSKI. Lectures on Lepton Nucleon Scattering and Quantum Chromodynamics
ISBN 3-7643-3079-1
5 DITA/GEORGESCU/PURICE. Gauge Theories: Fundamental Interactions and Rigorous Results
ISBN 3-7643-3095-3
6 FRAMPTON/GLASHOW/VANDAM. Third Workshop on Grand Unification, 1982
ISBN 3-7643-3105-4
7 FRÖHLICH. Scaling and Self-Similarity in Physics: Renormalization in Statistical Mechanics and Dynamics
ISBN 3-7643-3168-2
8 MILTON/SAMUEL. Workshop on Non-Perturbative Quantum Chromodynamics
ISBN 3-7643-3127-5
9 WELDON/LANGACKER/STEINHARDT. Fourth Workshop on Grand Unification
ISBN 3-7643-3169-0

10 FRITZ/JAFFE/SZASZ. Statistical Physics and Dynamical Systems: Rigorous Results
ISBN 3-7643-3300-6

11 CEAUSESCU/COSTACHE/GEORGESCU. Critical Phenomena: 1983 Brasov School Conference
ISBN 3-7643-3289-1

12 PIGUET/SIBOLD. Renormalized Supersymmetry: The Perturbation Theory of N=1 Supersymmetric Theories in Flat Space-Time
ISBN 3-7643-3346-4

13 HABA/SOBCZYK. Functional Integration, Geometry and Strings: Proceedings of the XXV Karpacz Winter School of Theoretical Physics
ISBN 3-7643-2387-6

14 SMIRNOV. Renormalization and Asymptotic Expansions
ISBN 3-7643-2640-9

15 Leznov/Saveliev. Group-Theoretical Methods for Integration of Nonlinear Dynamical Systems
ISBN 3-7643-2615-8

16 Maslov. The Complex WKB Method for Nonlinear Equations I. Linear Theory
ISBN 3-7643-5088-1

17 BAYLIS, Electrodynamics: A Modern Geometric Approach
ISBN 0-8176-4025-8